ŞE

Rehabilitating Older and Historic Buildings

Rehabilitating Older and Historic Buildings

LAW
TAXATION
STRATEGIES

STEPHEN L. KASS

JUDITH M. LaBELLE

DAVID A. HANSELL

Berle, Kass & Case
New York, New York

JOHN WILEY & SONS

A Ronald Press Publication

New York · Chichester · Brisbane · Toronto · Singapore

Library of Congress Cataloging in Publication Data:

Kass, Stephen L.
    Rehabilitating older and historic buildings.

    1. Historic buildings—Law and legislation—United States.
2. Real property and taxation—United States.
3. Tax incentives—Law and legislation—United States.
I. LaBelle, Judith M., 1948- . II. Hansell, David A., 1953- .
III. Title.

KF4310.K37 1985   344.73'094   85-18370
ISBN 0-471-80154-2   347.30494

Printed in the United States of America

10 9 8 7 6 5 4 3 2

Preface

The widespread rediscovery of the built environment, by both the public and the development community, has reflected increasing appreciation of the craftsmanship, design, and scale of thousands of older buildings in America's cities and towns, as well as the economic and planning benefits of rehabilitating historic structures for contemporary use. Significant tax incentives for rehabilitating historic and older buildings have accelerated this trend in the past several years. As a result of the increased level of rehabilitation activity, it is not surprising that few, if any, fields of law have grown as rapidly within the past decade as historic preservation, leading to increased demands for a concise guide to this developing field.

Because the legal issues relating to historic preservation and the rehabilitation process include such diverse subjects as federal, state, and municipal taxation; securities, partnership, and constitutional law; and zoning, environmental, and landmarks regulation, there are relatively few attorneys or accountants who feel confident advising a client concerning a proposed rehabilitation project. Similarly, private developers seeking to undertake rehabilitation projects, investment advisors counseling clients and prospective investors, and governmental officials responsible for approving and regulating such projects, often find it difficult to understand the overall legal, financial, and regulatory context within which they must function.

This book seeks to meet these needs by providing a guide to the process of rehabilitating historic and older buildings. It attempts to distill for lawyers, accountants, financial advisors, developers, government officials, and nonprofit organizations the principal components of the rehabilitation process and to set forth, as clearly as possible, the benefits and potential pitfalls of rehabilitation projects. This is neither a treatise nor an exhaustive exploration of the subject. Rather, this book is intended to provide interested readers with an overview of an increasingly complex and important field and to provide a foundation from which lawyers, accountants, and others can explore specific problems in greater detail.

As this volume is being completed, the Treasury Department has proposed fundamental revisions to the Internal Revenue Code, some of which would alter significantly the tax benefits and regulatory processes described herein. Because early enactment of such proposals appears unlikely, the authors anticipate periodic supplements in order to assist readers in evaluating the impacts of any legislative changes ultimately adopted by Congress. To the extent that future code amendments limit (or eliminate) tax considerations in connection with rehabilitation projects, an understanding of the underlying real estate, partnership, securities, environmental, and landmarking aspects of such projects is likely to become of even greater importance to project sponsors and to their advisors.

The authors wish to acknowledge and express their gratitude for the cooperation of the National Trust for Historic Preservation, Preservation Action, the Providence Preservation Society Revolving Fund, James Pickman & Associates, Inc., and the many other individuals and organizations who have provided us with material concerning rehabilitation projects or have reviewed portions of the text. The authors also wish to thank their colleagues at Berle, Kass & Case in New York City for sharing their professional experience and for their patience during the preparation of this volume. Stephen Warnath and Alan Epstein, law students who assisted in these efforts, deserve special thanks. Debra-Ann Chait also deserves special recognition for her skill and dedication in typing the manuscript.

STEPHEN L. KASS
JUDITH M. LaBELLE
DAVID A. HANSELL

New York, New York
September 1985

Contents

Rehabilitating Older and Historic Buildings

One

Federal Income Tax Incentives for Rehabilitation

Introduction

Of the variety of social and economic factors which have contributed to the current interest in rehabilitation, the most important are probably the federal income tax incentives now available for the rehabilitation of historic and certain other old buildings. To understand the impact of these incentives, it is useful to briefly trace their history.

For many years, the federal tax structure reflected a policy bias favoring new construction over the rehabilitation of older buildings. The tax code was structured so that rehabilitation of an old building entailed the forfeiture of substantial tax benefits. The fact that a new building could often be depreciated faster than one already in service resulted in preservation often being viewed as a less attractive economic option than demolition and new construction. The ability to deduct demolition costs as an expense in the year in which the demolition occurred provided further incentive to demolish rather than to rehabilitate. These and other tax provisions often reinforced economic incentives favoring replacement rather than preservation of older buildings, including the fact that rehabilitation of an older building often means foregoing an opportunity to construct a larger and more lucrative building on the same site. As investment decisions were often made based upon those considerations, the country lost innumerable structures important to its historic and architectural heritage.

Public awareness of the importance of preserving this heritage increased significantly as a result of activities surrounding the Bicentennial and a renewed focus on the nation's cities. As the loss of buildings continued and as preservation proponents gathered strength, pressures increased for changes in the Internal Revenue Code to support the preservation effort, and Congress began to respond.

1.1 EVOLUTION OF TAX TREATMENT OF OLDER BUILDINGS

Reform of the tax laws affecting existing structures began with the Tax Reform Act of 1976,[1] which provided two incentives intended to encourage rehabilitation of buildings listed on the National Register of Historic Places or situated in local historic districts. Owners of qualifying historic buildings were given a choice: They could amortize their rehabilitation expenditures over a period of 60 months[2] (a significantly shorter time than that previously allowed, which was tied to the life of the property improvements and generally averaged 25 to 30 years), or they could use the methods of depreciation available to owners of similar new construction.[3] Rather than being limited to the straight-line or 125-percent declining balance rates previously available, owners of substantially rehabilitated historic buildings could, depending on the type of project, use either the 150- or 200-percent declining balance rate for depreciation.

The 1976 Act also included two disincentives to demolition of certified historic structures, or of structures located in historic districts and certified as contributing to the significance of those districts.[4] The first disincentive required that such demolition costs be capitalized as part of nondepreciable land cost. Demolition costs relating to those structures could no longer be taken as current deductions.[5] (The Tax Reform Act of 1984 broadened this prohibition, and the code now denies a deduction for costs of, or losses incurred in connection with, the demolition of any building, not only certified historic structures.)[6] The second disincentive required that depreciation of any structure which replaced a demolished historic structure be taken using the straight-line method.[7]

---

[1] P.L. 94-455 (1976).

[2] Former I.R.C. § 191 (1976) (repealed in 1981)

[3] Former I.R.C. § 167(o) (repealed in 1981).

[4] *See* Section 2.2.

[5] Former I.R.C. § 280B.

[6] I.R.C. § 80B, as amended by P.L. 98-369, § 1063.

[7] Former I.R.C. § 167(n).

Two years later, in the Revenue Act of 1978, Congress provided a 10-percent investment tax credit as a further incentive for owners to rehabilitate commercial and industrial buildings which were more than 20 years old and which had not been rehabilitated within the preceding 20 years.[8] The tax credit was intended to spur private investment in rehabilitation projects by providing an incentive of greater value than the depreciation deduction alone. Because a tax credit reduces tax liability dollar for dollar, it is worth more to the taxpayer than a deduction, which only reduces taxable income by a variable. percentage reflecting the taxpayer's marginal bracket.[9] While making the investment tax credit available for qualified rehabilitation projects, however, the 1978 Act limited the manner in which the credit could be coupled with the incentives provided in 1976. An owner taking the investment tax credit for such a project could also use accelerated depreciation, but could not take advantage of the 60-month amortization period.[10]

The shift away from the incentives favoring demolition and new construction over rehabilitation continued with the enactment of Section 212 of the Economic Recovery Tax Act of 1981 (ERTA).[11] With ERTA's passage, Congress intended to accomplish a fundamental change in the manner in which investments in real estate were taxed, including a further reduction in tax preferences for the demolition and replacement of existing buildings. In doing so, Congress rejected the assumption that new construction was the preferable, if not the only, road to urban economic revitalization. As the Senate Finance Committee noted:

> Investments in new structures and new locations . . . do not necessarily promote economic recovery if they are at the expense of older structures, neighborhoods, and regions. A new structure with new equipment may add little to capital formation or productivity if it simply replaces an existing plant in the older structure in which the new equipment could have been installed.[12]

ERTA replaced the amortization and depreciation incentives created by the 1976 Act and the ten-percent investment tax credit provided by

[8] Former I.R.C. § 48(g).

[9] For example, a taxpayer in the 30-percent tax bracket with $90,000 taxable income, has a federal tax liability of $30,000. If the taxpayer earned a $9,000 deduction, this would reduce his taxable income to $81,000, resulting in a tax liability after the deduction of $27,000 (assuming no change in bracket). In contrast, the dollar-for-dollar offset that a taxpayer receives for a $9,000 tax credit allows the taxpayer to subtract the full $9,000, resulting in a total tax liability of $21,000.

[10] Former I.R.C. § 48(g)(2)(B)(iv).

[11] P.L. 97-34 (1981). Citations to the regulations promulgated in connection with these provisions were also changed by ERTA. Earlier references to 36 C.F.R. Part 1208 became 36 C.F.R. Part 67. Standards for evaluating National Register eligibility were changed from 36 C.F.R. Part 1202 to 36 C.F.R. Part 60.

[12] S. Rep. No. 144, 97th Cong., 1st Sess. 72 (1981).

the 1978 Act with a three-tier system of tax credits available for the substantial rehabilitation of qualified buildings, including qualified historic buildings. Among the changes to the Internal Revenue Code made by ERTA, one of the most significant with regard to rehabilitation of most older buildings was the replacement of the previous depreciation system with the Accelerated Cost Recovery System (ACRS).[13]

Of even greater significance was the adoption of a new set of tax credits for rehabilitation of certain older structures. As explained in this chapter, ERTA created three categories of favored buildings and provided for specified investment tax credits (equal to a percentage of the qualified rehabilitation expenditures) in connection with the rehabilitation of each building category. To qualify for the investment tax credit, the rehabilitation must meet minimum standards which seek to encourage significant rehabilitation projects that at the same time retain the integrity of the building.

Amendments to the Internal Revenue Code since 1981 have adjusted the incentives for rehabilitating historic and older structures, but have not made any fundamental change in the types of incentives available. These modifications are discussed in the specific analysis of preservation tax benefits which follows.

1.2 THE REHABILITATION TAX CREDIT

Section 38 of the Internal Revenue Code allows credits against income for investment in certain depreciable property, known as Section 38 property. One category of investments eligible for the tax credit is the expense of rehabilitating certain older buildings. The credit allowed for these expenses, known as qualified rehabilitation expenditures, is determined by multiplying the portion of the basis of the property which is attributable to such expenditures by the "rehabilitation percentage."[14]

To determine whether an investment tax credit (ITC) will be available in connection with any particular rehabilitation project, several factors must be analyzed, including (1) the age of the building and the use to which it will be put after rehabilitation, (2) the taxpayer's adjusted basis in the building and the total rehabilitation costs to be incurred, (3) the timing of the rehabilitation process, and (4) the physical changes to be made to the building. In addition, as explained in detail in chapter 2,

---

[13] The Internal Revenue Service has adopted regulations relating to the investment tax credit for rehabilitated buildings, which do not incorporate changes made by ERTA and which generally apply only to expenditures incurred by December 31, 1981. Regulations governing subsequent expenditures had been proposed but not adopted at the time this book went to press. *See* 50 Fed. Reg. 26794 (June 28, 1985). References herein are made to the existing regulations, but persons contemplating rehabilitation projects should consult the proposed regulations as well.

[14] I.R.C. § 46(a)(3).

if the building is a "certified historic structure," additional requirements must be met, particularly with regard to the type of physical alterations made.[15]

1.2.1   The Rehabilitation Percentage

The rehabilitation percentage varies with the class of building, as follows:

Class of Building	Rehabilitation Percentage[16]
30 or more years old[17]	15
40 or more years old	20
Certified historic structure	25

The 10-percent investment tax credit otherwise available for a qualified investment in Section 38 property, and the energy credit which might otherwise be available under Section 46(a)(2)(C), are specifically disallowed with regard to the portion of the basis of any property which is attributable to qualified rehabilitation expenditures.[18] In other words, the same expenditure may not be used in computing more than one of these tax credits.

1.2.2   Qualified Rehabilitation Expenditure

"Qualified rehabilitation expenditures," the amount to which the rehabilitation percentage is applied to determine the size of the available investment tax credit, means any amount (1) properly chargeable to capital account, (2) which is incurred after December 31, 1981 in connection with the rehabilitation of a qualified rehabilitated building, (3) for property (or additions or improvements to property) which have a

---

[15] It is critical that a determination be made at the beginning of the planning process as to whether the building under consideration is a certified historic structure. If it is, a qualifying rehabilitation will result in a 25-percent ITC. However, if such a structure is rehabilitated in a manner which does not qualify under the applicable standards, *no* ITC will be available. *See* Section 2.2. If a building has not been certified, the decision whether to seek certification may depend on a comparison at the outset of the marginal benefit of the 25-percent (vs. 15 or 20 percent) credit with the incremental cost of a certified rehabilitation. If the building is in a registered historic district, however, it must be certified either as significant or nonsignificant to benefit from any credit, as explained in chapter 2.

[16] I.R.C. § 46(b)(4).

[17] *Id.*

[18] I.R.C.§ 46(b)(4)(B).

recovery period, under ACRS, of 18 years (15 years in the case of low-income housing).[19]

Examples of expenditures which may be aggregated in computing qualified rehabilitation expenditures are as follows:

☐ Expenditures for removal of existing interior walls, plumbing, electrical wiring, flooring, and the like.

Capital expenditures for the construction of permanent interior partitions and walls, the installation of heating or air-conditioning systems (including temperature control systems), and the replacement of plumbing, electrical wiring, flooring, and the like, in connection with a rehabilitation.

Many "soft costs" incident to rehabilitation, including legal and architectural fees (other than fees associated with the organization of a partnership or syndicate to conduct a rehabilitation).[20]

The ability to capitalize soft costs is, at present, somewhat uncertain. To the extent that they represent fees for services provided in connection with the acquisition or development of real property (for example, appraisal fees, title and mortgage insurance, and real estate commissions), they should be properly allocable to basis, and hence qualify for the ITC.[21] The now-rescinded regulations implementing the former sixty-month amortization provision for certified rehabilitation expenditures expressly allowed capitalization of such costs, but the current regulation is silent on the point.[22] It is unclear whether this reflects a decreasing willingness on the part of the IRS to allow capital expense treatment of these items.

Construction period interest and taxes, if properly capitalized in accordance with the specific statutory procedure for such items, should also qualify.[23]

Specifically excluded from the definition of qualified rehabilitation expenditures are the cost of acquiring the building (or any interest in the building) and any expenditures attributable to the enlargement of an existing building.[24] Expenditures are attributable to enlargement, and

---

[19] I.R.C. § 48(g)(2)(A), as amended by the Tax Reform Act of 1984 § 111(e)(5). *See* Section 1.3.2.

[20] Treas. Reg. § 1.48-11(c). Partnership organizational expenses must be amortized over a period of not less than 60 months. I.R.C. § 709.

[21] *See, e.g., Jackson E. Cagle, Jr.,* 63 T.C. 86, *aff'd.,* 539 F. 2d 409 (5th Cir. 1976).

[22] *Compare* Former Treas. Reg. § 1.191-2(e)(1) *with* Treas. Reg. § 1.48-11.

[23] To make them eligible for the credit, the taxpayer would have to elect to capitalize construction period expenses under I.R.C. § 266. Only interest paid prior to placing property in service may be capitalized. *See* Rev. Rul. 70-37, 1970-1 C. B. 3.

[24] I.R.C. § 48(g)(2)(B)(ii) and (iii).

therefore do not qualify, if the total volume of the building is increased. Expenses relating to the construction of a new floor in an existing building, for example, will not qualify. However, if the interior of the building is remodeled in a way which increases usable square footage of floor space without increasing the volume of the building (as through conversion to loft spaces), the expenses attributable to such remodeling will qualify.[25] Expenses incurred in connection with ancillary facilities such as a parking lot will not qualify.[26] And qualifying expenses must be incurred in connection with rehabilitation, not new construction.

Also excluded are any expenditures with regard to which accelerated methods of depreciation are used.[27] Therefore, in order to take advantage of the investment tax credit, an election must be made under Section 168(b)(3) of the Code to use the straight-line, rather than the accelerated, method of depreciation. However, the taxpayer may, within three years after a building shell is placed in service, elect to depreciate the shell on the straight-line method and still qualify for the credit,[28] and may also separately depreciate "substantial improvements" to property, which will be in this chapter.

Expenditures made by a lessee will also be excluded from qualified rehabilitation expenditures unless, on the date the rehabilitation is completed, the remaining term of the lease (without regard to renewal periods) is 18 years or longer.[29]

The exclusions from qualified rehabilitation expenditures just outlined apply whether the rehabilitated building is eligible for the 15-, 20-, or 25-percent investment tax credit. Additional provisions apply with regard to rehabilitation expenses incurred in connection with the rehabilitation of a certified historic structure, as explained in chapter 2.[30] Any expenditures attributable to the rehabilitation of a certified historic structure or building in a registered historic district are also excluded unless the rehabilitation is a certified rehabilitation; that is, a rehabilitation which the Secretary of the Interior has certified as being consistent with the historic character of the property or the district in which it is located.[31] This exclusion may be avoided with regard to a building within a registered historic district if the building was not individually designated as a certified historic structure and if the Secretary of the Interior certifies that the building is not of historic significance to the district. This "decertification" of the building must occur

---

[25] Treas. Reg. § 1.48-11(c)(7).

[26] *Id.* § 1.48-11(c)(5).

[27] I.R.C. § 48(g)(2)(B)(i).

[28] Technical Corrections Act of 1982 § 102(f)(4). In addition, "substantial improvements" to a building may be separately depreciated. *See* Section 1.3.2.

[29] I.R.C. § 48(g)(2)(B)(v).

[30] *See* Section 2.3.

[31] I.R.C. § 48(g)(2)(B)(iv); § 48(g)(2)(c).

prior to the commencement of the rehabilitation of the building unless the taxpayer can certify to the Secretary of the Treasury that, at the time rehabilitation began, the taxpayer in good faith was not aware of the requirement to obtain such certification.[32]

1.2.3 Qualified Rehabilitated Building

Investment tax credits are available only with regard to the rehabilitation of "qualified rehabilitated buildings," which are defined on the basis of the building's age and of the nature and cost of the rehabilitation. Except in the case of a certified historic structure, a building cannot qualify as a qualified rehabilitated building unless at least 30 years have elapsed between the date physical work on the rehabilitation begins and the date the building was first placed in service.[33] Since the building must be depreciable to qualify, it cannot be used by the taxpayer for personal purposes. Moreover, the 15- and 20-percent credits are available only for commercial buildings, not for those used for residential purposes. The 25-percent credit is, however, available for residential buildings. The rented portion of a multi-unit dwelling in which one unit is occupied by the owner would be eligible for the 25-percent credit. In addition, buildings used for transient lodging, such as inns or hotels, are not deemed to be used for residential purposes and thus may qualify for the 15- or 20-percent credit.[34]

In addition to satisfying these age and use tests, the building (and its structural components) must have been "substantially rehabilitated," must have been placed in service before the beginning of the rehabilitation, and 75 percent or more of the existing external walls must be retained in place as external walls in the rehabilitation process.[35]

1.2.4 Substantial Rehabilitation

A building will be treated as having been substantially rehabilitated only if the qualified rehabilitation expenditures during the 24-month period selected by the taxpayer (and ending with or within the taxable year for which the credit is sought) exceed the *greater* of (1) the pre-rehabilitation adjusted basis of building (and its structural components); or (2) $5,000.[36]

---

[32] I.R.C. § 48(g)(2)(B)(iv).

[33] I.R.C. § 48(g)(1)(B). The term *placed in service* refers to the earlier of the taxable years in which either depreciation on the building began or the property was available for its intended use. Treas. Reg. § 1.46-3(d)(1). *Cf.* Prop. Treas. Reg. § 1.168-2(l)(2), providing a consistent definition for ACRS purposes.

[34] I.R.C. § 48(a)(3).

[35] I.R.C. § 48(g)(1)(A)(iii).

[36] I.R.C. § 48(g)(1)(C)(i).

The adjusted basis of the building (and its structural components) is determined as of the later of the first day of the 24-month period selected by the taxpayer or the holding period of the building as defined in Section 1250(e).[37] For this purpose, land costs are excluded from the prerehabilitation basis.[38]

When a major portion of a building is rehabilitated independently, it may be treatable as a separate building in determining whether it is a qualified rehabilitated building. This approach was taken in the regulations proposed by the Service prior to the passage of ERTA in 1981,[39] and there would seem to be no reason why it should not continue to apply.[40] To be deemed a separate building under the regulations, a part of a building is considered a major portion of the building only if it is comprised of contiguous portions of the building and is clearly identifiable (for example, the first 5 stories of a 7-story building or the east wing of a building).[41] The determination is to be made "on the basis of all the facts and circumstances."

The treatment of expenditures incurred within the 24-month period but paid for after that time is not settled. Although one might assume that this would depend on whether the taxpayer uses the cash or accrual method of accounting, it appears that all taxpayers, regardless of their accounting method, will be treated as being on the accrual basis.[42] If so, all expenses incurred as well as paid during the 24-month period would be aggregated in determining whether the building was substantially rehabilitated.

The statute provides an alternative to the 24-month period for phased rehabilitation projects, which allows rehabilitation expenditures to be aggregated over a 60-month period. This option is not intended to be available for projects that unintentionally extend past their original completion date and incur costs after the conclusion of the 24-month period. To qualify under this provision, the architectural plans and specifications for the project must be completed before the rehabilitation begins and must set forth the phases in which the project will be completed.[43]

Although these objective criteria for satisfying the substantial rehabilitation test may aid in the administration of the tax law, the test has

---

[37] For these purposes, the determination of the beginning of the holding period is made without regard to any reconstruction by the taxpayer in connection with the rehabilitation.

[38] S. Rep. No. 97-144, 97th Cong., 1st Sess., at 177-78.

[39] Treas. Reg. § 1.48-11(b)(5).

[40] *See e.g.*, Private Letter Ruling 8226125 (March 31, 1982), which ruled that the first floor of a nonhistoric building was a "separate building" for this purpose.

[41] Treas. Reg. § 1.48-11(b)(5)(ii).

[42] *See* Treas. Reg. § 1.48-11(c)(3).

[43] I.R.C. § 48(g)(1)(C)(ii). For example of a phased rehabilitation approved by the IRS, *see* Private Letter Ruling 8338048 (June 20, 1983).

been criticized for several reasons. First, the criteria may encourage developers to make expenditures solely to meet the substantial rehabilitation test even though the expenditures are not otherwise necessary. Second, the substantial rehabilitation standards make rehabilitation projects more difficult in areas where building values are high. In cities such as New York, the puchase price of an old building often exceeds the amount which would ordinarily be spent to rehabilitate it, thus making it impossible to meet the substantial rehabilitation test and to qualify for the tax credit unless the transaction is restructured for this purpose. In some cases, developers have attempted to avoid this problem by creating a partnership to undertake the rehabilitation of the building and having the current owner transfer the building to the partnership as its contribution to capital. The partnership can then use the owner's existing adjusted basis, which, especially if the building has been depreciated, may be low enough to allow the rehabilitation to qualify under the substantial rehabilitation test.

This technique may, however, have drawbacks as well. The partnership receiving the property may, because of the antichurning rules discussed later in this chapter, be ineligible for liberalized depreciation of the building under the accelerated cost recovery system. Moreover, if the property is transferred to the partnership at less than its current market value, the partnership will forgo the opportunity to depreciate the increment between the transferring partner's basis in the property and its fair market value, regardless of what depreciation formula is applied.

Owners may wish to consider other ways to reduce the adjusted basis of a building so as to decrease the amount which must be spent to qualify as a substantial rehabilitation.[44] The purchaser of a property for rehabilitation may wish to allocate the purchase price as heavily as possible to the cost of land, which is not included in the substantial rehabilitation test, rather than to the cost of the building (though this will, of course, reduce the purchaser's basis for later depreciation of the building). The purchaser must, in any case, be able to substantiate any allocation which is made. It may be desirable for the seller, or the purchaser prior to beginning rehabilitation, to make a charitable contribution of a facade easement on the building, thereby reducing basis.[45] In some circumstances, it may be necessary to hold onto the building until its basis is reduced, through accelerated cost recovery depreciation deductions,[46] to the point at which rehabilitation becomes financially viable.

[44] For a more detailed discussion of these and other approaches, *see* Luxemberger, "Getting to 'Substantially Rehabilitated' in the Most Economical Way: One Aspect of the Rehabilitation Tax Credit," 1 *Preservation L. Rept.* 2036 (1982).

[45] *See* Section 1.4.

[46] *See* Section 1.3.2.

Alternatively, there may be ways to increase qualified rehabilitation expenditures to satisfy the substantial rehabilitation test. An owner may be able to capitalize as rehabilitation costs certain expenditures which might otherwise be immediately deducted as current expenditures. Or items which might have been eligible for the regular investment tax credit may instead be treated as "building and structural components" which are added to rehabilitation costs, and which would then be subject to the 15- to 25-percent rehabilitation credit, rather than the regular 10-percent ITC.

Each of these approaches, however, may involve other adverse tax consequences, or may simply not be appropriate or possible, in the circumstances of a given project. It is essential, therefore, that each of these approaches be thoroughly explored with the purchaser's tax advisor before they are employed.

1.2.5   Retention of External Walls

To qualify for the rehabilitation credit, seventy five percent of the external walls existing prior to the rehabilitation must be retained in place as external walls.[47] An "external wall" is defined as a wall with one face exposed to the weather or earth. The portion of a building which serves as a common wall or party wall is deemed to be an external wall.[48] The exterior wall is not deemed to have been retained in place if it is converted into an interior wall.[49] For example, if rehabilitation plans create an enclosed, unroofed interior space, former exterior walls which face such a space after the rehabilitation are not deemed to have been retained in place as external walls. However, the walls surrounding an uncovered interior courtyard, which are usable by the building's occupants, may be treated as exterior walls.[50]

Except with regard to certified historic buildings, the retention requirement can be met by retaining the structural framework that supports the exterior facade, which may be replaced or concealed.[51] In determining the area of existing external walls, the area of windows and doors are included and they may be replaced or enlarged.[52]

The Tax Reform Act of 1984 adopted an alternative test to the straight 75 percent of external walls requirement, which will make it somewhat easier for unconventionally shaped buildings to qualify. Under the new

[47] I.R.C. § 48(g)(1)(A)(iii).
[48] Treas. Reg. § 1.48-11(b)(4)(ii).
[49] *Id.* § 1.48-11(b)(4)(v).
[50] *Id.* § 1.48-11(b)(4)(vi).
[51] *Id.* § 1.48-11(b)(4)(iv).
[52] *Id.* § 1.48-11(b)(4)(i) and (iv).

test, a rehabilitated building will satisfy the existing walls requirement if—

1. at least 50 percent of the building's existing external walls are retained in place as external walls;
2. at least 75 percent of the building's existing external walls are retained in place as either external or internal walls; and
3. at least 75 percent of the building's internal structural framework is retained in place.[53]

The House Ways and Means Committee report discussing this porvision suggests that a building's internal structural framework will be deemed to include all load-bearing internal walls and any other internal structural supports present in a building before commencing rehabilitation, including the columns, girders, beams, trusses, spandrels, and all other members that are essential to the stability of the building.[54]

The requirement that at least 75 percent of the existing external walls be retained in place has also been criticized as being unnecessarily restrictive and difficult to apply. Most architecturally valued buildings have only one wall which was designed to face the world and which may therefore be worth preserving as an external wall, but the retention requirement applies to all buildings—historic and nonhistoric alike.

1.2.6 Recipients of the Investment Tax Credit

Generally, the taxpayer who owns the qualified rehabilitated building is eligible to take advantage of any available investment tax credit. The determination of ownership is made when the property is placed in service.[55] Consequently, the taxpayer taking the credit is not necessarily the taxpayer who owned the property during the rehabilitation. This provision is particularly significant in facilitating syndication of rehabilitation projects, as discussed in chapter 3, because it permits the credit to be passed through to limited partners who may not have been investors in the original purchase of the property. In addition, there are some circumstances in which the tax credits may be passed through to the lessee of a rehabilitated building.

The Treasury regulations governing the repealed 60-month amortization option for older buildings, and the proposed regulations govern-

---

[53] I.R.C. § 48(g)(1)(E), as amended by P.L. 98-369, § 1043 (1984).

[54] H. Rept. No. 98-432, 98th Cong., 2d Sess. (1984), reprinted in CCH Standard Federal Tax Reports § 556 (1985).

[55] Treas. Reg. § 1.48-11(c)(3).

ing the investment tax credit, establish a formula for determining a transferee's basis in qualified rehabilitation expenditures in cases in which a transfer is made after rehabilitation has occurred but before a building has been placed in service. A transferee who wishes to take advantage of the credit may treat as paid or incurred by the transferee the lesser of (1) the transferor's basis in the certified or qualified rehabilitation expenditures, or (2) that portion of the transferee's costs of acquisition which were attributable to such expenditures.[56]

There are several situations in which the ability to pass along rehabilitation expenditures to a transferee may be especially helpful. One is the case of a condominium conversion. Purchasers of condominium units for depreciable purposes are eligible for the rehabilitation credit for those expenditures which relate to their individual units and to their proportionate shares of common areas.[57] Qualified rehabilitation expenditures incurred by a condominium developer may be taken into account by purchasers for purposes of calculating the amount of the credit for which they are eligible, so long as the units are acquired prior to their being placed in service (i.e., offered for rent as commercial units or used in a trade or business, whichever is earlier).

A lessor of property eligible for the rehabilitation tax credit may elect to treat its lessee as a purchaser of the property.[58] If he does so, the lessee is treated as the actual owner and may take the tax credit. A number of conditions apply to such elections,[59] including that the lessee be the original user of the property. If the lease term is less than 15 years, the lessor may only pass through a fraction of the credit, equal to the ratio between the lease term and the class life of the property (typically 18 years).

Passing the tax credit through to a transferee may also be important when rehabilitation property is transferred to a partnership. This must be done before the property is placed in service if the partnership is to be eligible for the credit, and all partners who wish to share the benefits of the credit must also have joined prior to that time. Similarly, if property is to be transferred from a partnership to other entities or individuals, this must occur before it is placed in service if the transferees are to be eligible for the credit.[60] Moreover, the allocation of profits among the partners may not be altered for five years after the property is placed in service, or a disposition of Section 38 property

---

[56] Former Treas. Reg. § 1.191-1(c)(2)(iii); Treas. Reg. § 1.48-11(c)(3)(ii)(B).

[57] See, e.g., Private Letter Rulings 8347047 (August 22, 1983), 8340021 (June 28, 1983), and 8305104 (November 3, 1982).

[58] I.R.C. § 48(d); Treas. Reg. § 1.48-11(d)(1).

[59] Treas. Reg. § 1.48-4.

[60] See Private Letter Ruling 8506036 (November 9, 1984).

will be deemed to have occurred and some or all of the credit taken will be recaptured.[61]

1.2.7 Limit on Tax Offset; Carrybacks and Carryforward

As noted, the rehabilitation tax credit is available in the year the rehabilitated building is placed in service.[62] As a general matter, property is deemed to have been placed in service when it is "placed in a condition or state of readiness and availability for a specifically assigned function, whether in a trade or business, in the production of income, in a tax-exempt activity, or in a personal activity."[63] This is often demonstrated by an objective event such as the receipt of a certificate of occupancy from the appropriate governmental authority. However, the IRS will not necessarily accept such an event as determinative of when a building is placed in service.[64]

The amount of the investment tax credit which a taxpayer may claim in any taxable year may not exceed $25,000 plus 85 percent of the taxpayer's liability for tax in excess of $25,000.[65] The taxpayer may receive the full benefit of the available credit even if the taxes for which the taxpayer would otherwise be liable in the year the property is put in service are less than the amount of the credit for which he is eligible. The excess credit may be carried back to the three taxable years preceding the year in which the unused credit accrued and applied to the taxes paid on those years (starting with the most recent).[66]

If the tax paid during the three prior years does not fully offset the unused credit, the remainder may be carried forward and offset against taxes which would otherwise be due during the 15 years following the year in which the credit became available.[67] The ability to maximize the tax offset power of the credit in this manner may be an important incentive in attracting syndication investors to a rehabilitation project.

---

[61] Treas. Reg. § 1.47-6.

[62] *Id.* § 1.46-3(d)(4)(i).

[63] *Id.* § 1.46-3(d)(1); *see* Treas. Reg. § 1.48-11(d)(2).

[64] *See e.g.,* Private Letter Ruling 8415003 (December 15, 1983), in which the IRS ruled that the installation of drywall and water and electrical meters, not the issuance of a temporary certificate of occupancy, determined when an apartment was "in a condition or state of readiness and availability for occupancy." The IRS found that the taxpayer had failed to demonstrate a sufficient connection between the issuance of a certificate and the time an apartment was ready for occupancy.

[65] I.R.C. § 38(c).

[66] I.R.C. § 39.

[67] Note that any recapture of the tax credit will result in an adjustment to the carryback or carryover. *See* I.R.C. § 47(a)(6).

1.2.8 Recapture of Tax Credits

After claiming an investment tax credit, the taxpayer must hold the rehabilitated building for five full years after the property is placed in service in order to avoid being subject to a recapture of the credit.[68] The percentage of the credit which is recaptured is determined by the amount of time which has elapsed since the property was placed in service, and is determined according to the following schedule:

If disposition of property occurs within:	The percentage of the credit recaptured is:
one full year after placed in service	100
one to two full years after placed in service	80
two to three full years after placed in service	60
three to four full years after placed in service	40
four to five full years after placed in service	20

To determine the taxpayer's tax liability for the taxable year in which the disposition is made, one must first determine the amount of the credit which has been taken in all prior taxable years as a result of qualified investment in the property. The amount of the credit which will be recaptured is determined by multiplying the applicable recapture percentage by the total prior credit taken. This recapture amount is added to the taxpayer's current tax liability.

However, current tax liability is adjusted only with regard to credits actually taken. If the taxpayer had excess credits which could not be fully used in the year they became available and which have been carried back or carried over, such carrybacks and carryovers are adjusted.[69]

---

[68] I.R.C. § 47(a)(5).

[69] I.R.C. § 47(a)(5)(D). This section as currently written is effective generally for agreements entered into on or after July 1, 1982, or as to property placed in service after that date.

Any portion of the credit which has not yet been taken will not affect current tax liability on disposition of the property.

Most dispositions and changes in the nature of the use of the building result in recapture of the tax credit previously taken for rehabilitation. This includes sale, like-kind exchange, and conversion of commercial property to rental or personal use.[70] It also includes gifts of rehabilitated property. Transactions which do not trigger recapture include transfer by reason of death, transfer between spouses incident to divorce, and certain sale-and-leaseback arrangements.[71] While recapture generally applies to casualty losses, replacement property will usually be treated as eligible for the credit.

1.3 DEPRECIATION

Renovation expenditures will be treated as qualified rehabilitation expenditures (and thus be counted when determining whether the substantial rehabilitation test has been met and the ITC will be available) only if an election has been made under Section 168(b)(3) to use the straight-line method of depreciation.[72] Thus, to obtain an investment tax credit for rehabilitation expenditures, the taxpayer must forego the use of the accelerated forms of depreciation (either the 175-percent or, in the case of low-income housing, the 200-percent declining balance methods) which would otherwise be available. Pursuant to the Tax Reform Act of 1984, the taxpayer may elect to apply the straight-line method over a recovery period of 18, 35, or 45 years.[73]

1.3.1 Adjustment to Basis

The amount of the depreciation deduction which a taxpayer may take to reflect the decreasing useful life of a building is based on the taxpayer's adjusted basis in the property. The higher the adjusted basis, the larger the amount of the deduction which may be taken. When a taxpayer receives an investment tax credit for a qualified rehabilitation, the taxpayer's basis in the rehabilitated property must be adjusted. In the case of buildings which qualify for the 15- or 20-percent credit, 100 percent of the credit must be subtracted from basis.[74] For a certified historic structure, basis is reduced by 50 percent of the amount of the

[70] Treas. Reg. § 1.47-2.
[71] *Id.* § 1.47-3.
[72] I.R.C. § 48(g)(2)(B)(i).
[73] I.R.C. § 168(b)(3).
[74] I.R.C. § 48(q)(3).

credit.[75] Thus, the rehabilitation of a certified historic structure may receive the benefit not only of an investment tax credit which is 5 to 10 percent greater than that allowed for other rehabilitation projects, but also of a smaller adjustment to basis, which results in higher depreciation deductions. Moreover, depreciation deductions offset ordinary income, while reductions in basis will result in larger capital gains upon disposition of the property.

Prior to the Tax Equity and Fiscal Responsibility Act of 1982 (TEFRA), no adjustment to basis was required with regard to certified historic structures. Pursuant to the transitional rules provided, certain projects may still be exempt from the requirement to adjust basis.[76]

1.3.2 Accelerated Cost Recovery System

In order to spur productive investment in personal and real property, ERTA established the Accelerated Cost Recovery System (ACRS). The purposes of ACRS were to allow faster write-off of investments and to establish mandatory "audit-proof" depreciation classes for various types of property as a tax simplification device. As originally enacted, ACRS provided a class life of 15 years for most depreciable real property. The Tax Reform Act of 1984 increased the period to 18 years for all categories of property but low-income housing.[77] The 1984 changes apply to property placed in service after March 15, 1984.

Under ACRS, the taxpayer still has the choice of several depreciation methods. Property may be depreciated using the straight-line method over the recovery period of 18 years, or alternatively, over periods of 35 or 45 years.[78] Alternatively, a property may be depreciated by means of the 175-percent declining balance method or, in the case of low-income housing, the 200-percent declining balance method, and the taxpayer may switch to the straight-line method at the time it becomes advantageous to do so.[79] The deduction available in the year the property

---

[75] I.R.C. § 48(q)(1).

[76] A taxpayer is not required to subtract any of the ITC from basis if—

    a. a contract for rehabilitating the building was entered into after August 13, 1981 and was binding on July 1, 1981 and at all times thereafter, and the building is placed in service prior to January 1, 1986; or

    b. rehabilitation work was begun between December 31, 1980 and July 1, 1982 *and* the building is placed in service prior to January 1, 1986; or

    c. a Securities and Exchange Commission filing and a Department of Housing and Urban Development Section 8 application filing were made before July 1, 1982 and the building was placed in service prior to July 1, 1984. P.L. 97-248, § 205(c)(1).

[77] I.R.C. § 168(b)(3), as amended by P.L. 98-369 § 111.

[78] *Id.*

[79] I.R.C. § 168(b)(2)(A).

is placed in service depends upon the number of months in that year that the property was in service; the so-called half-year convention does not apply.[80]

Component depreciation of real property is not permitted under ACRS, but a substantial improvement may be treated as a separate building for depreciation purposes.[81] Substantial improvements are those made at least three years after the building was first placed in service, and which are additions to the capital account of a building which exceed 25 percent of the adjusted basis of the building within any 24-month period. Once the 25-percent figure is attained, a "rehab-basis account" is maintained with respect to the improvement, on which depreciation deductions may be calculated.

ACRS also prescribes a set of anti-churning rules to prevent taxpayers from converting pre-1981 property into ACRS-eligible property for which more favorable treatment may be available.[82] Generally, ACRS treatment will not apply to property if—

1. It was owned by the taxpayer or a "related person" at any time during 1980;
2. The taxpayer leases the property to a person or person related to a person who owned the property at any time during 1980; or
3. It was acquired through certain nontaxable exchanges or property, to the extent that the basis includes an amount representing the adjusted basis of other property owned by the taxpayer or a related person during 1980.

For purposes of the foregoing tests, categories of related persons are as defined in code Sections 52(a) and (b), 267(b) and 707(b)(1), except that common control requirements are reduced from 50 percent to 10 percent.[83]

1.3.3 Recapture under ACRS

Depreciation taken on real property may be recaptured upon the sale, exchange, or other disposition of that property. If ACRS property is depreciated on a straight-line basis, whether over 18, 35, or 45 years, there is no recapture.[84] If other methods are used, however, depreciation

---

[80] I.R.C. § 168(b)(2). For a determination of when a building is "placed in service" for depreciation purposes, see Rev. Rul. 76-238, 1976-1 C.B. 55.

[81] I.R.C. § 168(f)(1)(C); see also Prop. Treas. Reg. § 1.168-2(e).

[82] I.R.C. § 168(e)(4)(B).

[83] I.R.C. § 168(e)(4)(D).

[84] I.R.C. § 1245(a)(5)(C).

is generally fully recaptured, to the extent of gain on the disposition.[85] The two exceptions to this rule related to residential rental property and low-income housing.[86] Recapture on the former is limited to the excess of accelerated over straight-line depreciation. Recapture on the latter is similarly limited, and is further limited by the continued application of phase-out provisions which reduce the amount recaptured over time.[87]

The Tax Reform Act of 1984 provides that all recapture must be taken into account in the year of disposition even if the sale is on an installment basis.[88] This may create serious cash flow problems, since the tax on the recaptured credit may exceed the proceeds received from the disposition prior to the time that the tax is due. On the other hand, certain types of dispositions do not trigger recapture.[89] These include gifts, death transfers, certain tax-free transactions, like-kind exchanges, and involuntary conversions.

1.4   PRESERVATION EASEMENTS

Under the terms of the Tax Treatment Extension Act of 1980, certain donations of easements affecting historic buildings or districts are eligible for charitable contribution deductions under Section 170 of the Internal Revenue Code. The most common application of this provision involves gifts of facade easements, which restrict future alteration or demolition of building facades and thereby preserve an important part of the character of a historic structure. Alone or in tandem with the rehabilitation tax credit, facade easements can contribute significantly to making restoration of historic buildings financially attractive. Investors should be aware, however, that the IRS has become increasingly aggressive in challenging deductions relating to these donations, and that strict compliance with the substantiation requirements discussed next may be essential to a determination of validity. In addition, all of the requirements and, where possible, the precise language of the proposed IRS regulations should be incorporated in the drafting of easement donation agreements.

The IRS first ruled on the deductibility of open space easements in gross in 1964;[90] subsequent regulations provided that easement gifts

---

[85] I.R.C. § 1245(a).

[86] I.R.C. § 1245(a)(5)(A) and (D).

[87] I.R.C. 1250(a)(1)(B).

[88] I.R.C. § 453(i), as amended by Tax Reform Act of 1984 § 112.

[89] I.R.C. § 1245(b).

[90] *See* Rev. Rul. 64-205, 1964-2 C.B. 62.

were not subject to limitations otherwise applicable to donations of partial interests in property.[91] The Tax Reform Act of 1976 codified a deduction for gifts of a "lease on, option to purchase or easement with respect to real property of not less than 30 years' duration" given to a conservation organization.[92] All of these earlier attempts to recognize, for tax purposes, the legitimate interests motivating preservation easements were superseded in the 1980 legislation.

The 1980 Act created a federal income, gift, and estate tax deduction for a "qualified conservation contribution."[93] As defined in Section 170(h) of the Code, the term qualified conservation contribution means a gift (a) of a "qualified real property interest," (b) to a "qualified organization," (c) to be used exclusively for conservation purposes. Each of these conditions, discussed in greater detail in this chapter, must be met to qualify for the deduction, and the Treasury has proposed regulations providing additional elaboration on the statutory requirements.[94] In principle, however, it is clear that appropriate facade easements do fall within the terms of the statute's definition, and hence do qualify for a charitable deduction.

1.4.1   Qualified Real Property Interest

To receive a tax deduction, the easement which is donated must be "a restriction (granted in perpetuity) on the use which may be made of the real property."[95] Moreover, the proposed regulations make clear that the easement must be enforceable.[96] This entails the rights of a donee to inspect the property to determine whether the terms of conveyance have been complied with, to enforce the easement restrictions by resort to appropriate legal proceedings, and to compel restoration of the property to its condition prior to the donation. All of these powers should be affirmatively stated in the document creating the easement. In addition, it may be advisable for the donor to get a legal opinion that the easement will be enforceable in perpetuity under state and local law.

Compliance with the perpetuity requirement may be somewhat problematic. State law often provides procedures for relaxing or terminating property restrictions which impair the conveyance of marketable title or are otherwise determined by courts to constrain unduly the use of

---

[91] Treas. Reg. § 1.170A-7(b)(1)(ii).

[92] P.L. 94-455, § 2124(e) (1976).

[93] I.R.C. § 170(f)(3)(B)(iii).

[94] *See* 48 Fed. Reg. 22940 (May 23, 1983).

[95] I.R.C. § 170(h)(2)(C).

[96] Prop. Treas. Reg. § 1.170A-13(g)(4)(ii).

property.[97] Many states bar the original grantee from assigning an easement.[98] Some states have recording statutes requiring that restrictions
on the use of property be re-recorded at periodic intervals.[99] As a legal
matter, any of these makes it impossible to guarantee that the easement
will be perpetual. Some, but not all, of these assignment-restricting and
recording statutes contain exceptions for easements granted for conservation purposes.[100] Absent such exception, however, there is an un-
resolved conflict between the property right which can be created under
state law and the requirements of the federal tax laws.

That conflict is averted under the Uniform Conservation Easement
Act, adopted by the National Conference of Commissioners on Uniform
State Laws in 1981.[101] The Uniform Act removes obstacles to the creation
of perpetually enforceable easements in gross for conservation and historic preservation purposes. However, it limits the categories of easement holders possessing the right to permanent protection to governmental bodies and charitable organizations devoted, *inter alia*, to
conservation purposes. Versions of the Uniform Act have been adopted
in Nevada, Wisconsin, Arkansas, Oregon, and New York, and additional
states can be expected to follow suit.

If the terms of the easement permit future development of land or
alteration to buildings within a registered historic district, the proposed
regulations require that the development or alteration conform to local,
state, or federal standards applicable to construction or rehabilitation.[102]
This would presumably include the Secretary of the Interior's Standards
for Rehabilitation, discussed in chapter 2. Although as currently drafted
the regulation does not apply to National Register properties, it would
be prudent to include the requirement in drafting any easement to
comply with these provisions.

Another requirement of the proposed regulations is that the easement
prohibit subsequent transfer from the "qualified organization" to which
it is originally granted unless two conditions are met. The subsequent
transferee must also be a qualified organization, under the rule ex-

---

[97] For example, New York's "cy pres" statute permits a court to remove restrictions from
dispositions for "religious, charitable, educational or benevolent purposes" if a determination is made that "circumstances have so changed . . .as to render impracticable or
impossible a literal compliance" with the terms of those restrictions. EPTL § 8-1.1(c).

[98] *See* Netherton, "Restrictive Agreements for Historic Preservation," 12 *Urb. Law.* 54
(1980).

[99] *See* Madden, "Tax Incentives for Land Conservation: The Charitable Contribution Deduction for Gifts of Conservation Easements," 2 *B.C. Envtl. Aff. L. Rev.* 105, 107 (1983).

[100] *See e.g.,* Mass. Gen. Laws Ch. 184 § 26(c).

[101] *See* 12 *Uniform Laws Annotated* 57 (Supp. 1984).

[102] Prop. Treas. Reg. § 1.170A-13(d)(5)(i).

plained in the next section, and future transfers must be contingent upon assurances that the conservation purposes for which the gift was made will continue to be carried out.[103]

The easement must also give the donee the right to receive a fair market value for the property interest, equal to a minimum ascertainable share of the fair market value of the entire property.[104] This represents a significant limitation on the donor's remaining rights in the property, including his right to dispose of it. If the restriction is extinguished for any reason, upon subsequent sale or exchange of the property the donee would be entitled to a share of the proceeds equal to the proportionate value of the original restriction. This would be true even if the easement is lifted against the donor's wishes, as, for example, by order of a court upon application of another party. Moreover, any proceeds received by the donee must be used in a manner consistent with the purposes of the original grant.[105]

A final issue deals with public benefits from the conservation interests which the easement is designed to protect. According to the proposed regulations, public access limitations, if necessary to protect those conservation interests, will not in themselves render a donation nondeductible.[106] On the other hand, in the case of a facade or interior easement to protect a building's appearance, there must be substantial and regular opportunity for public viewing of the protected features.[107] Thus, if the subject property is to be partially or totally closed to the public, the conveying instrument should clearly indicate why such restrictions are necessary to protect the historic qualities of the building.

1.4.2 Qualified Organization

To achieve deductible status, the donee of a qualified real property interest must be a qualified organization, of which there are three types. The organization may be a governmental unit, a publicly supported charity as defined in Section 509(a)(2) of the Internal Revenue Code, or an organization controlled by a publicly supported charity as defined by Section 509(a)(3) of the Code. It is beyond the scope of this book to describe the IRS's public support requirements in detail, but it should be noted that, in general, entities such as school and religious organizations will not qualify as recipient organizations.[108] Donors and donees should also secure professional advice on whether a donated easement

---

[103] *Id.* § 1.170A-13(c)(2).

[104] *Id.* § 1.170A-13(g)(5)(ii).

[105] *Id.* § 1.170A-13(c)(2).

[106] *Id.* § 1.170A-13(e)(2).

[107] *Id.* § 1.170A-13(d)(5)(iii)(B).

[108] *See e.g.,* B. Hopkins, *The Law of Tax-Exempt Organizations* (1983).

must be treated as a private gift which may jeopardize the donee's "publicly supported" status.

The proposed regulations require that a donee organization "have the resources to enforce the restriction" and "be able to demonstrate a commitment to protect the conservation purposes of the donation."[109] The regulations suggest that an established group dedicated to conservation purposes would automatically meet both prongs of this test. Presumably, a governmental unit with revenue-raising powers would as well. The regulations do not require that funds actually be set aside to enforce the easement restrictions. As a practical matter, however, it may be useful to establish an endowment for meeting future costs of maintaining and enforcing the easement. Methods for doing so include requiring the donor to contribute a percentage of the value of the easement, or of the unrestricted value of the property; requiring a contribution based upon the salable or rentable floor area of the building; or capitalizing, on an organization-wide basis, the costs of administering an entire easement program and assessing all donors proportionately.[110]

1.4.3 Conservation Purposes

Among the allowable conservation purposes for which an easement may be granted under the statute is "the preservation of an historically important land area or a Certified Historic Structure."[111] The definition of a "certified historic structure" for this purpose is substantially similar to that applicable to the rehabilitation tax credit, but with two significant differences. First, the building must actually be listed on the National Register or certified as significant by the due date (including extensions) of the taxpayer's return for the year of transfer.[112] Second, both depreciable and nondepreciable structures qualify, thus making private residences eligible for easement donations.[113]

Lack of certification, however, does not necessarily exclude a building within a historic district from eligibility for protection through an easement donation. Under the proposed regulation, the term "historically important land area" includes—

1. An independently significant land area that "substantially" meets National Register criteria (such as an archaeological site or Civil War battlefield);

---

[109] Prop. Treas. Reg. § 1.170A-13(c)(i).

[110] Coughlin, "Historic Preservation Easements," reprinted in *Historic Preservation Law: Tax Planning for Old and Historic Buildings* at 344–45 (American Law Institute, 1984).

[111] I.R.C. § 170(h)(4)(A)(iv).

[112] I.R.C. § 170(h)(4)(B)

[113] Prop. Treas. Reg. § 1.170A-13(d)(5)(iii).

2. Any building or land area within a registered historic district (except buildings that cannot reasonably be considered as contributing to the significance of the district); and

3. A land area adjacent to a Register property outside a registered historic district "in a case where the physical or environmental features of the land area contribute to the historic or cultural integrity of the structure."[114]

Thus, under (2), the protection of an uncertified but significant building within a registered district is an acceptable "conservation purpose."[115] It is not entirely clear how the "significance" requirement for this purpose relates to the certification of significance for purposes of the rehabilitation tax credit, but it would certainly be advisable to obtain such a certification where possible, even if the proposed regulation appears not to require it.

A gift to achieve a statutorily enumerated conservation purpose will not be permitted if it would also result in or allow the destruction of other significant conservation purposes.[116] The proposed regulations illustrate this provision with the example of preservation of farmland for purposes of flood prevention and control, which would not be eligible for a deduction if the use of pesticides in the operation of the farm could damage a significant natural ecosystem. Uses which are destructive of conservation interests are permissible only if necessary for the protection of conservation interests which the contribution is intended to further.[117] Thus, the regulations indicate, a deduction for an easement to preserve an archaeological site would not be disallowed because site excavation might impair a scenic view. How the IRS will balance competing considerations of this sort in practice remains to be determined.

1.4.4 Valuation of Easement Donations

Except in the rare instance in which a market place value for the easement can be established, its value is to be determined by taking the difference between the fair market value of the land before the donation and its fair market value afterward.[118] This is consistent with prior revenue rulings by the IRS,[119] and with judicial decisions on the issue

---

[114] *Id.* § 1.170A-13(d)(5)(ii).

[115] *See* 3 *Preservation L. Rept.* 1025 (1984), with respect to Private Letter Ruling 8410034, upholding the deductibility of an easement to protect a noncertified structure.

[116] Prop. Treas. Reg. § 1.170-13(e)(3).

[117] *Id.* § 1.170A-13(e)(4).

[118] *Id.* § 1.170A-13(h)(3)(i).

[119] Rev. Ruls. 73-339, 76-376.

of valuation of easements.[120] In making this determination, the taxpayer must consider not only current uses of the property but also future uses permitted by the conservation restriction, adjusted for the likelihood that those future uses will actually be undertaken.

When the easement affects only a portion of a contiguous parcel of land owned by one taxpayer or members of the taxpayer's family, the amount of the deduction must be determined by calculating the difference between the fair market value, before and after the gift, of the entire tract, not merely the piece burdened by the easements.[121] Thus, if a benefit will accrue to one part of the land as a result of the conveyance, that must be offset against the diminution in value of another part. If the net result of the transfer is a financial benefit to the landowner, no deduction is allowed. Easement donations are frequently challenged by the IRS on this ground.

Both donors and donees of easements must comply with the new procedural requirements governing charitable contributions of appreciated property under Section 155 of the Tax Reform Act of 1984.[122] If the claimed value of the easement exceeds $5,000, the taxpayer must have an appraisal done by a "qualified appraiser" who is not the taxpayer, a party to the transaction, or a person related to either. The taxpayer must attach an appraisal summary to his tax return, providing information which the IRS is to specify by regulation. Moreover, if the donee charity sells, exchanges, or disposes of the easement within two years of receipt, it must file an information report with the IRS, identifying the donor, describing the property, and giving the date of contribution, date of disposition, and the amount received upon disposition.

There is considerable indication that the IRS is aggressively auditing returns claiming deductions for easement donations, and challenging allegedly inflated appraisals.[123] To minimize conflict, it is important that easement values be accurately appraised. To assist taxpayers in this

---

[120] *See e.g., Robert H. Thayer & Virginia Thayer* v. *Comm'r.,* 1977 T.C. Memo 370.

[121] Prop. Treas. Reg. § 1.170A-13(h)(3)(i).

[122] The new rules are codified at I.R.C. §§ 6050L, 6652, 6659, 6660 and 6678.

[123] *See e.g., William B. Akers* v. *Comm'r.,* 1984 T.C. Memo 490, in which the IRS disallowed an easement deduction entirely, and the Tax Court reduced the value of the donated easement to $114,000 from the claimed value of $789,000. *See also* Penico, "Valuation of Preservation Easements," 3 *Preservation L. Rept.* 2069 (1984). In public information release I.R. 84-125 (December 10, 1984), however, the Commissioner of Internal Revenue stated that the IRS would review donated easements on a case-by-case basis, and that deductions would be allowed if a taxpayer's appraiser used the before-and-after method and applied sound appraisal practices in calculating the value of a donation.

A useful source of current information on easement donations and valuation is the *Conservation Tax Program,* a series of publications of the Land Trust Exchange, which may be contacted at P.O. Box 364, 2 Mt. Desert Street, Bar Harbor, ME 04609 (207-288-9751).

respect, the National Trust for Historic Preservation has published a pamphlet entitled *Appraising Easements: Guidelines for Valuation of Historic Preservation and Land Conservation Easements,* which prospective donors may wish to consult.

1.4.5. Interaction of Rehabilitation Tax Credit and Preservation Easement Deduction

To maximize tax benefits for rehabilitation of old buildings, owners may wish to take advantage of both rehabilitation tax credits and deductions for donation of facade easements. Timing may, however, be of central importance. The proposed regulations require that the basis of property be reduced in the case of an easement donation, and in an amount proportionate to the amount of predonation basis allocable to the easement.[124] If an easement on a certified historic structure is conveyed after rehabilitation has been done and after the building has been placed in service, the reduction in basis attributable to the rehabilitation may be treated as a disposition of Section 38 property, which would trigger a recapture of part of the investment tax credit.[125]

Timing of the gift may have other implications as well. If the donation is made while the property remains unrehabilitated, it may reduce adjusted basis for purposes of meeting the substantial rehabilitation test of eligibility for the investment credit.[126] Moreover, the timing of the gift may determine the allocation of basis reduction between the land and building basis account and the rehabilitation basis account, which in turn may affect the amount of the investment credit for which the taxpayer is eligible.[127] These are only a few of the many complex issues that may arise in attempting to coordinate use of these two tax provisions.

1.5   SUMMARY OF DIFFERENCES IN TREATMENT OF HISTORIC AND NONHISTORIC OLD BUILDINGS

To summarize the foregoing discussion, it may be useful to note the major differences in tax treatment between rehabilitations of historic and nonhistoric structures.

1.   The rehabilitation of a certified historic structure (or a structure located in a historic district which is certified as contributing to the

---

[124] Prop. Treas. Reg. § 1.170A-13(h)(3)(iii).

[125] Treas. Reg. § 1.47-1(a)(1).

[126] *See* Section 1.2.4.

[127] For general discussion of these issues, *see* Luxemburger, "Preservation Easements Rehabilitated Buildings: Preservation Easements and Mortgage Property," 10 *Real Estate Tax* 273 (1983).

significance of that district) is, if the rehabilitation itself is certified, eligible for a 25-percent rehabilitation tax credit. On the other hand, rehabilitation of a 30-or-more-year-old building (which has received a certification of nonsignificance if it is located within a historic district) is eligible for a 15-percent tax credit, and a 40-or-more-year-old building under the same circumstances is eligible for a 20-percent credit.

2. To qualify for a tax credit, a rehabilitation of a certified historic structure, or of any structure within a historic district which has not been certified as not contributing to the significance of that district, must in turn be certified as meeting the Secretary of the Interior's Standards for Rehabilitation. On the other hand, rehabilitation of a noncertified structure, or of a structure within a historic district which has been certified as nonsignificant, need not be a certified rehabilitation, but must meet the other requirements for the tax credit. Thus, although the potential tax benefit for a certified rehabilitation is greater, such a rehabilitation must overcome an additional, and in some instances perhaps costly, statutory hurdle.

3. The rehabilitation tax credit applies to a certified rehabilitation of an appropriately certified historic structure regardless of age, while a noncertified rehabilitated building must be at least 30 years old to qualify.

4. Residential rental property is eligible for the rehabilitation tax credit only if it is a certified historic structure.

5. When a rehabilitation tax credit is taken for a noncertified rehabilitation of a noncertified structure (or, if the structure is located in a historic district, a structure which has been certified as noncontributing), the basis of the building must be reduced on a dollar-for-dollar basis. On the other hand, when the credit is taken for a certified rehabilitation, the basis of the building is reduced by only 50 percent of the amount of the credit.

6. To qualify for a charitable contribution deduction under the proposed regulations, an easement generally must be donated to protect a certified historic structure or a building within a historic district, unless the building cannot reasonably be considered as contributing to the significance of the district.

1.6   TAX TREATMENT OF PROPERTY LEASED TO THE TAX-EXEMPT ORGANIZATIONS

Prior to the Tax Reform Act of 1984, property used but not owned by a tax-exempt organization was generally eligible for the rehabilitation tax credit on the same basis as other property.[128] In addition, taxpayers could use ACRS to depreciate property used but not owned by tax-

---

[128] I.R.C. § 48(a)(4).

exempt organizations. The combined effect of these two provisions was to encourage a large volume of transactions in which tax-exempts would sell property to taxable entities and then lease it back, thereby reaping the benefits of investment credits and depreciation deductions which would otherwise have been unavailable.

The significant revenue losses occasioned by these transactions led Congress to tighten up the tax treatment of property used by tax-exempt organizations. A new set of rules, described by one commentator as "grotesquely intricate,"[129] were adopted in the Tax Reform Act of 1984.[130] Under the new provisions, the involvement of a tax-exempt organization in a historic rehabilitation may bar or at least limit the eligibility of the project for federal tax incentives. There may still be substantial advantages from the involvement of tax-exempt entities in historic preservation work (such as the availability of tax-exempt financing[131] or charitable contributions[132] to support the work), but both the benefits and detriments of tax-exempt participation should be considered carefully in deciding how best to structure a preservation or rehabilitation.

1.6.1 Tax-Exempt Property

The new statutory provisions apply to a category of property known as tax-exempt use property. Real property becomes tax-exempt use property to the extent that it is committed to "disqualified" uses. A lease of property to a tax-exempt entity is a disqualified use if:

1. The tax-exempt entity participates in financing the property through tax-exempt obligations, whether before or after financing is committed;
2. The lease governing the tax-exempt entity's use contains a purchase or sale option or the equivalent;
3. The lease has a term exceeding twenty years; or
4. The lease is entered into after a sale, lease, or other transfer of the property from the tax-exempt entity, and the property had been used by the entity prior to the transfer.[133]

A service contract arrangement governing use of a particular property may be treated as a "disqualified lease" for this purpose, depending upon the degree to which it possess the characteristics of a lease. The

---

[129] J. Eustice, *The Tax Reform Act of 1984: A Selective Analysis* at 5-40 (1984).

[130] P.L. 98-369, § 31.

[131] *See* Section 4.2.1.

[132] *See* Section 4.3.8.

[133] I.R.C. § 168(j)(3)(B).

determination will be made by evaluating a number of factors, including whether the tax-exempt entity has physical possession, control, or a significant possessory or economic interest in the property; whether the service provider bears any economic risks in the event of nonperformance; whether the service provider concurrently uses the property for the benefit of parties unrelated to the tax-exempt entity; and whether the total contract price substantially exceeds the rental value of the property.[134]

Notwithstanding these provisions, depreciable real property will not generally be treated as tax-exempt use property unless the tax-exempt entity uses more than 35 percent of the property in a disqualified manner.[135] Moreover, property will not be deemed tax-exempt use property if it is predominantly used by a tax-exempt entity in an unrelated trade or business which produces income taxable as unrelated business income.[136] Nor will it be treated as tax-exempt use property if the use is subject to a short-term lease, generally three years or less.[137]

The new amendments expand the definition of "tax-exempt entity" in one potentially significant respect. For purposes of the limitations on depreciation and investment credits, tax-exempt entities are deemed to include "foreign persons or entities."[138] Not in all cases, however: Foreign investors are not treated as tax-exempt entities with respect to a property if 50 percent or more of the income derived by the foreign investor from the use of the property is subject to United States income tax.

1.6.2 Consequences of Tax-Exempt Participation

Several significant results flow from a determination that any portion of a rehabilitated or to-be-rehabilitated building represents tax-exempt use property. First, the rehabilitation credit will not apply to expenditures attributable to the rehabilitation of any part of a building which is tax-exempt use property if any portion of either the rehabilitation expenditures themselves or the cost of acquiring the building (or any interest in it) is financed by tax-exempt obligations. This restriction is, however, subject to the 35 percent threshold. If a rehabilitated building becomes tax-exempt use property within 5 years after the rehabilitation occurs, this will be treated as a disposition and the credit will be recaptured.

[134] I.R.C. § 7701(e)(1).
[135] I.R.C. § 168(j)(3)(B)(iii).
[136] I.R.C. § 168(j)(3)(D).
[137] I.R.C. § 168(j)(3)(C).
[138] I.R.C. § 168(j)(4)(iii).

Second, the straight-line method, disregarding salvage value, must be used to calculate depreciation or amortization deductions for tax-exempt use property.[139] A 40-year recovery period must be used with respect to most depreciable real property (so-called 18-year property).[140] In any event, the recovery period cannot be less than 125 percent of the term of the lease, a limitation which will be significant only in the case of extremely long-term leases. First-year deductions are computed based upon the number of months of the taxable year in which the property is in service.

Finally, the rules contain restrictions to prevent the use of partnerships or other pass-through entities to circumvent their application.[141] If property which is otherwise not tax-exempt use property is owned by a partnership in which a tax-exempt entity is a member, the tax-exempt's proportionate share of the property will be treated as tax-exempt use property, unless the allocation of partnership items (income, deduction, gain, loss credit, and basis) is a "qualified allocation." An allocation is qualified if (1) the tax-exempt entity receives the same distributive share of all partnership items, (2) that share remains the same throughout the tax-exempt entity's participation in the partnership, and (3) the allocation has a substantial economic effect.[142]

1.7 EFFECTIVENESS OF REHABILITATION INCENTIVES

Evidence indicates that the federal tax incentives for rehabilitation have successfully increased private investment in historic preservation. The National Park Service reported in 1983 that since the passage of the Tax Reform Act of 1976, more than $3.95 billion had been privately invested in over 6,935 rehabilitation projects.[143] A 64-percent increase in rehabilitation activity had occurred in the first three quarters of fiscal year 1983 over the same period a year earlier.

When a property owner is given final certification for preservation tax benefits, the owner is asked to fill out a data sheet indicating the particular tax treatment chosen and whether the owner would have undertaken the rehabilitation project if the preservation tax incentives had not been available. Of those who responded after the passage of ERTA in 1981, 78 percent indicated that they were able to take advantage

---

[139] I.R.C. § 168(j)(1).

[140] I.R.C. § 168(j)(1)(B)(III).

[141] I.R.C. § 168(j)(8).

[142] See Section 3.2.2.

[143] National Park Service Information Update, July 12, 1983.

of the 25-percent rehabilitation credit. Another 14 percent had been able to use the 5-year amortization option which had been available under the Tax Reform Act of 1976. Eight percent had donated a conservation easement on their property. Of the same sample, 63 percent said they would not have undertaken the preservation project had tax incentives not been available; only 17 percent indicated they would have gone ahead with the project in any event.[144]

[144] National Park Service, "Tax Incentives for Rehabilitating Historic Structures: Facts and Figures 1977–1983" (January 1984).

Two

Historic
Preservation
Certifications

Introduction

Eligibility for the federal tax incentives discussed in chapter 1 requires compliance with the certification procedures established by the National Park Service (NPS). While the Internal Revenue Code vests the Secretary of the Interior with responsibility for the certification program, the Secretary has delegated virtually all operating authority to the director of the NPS. Within the NPS, the associate director for cultural resources has responsibility for administration of the tax incentive program, but the Technical Preservation Services Division has day-to-day responsibility for the program. In addition, state historic preservation officers (SHPOs) play an important role in reviewing certification applications and in making recommendations to the NPS.

The certification process involves two steps. First, to be eligible for the 25-percent ITC, a building must qualify as a "certified historic structure." Any building individually listed on the National Register of Historic Places comes within this definition, as does any building which is certified as contributing to the historical significance of a registered historic district.[1] To be eligible for the 15- or 20-percent ITCs, a building

[1] 36 C.F.R. § 67.2.

within a registered historic district must be certified as not contributing to the significance of that district. Second, the rehabilitation project undertaken on the certified historic structure must itself be certified as meeting NPS requirements. This chapter will first describe the process of listing on the National Register and then discuss the requirements for certification of both historic structures and rehabilitation activities.

2.1   THE NATIONAL REGISTER OF HISTORIC PLACES

The National Register of Historic Places is the official list of this nation's historic and cultural resources. The National Historic Preservation Act,[2] discussed in greater detail in chapter 6, authorizes the Secretary of the Interior to expand and maintain the National Register of buildings, districts, sites, structures, and objects which because of their significance in history, architecture, archaeology, engineering, and culture are worthy of preservation.[3] The property may be significant on a national, state, or local level.[4]

Administered by the Keeper of the National Register within the NPS, the Register is intended to be a comprehensive national inventory of historic properties. It is published in its entirety in the Federal Register every February. Updates are usually listed on the first Tuesday of each month.[5]

A national inventory serves several purposes. The most important result of listing is that any action by a federal agency that affects listed or eligible property is subject to the review and comment procedures of Section 106 of NHPA, discussed more fully in this chapter. The Reg-

[2] 16 U.S.C. § 470 et seq.

[3] 16 U.S.C. § 470a(a)(1)(A). Original authorization to survey "historic and archeological sites, buildings, and objects for the purpose of determining which possess exceptional value as commemorating or illustrating the history of the United States" was granted under the Historic Sites Act of 1935, 16 U.S.C. § 462(b).

[4] 36 C.F.R. § 800.2(d). An example of a site that the Secretary of the Interior found to be eligible for the National Register without being of national significance is Hawaii's Moanatua Valley. The Advisory Council on Historic Preservation found that:

the historical and cultural significance of the [Moanatua] [V]alley stems from Hawaiian folklore and tradition and continues into the 20th century. The valley contains Kamanui, the valley of the great power, and Waolani, the valley of the spirits which was, in tradition, "the dwelling place of the gods." The forest of the valley retains a traditional natural state associated with the legend and history of the area.

The valley was the property of the royal house of Oahu, the scene of battles and other exploits which are extolled in the ancient Hawaiian chants, the Kahikilaulani. . . .

Stop H-3 Ass'n v. Coleman, 533 F.2d 434, 436 n.1 (9th Cir.), cert. denied, 429 U.S. 999 (1976). Other examples from throughout the country can be located by consulting the Federal Register.

[5] 36 C.F.R. § 800.2(d).

ister also functions as a central reference tool for anyone contemplating action that may affect a building suspected of having significance. It can be consulted in designation proceedings to suggest, by comparison, whether other properties should be considered for protection. Finally, inclusion in the Register makes a property eligible for financial assistance in the form of tax benefits, grants and loans.

2.1.1 Eligibility

Several broad categories of property are eligible for inclusion in the Register: "Districts, sites, buildings, structures, and objects significant in American history, architecture, archaeology, engineering and culture"[6] may be considered. These terms are fully defined and examples of each are provided in the regulations:[7]

(a) *Building.* A building is a structure created to shelter any form of human activity, such as a house, barn, church, hotel, or similar structure. Building may refer to a historically related complex such as a courthouse or jail or a house or a barn.

· · · · ·

(d) *District.* A district is a geographically definable area, urban or rural, possessing a significant concentration, linkage, or continuity of sites, buildings, structures, or objects united by past events or aesthetically by plan or physical development. A district may also comprise individual elements separated geographically but linked by association or history.[8]

· · · · ·

(j) *Object.* An object is a material thing of functional, aesthetic, cultural, historic, or scientific value that may be, by nature or design, movable yet related to a specific setting or environment.

· · · · ·

(l) *Site.* A site is the location of a significant event, a prehistoric or historic occupation or activity, or a building or structure, whether standing, ruined, or vanished, where the location itself maintains historical or archaeological value regardless of the value of any existing structure.

· · · · ·

(p) *Structure.* A structure is a work made up of interdependent and interrelated parts in a definite pattern of organization. Constructed by man, it is often an engineering project large in scale.[9]

---

[6] 16 U.S.C. § 470a(a)(1)(A).

[7] 36 C.F.R. § 60.3.

[8] Districts composed of geographically separated elements are known as thematic districts.

[9] It should be noted that a "structure," for National Register purposes, is not synonymous with a "structure" for rehabilitation tax credit purposes and that a given structure could qualify for the former but not the latter. A historic ship would be an example of such a structure.

Clearly, although one tends to think of landmarks in terms of old buildings and battlefields, for National Register purposes this is too limited a view. Almost any type of physical property is potentially eligible for inclusion in the National Register.

The regulations, however, do identify several types of property that are to be less favored when nominated for inclusion in the National Register. Cemeteries and individual graves, or birthplaces of historical figures; religious, commemorative, relocated, or reconstructed properties; and properties whose significance can be traced to the past 50 years are not, as a general rule, eligible for listing.[10]

2.1.2 The Standard of Evaluation

In evaluating the "significance" of nominated property, several criteria are applied:

The quality of significance in American history, architecture, archaeology, engineering, and culture is present in districts, sites, buildings, structures, and objects that possess integrity of location, design, setting, materials, workmanship, feeling, and association, and

(a) that are associated with events that have made a significant contribution to the broad patterns of our history; or

(b) that are associated with the lives of persons significant in our past; or

(c) that embody the distinctive characteristics of type, period, or method of construction, or that represent the work of a master, or that possess high artistic values, or that represent a significant and distinguishable entity whose components may lack individual distinction; or

(d) that have yielded, or may be likely to yield, information important in prehistory or history.[11]

Given the number and type of criteria to be applied, the determination of a property's significance necessarily contains an element of subjectivity. For purposes of the Register, properties of national, state, and local significance are treated equally.

---

[10] 36 C.F.R. § 60.4. However, broad categories of exceptions provide numerous havens within the general rule to protect specific properties. Sherfy and Luce, *How to Evaluate and Nominate Potential National Register Properties that Have Achieved Significance within the Last 50 Years* (Washington, D.C.: Heritage Conservation and Recreation Service, U.S. Department of Interior, 1979).

[11] 36 C.F.R. § 60.4.

2.1.3　The Nomination Process

The nomination process begins by submitting a standard nomination form[12] to the SHPO.[13] The nomination form may be completed by one of a number of interested parties including the property owner, the municipality within which the property is located,[14] or a preservation group. Although most nominations originate at the local level, the SHPO may also initiate the process.[15]

The SHPO must respond to the applicant submitting the nomination form within 60 days, providing a technical opinion indicating whether the form adequately documents the property and whether the property appears to meet the National Register criteria.[16] It is the applicant's responsibility to provide additional documentation, if necessary. If documentation is adequate, the SHPO must schedule the property for consideration at the earliest possible meeting of the State Review Board, and must give the applicant at least 75 days' notice of that meeting.

The state review board conducts the initial review of a nomination.[17] State review boards meet at least three times a year to evaluate nominations. The property owner, community officials, and the general public are notified[18] of the review process and their input regarding the nomination is solicited as part of this evaluation.

The review board will determine whether the property meets National Register criteria based on its consideration of the nomination form and supporting documents, comments presented to it, and possibly an on-site investigation by the review board or its staff. If the review board determines that the property meets the designation criteria, that rec-

---

[12] The nomination form is available from the Department of Interior, participating federal agencies, and state historic preservation offices. 36 C.F.R. § 60.5. Completed nomination forms must be "adequately documented" and "technically and professionally correct and sufficient." To meet these requirements the forms and accompanying maps and photographs must be completed in accord with requirements and guidance in the NPS publication, *How to Complete National Register Forms* and other NPS technical publications on this subject. 36 C.F.R. § 60.3(i).

[13] The State Historic Preservation Officer is the person in each state selected to administer the state's historic preservation program including identifying, nominating, and administering applications for listing historic properties in the National Register. 36 C.F.R. § 60.3(m); *see generally* 36 C.F.R. Part 61.

[14] To nominate property the local government must be certified by the Secretary of Interior and by the SHPO. 16 U.S.C. § 470a(c); 36 C.F.R. § 61.5(d)(1).

[15] 36 C.F.R. § 60.6(a).

[16] 36 C.F.R. § 60.11(a).

[17] A review board must consist of no fewer than five members, a majority of whom must be recognized professionals in the fields of history, prehistoric and historic archaeology, architectural history, architecture, and historic architecture. Members are designated by the SHPO unless state law provides for membership by other means. 36 C.F.R § 60.3(o).

[18] The review process is outlined in 36 C.F.R. § 60.6(b),(c),(d).

ommendation is forwarded to the SHPO for review. If the board does not so find, then the nomination is disapproved and is not sent to the SHPO for any further deliberation.

The review board's favorable decision to list the property on the Register, if concurred in by the SHPO, is forwarded to the Keeper of the National Register. Even if the SHPO does not agree, the nomination may reach the federal level by several avenues. The SHPO may still submit the nomination with relevant material, the opinion of the state review board, and his dissenting opinion. If he chooses not to do so, he must advise the applicant of that fact within 45 days. Nominations disputed by the SHPO and the state review board are substantially reviewed by the Keeper of the Register. Or, the property owner or the local government may appeal the SHPO's action to the Secretary of the Interior.[19]

A federal agency may nominate a property to the National Register without going through the state nominating process. By law, each federal agency must establish a method to identify, inventory, and nominate property which it owns or controls that appears to be eligible for National Register listing.[20] These nominations are made to the Secretary of the Interior rather than to the SHPO, but the SHPO and the affected community must be allowed to comment. Each federal agency must also appoint a "preservation officer" to implement the agency's program.

Applications that have made their way through the state or federal agency nominating process and have thus far been approved are forwarded to the Keeper of the National Register, where the nominated property is once again reviewed for compliance with the National Register criteria. Notice that property has been nominated for National Register listing is provided in the Federal Register[21] Any person or organization that supports or opposes a nomination may petition the Keeper to accept or reject the nomination.[22] If conformance is found, the nomination is approved and the property designated. If there is insufficient information upon which to base a decision, the application is sent back to the state level for supplementation.

Unless there is a dispute between the SHPO and the state review board, or unless the Keeper of the Register receives a petition from a person or organization objecting to the state's approval of the nomination and requesting the Keeper to substantially review the nomination, review by the Keeper will not typically amount to a *de novo* fact finding. More frequently the Keeper only considers the information in the application.

[19] 16 U.S.C. § 470a(a)(5).
[20] 16 U.S.C. § 470h-2(a)(2). *See* Wilson v. Block, 708 F.2d 735, 754 (D.C. Cir. 1983).
[21] 36 C.F.R. § 60.6(q).
[22] *Id.* at § 60.6(t).

Moderate procedural safeguards are provided for this process. A 15-day comment period is required, and notice for the comment period is published in the National Register.[23] This period may be waived if necessary, usually at the property owner's request. However, the form of the review is not specified in the regulations, nor do they require written findings evaluating the property's characteristics vis-a-vis the criteria.

The property is officially designated for listing upon the signature of the Secretary of Interior. Notice of the listing is provided to the owner, the SHPO, and the members of Congress representing the district where the property is located. The SHPO, in turn, is required to inform the owners and the chief elected official of the relevant municipality of the listing.[24]

The failure to nominate property may be appealed to the Keeper of the Register by any individual or by the local government with an interest in the property.[25] However, the preamble to the regulations implementing the 1980 amendments states that appeal is discouraged "in cases where there is a general consensus of opinion among involved State and/or Federal officials that a property is not eligible."[26]

2.1.4 Owner Consent

A property may not be listed on the National Register if its owner objects.[27] This owner consent provision was enacted when the tax laws provided significant disincentives for some private actions affecting Register buildings. Its enactment also reflected the resistance to Register listing caused by the fact that such listing may trigger the application of state regulation of actions affecting the property.

An owner is defined as "those individuals, partnerships, corporations or public agencies holding fee simple title to property."[28] Ownership does not include holding easements or less than fee interests, including leaseholds of any nature. After receiving notification that the state has initiated nomination procedures, the owner who wishes to object must submit to the SHPO a notarized statement certifying ownership of the property, in whole or in part, and the fact of his objection.[29] In the case of properties within a historic district, a majority vote of the owners within the proposed district can block the designation.[30] When the owner

---

[23] *Id.* at § 60.13(a).

[24] *Id.* at § 60.6(u).

[25] *Id.* at § 60.12.

[26] 48 Fed. Reg. 46306 (1983).

[27] 16 U.S.C. § 470a(a)(6).

[28] 36 C.F.R. § 60.3(k).

[29] *Id.* at § 60.6(g).

[30] *Id.* An owner cannot accumulate voting power based on property units owned or shares of a property. Each owner gets one vote. It is also irrelevant that the particular property in a historic district does not contribute to the significance of the district.

or majority of owners objects to the nomination, it is still submitted to the Keeper of the Register, but only for a determination of eligibility.[31] If the Keeper determines that this property is eligible, the SHPO notifies the appropriate elected local offical.

A property that is eligible for listing, but is not listed because its owner objects, is nevertheless afforded the procedural protection of Section 106 of the Act, discussed in chapter 6.[32] Many state environmental and preservation statutes also apply to Register-eligible property as well as to property that is actually listed. However, the owner of eligible but nonlisted property will not be eligible for the tax benefits or grants made available to registered property.

It is not clear whether the owner-consent provisions of the 1980 amendments apply to lands owned or controlled by the federal government. This question has particular significance to western states with enormous expanses of federal land, where frequent differences of opinion occur between the SHPO and federal agencies over the historic significance of sites on federal lands. The Reagan administration has taken the position that it has the right to withhold consent. A 1982 letter to all SHPOs from the Keeper of the Register directed that nominations for listing of property under federal ownership or control include a letter from the relevant federal agency preservation officer approving of the listing.[33]

2.1.5 Removal from the Register

The grounds for removal of property from the National Register are articulated in the regulations.[34] Generally, it is not a simple matter to change the status of property that has been determined to have met the criteria for listing—it must be shown that it presently does not, or that it never did, meet the criteria for eligibility. Among the grounds for removal are errors in the original determination due to substantive, technical, or procedural flaws, or the availability of new information pertaining to eligibility. Alterations made after December 12, 1980 that changed the character of the property to the extent that it no longer meets the criteria are also a basis for removal.

There may be instances in which delisting from the Register is necessary to allow an owner to benefit from any of the rehabilitation tax credits discussed in chapter 1. When preparations were being made for recent full-scale renovation of the National Press Building in Washington, D.C., for example, it was discovered that the facade of the building had been so badly stripped that a certified rehabilitation was impossible.

---

[31] 36 C.F.R. § 60.6(n), (s).

[32] *See* Section 6.1.1.

[33] *See* 3 *Preservation L. Rept.* 1006 (1984).

[34] 36 C.F.R. § 60.15.

Only by removing the building from the Register was the owner able to take advantage of the 20-percent tax credit, for which certification of restoration work is not required. Owners may, in this and other situations, weigh the benefits of the enhanced credit for a certified rehabilitation against the possibly increased cost or difficulty of undertaking such a rehabilitation in deciding whether to seek listing or delisting of a building.

2.2 CERTIFICATIONS OF HISTORIC SIGNIFICANCE

In addition to buildings individually listed on the National Register, any building which is certified as contributing to the historic significance of a registered historic district is eligible for the 25-percent rehabilitation tax credit.

There are two ways to establish that a particular district qualifies as a Registered Historic District, within which non-National Register properties may be certified for tax benefits. First, the district itself may be listed on the National Register. Second, the district may have been designated as historic under a certified state or local preservation statute. The process for certification of such local statutes is described in chapter 7.[35] In the latter case, the district must, in addition to local designation, be certified by the NPS as "meeting substantially all of the requirements for listing of districts in the National Register."[36]

An owner contemplating rehabilitation may also request preliminary certification that a building not yet on the National Register appears to meet National Register criteria, or that a building located within a potential historic district appears to contribute to the significance of that district. Preliminary certification is not binding, and does not guarantee that the NPS will ultimately issue a final certification for the building in question. Thus, owners who proceed with rehabilitation on the strength of a preliminary certification do so at their own risk.[37] Nevertheless, preliminary certification often provides useful guidance, particularly when a rehabilitation project is to be syndicated.

Requests for preliminary determinations of significance, moreover, must be accompanied by assurances of future state action consistent with the purported historic significance of the building in question. Thus, for instance, if the building does not fall within the existing bound-

---

[35] See Section 7.5.1.

[36] 36 C.F.R. § 67.2.

[37] 36 C.F.R. § 67.4(d). The taxpayer may assume that the listing will occur and take the tax benefits pending the listing without penalty if the listing occurs within 30 months of the first deduction. See C. Duerksen, ed., A Handbook on Historic Preservation Law at 155 (1983).

aries of a historic district, written assurances must be provided from the appropriate local official that the district is to be expanded. Likewise, if a certification request is made for a building situated within a geographic district but outside that district's established period of significance, a preliminary determination will be issued only if a local official provides written assurance that the district documentation is being revised to expand the historical timeframe upon which its significance is based.[38] If the district in question has been nominated but not yet registered as historic, the state must verify in writing that the nomination is being revised so that the building will fall within the district's area and period of significance.

The owner of a building within a registered district does not have the option either to seek certification and the 25-percent tax credit or to forego certification and undertake a rehabilitation which is eligible for only the 15- or 20-percent tax credit. In order to be able to obtain the lesser tax credits he must obtain a certification that the building is not of significance to the district within which it is located. Otherwise, to obtain any tax credit he must, as described next, undertake a fully certified and approved rehabilitation.[39]

2.2.1 Standards for Evaluating Historic Significance

In reviewing buildings located within registered historic districts to determine if they contribute to the significance of their districts, the NPS applies its "Standards for Evaluating Significance within Registered Historic Districts."[40] A building will be deemed as contributing, and hence as likely to be certified, if by location, design, setting, materials, workmanship, feeling, and association it adds to the district's sense of time, place, and historical development. A building will not be found to contribute, and therefore will be certified as nonsignificant, if it fails to add to the district's sense of time, place and historical development. This also applies when a building has been so altered or has so deteriorated that its overall integrity and the characteristics just listed have been irretrievably lost.

Buildings constructed within the past 50 years generally will not be considered to contribute to the significance of a historic district unless strong justification of their historical or architectural merit is provided, or unless the historical attributes of the district itself are considered to be less than 50 years old. An owner seeking a certification of nonsignificance may submit a condemnation order as evidence of physical deterioration, but that alone will not be considered sufficient evidence

[38] 36 C.F.R. § 67.4(d); 67.5(c).
[39] I.R.C. § 46(a)(2)(F)(i); 48(g)(2)(B)(iv)(II).
[40] 36 C.F.R. § 67.5.

of loss of integrity. It may also be necessary to submit a structural engineer's report to help substantiate physical deterioration or structural damage.[41]

The owner of a deteriorated or partially demolished historic building who wishes to undertake a certified rehabilitation may have trouble obtaining a certification of significance. Even if the rehabilitation plans call for the most accurate possible reconstruction of the building, the NPS may find that prior to the reconstruction, the time at which its determination must be made, the building does not retain sufficient structural integrity or historical significance.

2.2.2 Certification Procedure

A request for a certification of historic significance, or for certification of rehabilitation, must be made by means of a Historic Preservation Certification Application (approved OMB Form No. 1024–0009), which can be obtained from NPS regional offices or from appropriate state officials and which is reproduced in Appendix C of this book. Part 1 of the application pertains to certification of historic significance or nonsignificance, while Part 2 is used to request certification of rehabilitation projects. Although Parts 1 and 2 need not be submitted together, Part 2 of the application will not be processed until an adequately documented Part 1 has been received and acted upon, unless the building in question is already a certified historic structure.

Generally, only the fee simple owner of a building may submit an application for certification.[42] If, in appropriate circumstances, the application is made by another individual or entity, the application must be accompanied by a written statement from the owner indicating that he is aware of the application and has no objection to the request for certification. In certain situations, the NPS may undertake a certification determination at the request of a state official or on its own initiative, but only if notice is provided to the owner.

For purposes of the 15- and 20-percent rehabilitation tax credits (for nonhistoric buildings), buildings situated within registered historic districts are presumed to contribute to the significance of those districts unless certified as nonsignificant before the project is initiated. If an owner begins or completes demolition or substantial alteration without knowledge that certification is required, he may still request certification that the building was not of historic significance before the work was undertaken. If he does so, however, the owner must certify to the Secretary of the Treasury that at the time demolition or alteration was initiated, he was in good faith unaware of the certification requirement.[43]

---

[41] *Id.* § 67.5(b).

[42] *Id.* § 67.3(a)(1).

[43] I.R.C. § 48(g)(2)(B)(iv)(III).

The last point may be of critical importance for any purchaser of a building within a National Register district. It is entirely possible that a purchaser may buy property within a registered or nominated district without knowledge that the property has been or will be deemed "contributing," or, for that matter, without knowledge that the property falls within the boundaries of such a district at all. Accordingly, any buyer of urban property who intends to take advantage of the investment tax credit for rehabilitation (even at 15 or 20 percent) should, as a routine precaution, investigate the status of the property he intends to purchase, as well as the possibility that it is located within a historic district, to determine whether these restrictions will apply.

2.2.3 Application Requirements

An application for certification of significance or nonsignificance must, at a minimum, provide the following documentation:[44]

1. The address of the property, and the name if one exists.
2. The nature of the certification request.
3. The name, title, address, and telephone number of the authorized project contact to whom all inquiries should be addressed.
4. The name, address, and telephone number of the building owner.
5. The name of the historic district in which the building is located.
6. A brief description of the physical appearance of the building, both exterior and interior, including architectural details, alterations, and date of construction.
7. A brief statement of the significance of the building, summarizing its significance to the distinctive historical and visual character of the district.
8. Current photographs of the building, as well as photographs prior to alteration if rehabilitation has been completed.
9. Sketch maps indicating the building's location within the district.
10. The signature of the owner requesting or concurring in the request for an evaluation.

If a building has been or is to be moved into a historic district, additional documentation must be provided in the application. Information must be submitted discussing the effect of the move on the building's appearance and on the distinctive historical and visual character of the district, describing the new setting and general environment of the proposed site, and identifying the method of moving to be used.

A single application may generally be submitted to request certification of the historic significance of groups of related buildings within

---

[44] 36 C.F.R. § 67.4(c).

a historic district. However, documentation must be offered for every building proposed for certification. In addition, buildings whose functional relationship contributes to their historical significance (for instance, a mill complex or an industrial plant) may be treated as a single unit for purposes of certification as contributing or noncontributing to the historic district.

2.2.4 Processing of Applications

The manner in which certification applications are processed depends on whether the state in which the property is located has elected to participate in the review of certification requests. In participating states, completed applications are sent to the appropriate state official, generally the SHPO. Thirty-day review periods are permitted at the state and federal levels respectively. In the case of "qualified states," federal review is shortened to 15 days and the entire review period to 45 days. Qualified states are those which have elected to participate in the certification process, and which the NPS has determined provide review of certification requests by professionally qualified staff, in accordance with established guidelines, and within the specified 30-day timeframe.[45] In nonparticipating states, certification requests should be sent directly to the appropriate NPS regional office. Review of completed requests should generally be completed within 60 days[46] and federal review within an additional 60 days.

Failure to process certification applications in a timely manner appears to have been the primary complaint voiced in regard to the entire certification program.[47] Increased processing time can increase the cost of rehabilitation projects, and may deter property owners from participating in the program at all. At least one NPS regional office was unable to comply with the time limits that were subsequently shortened by the 1984 amendment of the certification regulations.[48] Whether the new time limits, which are not mandatory, can be enforced remains to be seen.

The owner or his duly authorized representative may appeal from any grant or denial of certification. Appeals must be made in writing and sent to the Chief Appeals Officer, Cultural Resources, National Park Service, U.S. Department of the Interior, Washington, DC 20240. The chief appeals officer must receive the appeal within 30 days of receipt

---

[45] *Id.* § 67.11(b).

[46] *Id.* § 67.3(b)(4).

[47] "Federal Taxation and the Preservation of America's Heritage: A Report Prepared for the Advisory Council on Historic Preservation by the National Trust for Historic Preservation" at IV–6 (1983).

[48] "Historic Tax Certification Backlogs Not a Problem, Says Interior Department," 11 *Housing & Development Reporter* 182 (1983).

of the contested decision by the complaining party. The appellant may request a meeting to discuss the appeal, and the appropriate state official will be notified that the appeal is pending. The chief appeals officer will review the appeal and the written record of the decision in question, and will notify the appellant of his decision within 30 days, if circumstances permit.[49]

The denial of a preliminary determination of significance is not automatically appealable, although an appeal may be granted on a case-by-case basis at the discretion of the chief appeals officer. If an appeal is not permitted, the owner must seek recourse to the usual process of nomination for inclusion in the National Register.

2.3   CERTIFICATIONS OF REHABILITATION

Before an owner may take advantage of the tax benefits for rehabilitation of an historic building, he must also obtain a determination that the project undertaken qualifies as a "certified rehabilitation." This is true whether the property is listed on the National Register or has been certified by the NPS as contributing to the historic significance of the registered district in which it is located. The purpose of this second-stage certification is to insure that the historic character of a building is preserved in the process of rehabilitation.

Owners contemplating rehabilitation projects should bear in mind two general factors which are of primary concern to the NPS. First, it is essential to identify the historic materials and features, and the unique craftmanship they represent, which make their buildings worth preserving. Second, they must insure that, to the greatest extent possible, these attributes will be retained, protected, and repaired in the process of rehabilitation. Removal or replacement, even with materials and features which appear to be historic, may jeopardize certification. Integrating efficient contemporary uses with preservation of historic features will often be a challenging task.

The definition of "rehabilitation project," for certification purposes, is extremely broad. It includes "all work on the significant interior and exterior features of the certified historic structure(s) and its setting and environment," as well as "related demolition, construction, or rehabilitation work which may affect the historic qualities, integrity, or setting of the certified historic structure(s)."[50] All elements of the rehabilitation must meet the Standards for Rehabilitation, discussed in this chapter, and no portion of a project may be exempted. Thus an owner may not choose to undertake one portion of a project in accordance with the Standards for Rehabilitation to obtain the 25 percent tax credit, and

[49] 36 C.F.R. § 67.10(a).
[50] *Id.* § 67.6(b).

forego any tax credit on the remainder of the project in order to avoid the application of the standards to that part of the project. If the NPS believes that a project submitted for certification does not in fact include the entire rehabilitation undertaken, it may choose to deny a rehabilitation certification, or alternatively, to withhold its decision until the Internal Revenue Service, by means of a private letter ruling, determines the proper scope of the rehabilitation project to be reviewed. In making their determinations, the NPS and the IRS will consider all the facts and circumstances surrounding a particular application, but will give particular attention to these issues: (1) whether previous demolition, construction, or rehabilitation work, irrespective of ownership or control at the time, was in fact undertaken as part of the rehabilitation project for which certification is sought; and (2) whether property conveyances, reconfigurations, ostensible ownership transfers, or other transactions were intended to limit the scope of a rehabilitation project for review purposes without substantially altering the beneficial ownership or control of the property.

In deciding whether to certify a rehabilitation project, the NPS will evaluate the condition of a structure after rehabilitation in relation to its condition prior to the commencement of rehabilitation. In addition, the project will be evaluated in the broadest possible context. Thus, for instance, if rehabilitation involves a number of structures which have been judged to have been functionally related to serve an overall purpose (again, a mill complex or an industrial plant are the typical illustrations), the NPS will consider certification based upon the merits of the overall project rather than on its individual components. Similarly, if a building in a historic district is to be rehabilitated, review will encompass the effects of the alteration on both the historic structure itself and on the district in which it is situated.

2.3.1　Standards for Rehabilitation

As guidelines in evaluating rehabilitation projects, the NPS applies its Standards for Rehabilitation.[51] The regulations call for the standards to "be applied taking into consideration the economic and technical feasibility of each project."[52] Nevertheless, a project must comply with each of the ten standards, or it will not be approved. The standards are as follows:

1. Every reasonable effort shall be made to provide a compatible use for a property which requires minimal alteration of the building, structure, or site and its environment, or to use a property for its originally intended purpose.

[51] *Id.* § 67.7(a).
[52] *Id.*

2. The distinguishing original qualities or character of a building, structure, or site and its environment shall not be destroyed. The removal or alteration of any historic material or distinctive architectural features should be avoided when possible.

3. All buildings, structures, and sites shall be recognized as products of their own time. Alterations that have no historical basis and which seek to create an earlier appearance shall be discouraged.

4. Changes which may have taken place in the course of time are evidence of the history and development of a building, structure, or site and its environment. These changes may have acquired significance in their own right, and this significance shall be recognized and respected.

5. Distinctive stylistic features or examples of skilled craftsmanship which characterize a building, structure, or site shall be treated with sensitivity.

6. Deteriorated architecture features shall be repaired rather than replaced, wherever possible. In the event replacement is necessary, the new material should match the material being replaced in composition, design, color, texture, and other visual qualities. Repair or replacement of missing architectural features should be based on accurate duplications of features, substantiated by historic, physical, or pictorial evidence rather than on conjectural designs or the availability of different architectural elements from other buildings or structures.

7. The surface cleaning of structures shall be undertaken with the gentlest means possible. Sandblasting and other cleaning methods that will damage the historic building materials shall not be undertaken.

8. Every reasonable effort shall be made to protect and preserve archeological resources affected by, or adjacent to any project.

9. Contemporary design for alterations and additions to existing properties shall not be discouraged when such alterations and additions do not destroy significant historical, architectural, or cultural material, and such design is compatible with the size, scale, color, material, and character of the property, neighborhood or environment.

10. Whenever possible, new additions or alterations to structures shall be done in such a manner that if such additions or alterations were to be removed in the future, the essential form and integrity of the structure would be unimpaired.

To assist in applying and implementing the Standards, the NPS has adopted "Guidelines for Rehabilitating Historic Structures."[53] These re-

---

[53] National Park Service (revised 1983).

flect a general hierarchy from least to most intrusive means of rehabilitating structures in order to accommodate efficient contemporary
uses. Not surprisingly, they represent a strong preference for the most
minimal rehabilitative activity which can possibly be undertaken. Building owners are instructed initially to identify, retain, and preserve those
architectural materials and features which are important in defining
the historic character of a structure. Once this is done, work should be
done as necessary to protect and maintain the materials and features
which have been identified, but with the least degree of intervention
possible. Protective work to be undertaken includes such things as rust
removal, caulking, limited paint removal, reapplication of protective
coatings, or installation of fencing or alarm systems.

If protective maintenance is not adequate to preserve historic features, then repair work should be undertaken. Again, the guidelines call
for the least intrusion possible. Patching, splicing, reinforcing, and the
like are preferable to more substantial alteration. Also, use of the original material is always desirable, although substitutes are acceptable if
the form and design of the substitute convey the same visual appearance
as the remaining parts of the original material or feature.

If repair is impossible, the next-best option is replacement of an entire
character-defining feature. The guidelines emphasize, however, that removal and replacement are never recommended when a feature, however damaged or deteriorated, can be repaired and preserved. If replacement is essential, use of the original material is, again, preferable
but not essential. If an entire interior or exterior feature is missing and
irretrievable (for instance, a cast iron facade, or a principal staircase)
and adequate historical, pictorial, and/or physical documentation exists
so that the original feature can be accurately reproduced, design and
construction of a new feature based upon that information may be
appropriate. A new design, which is compatible with the remaining
features of the building, may also be acceptable.

Finally, though least desirable, physical alterations or additions to
historic buildings may be necessary to allow continued use, to meet
health and safety code requirements, or to meet energy conservation
objectives. These must not radically change, obscure, or destroy character-defining spaces, materials, features, or finishes of the structure.
Interior alterations are generally preferable to exterior, and interior
solutions are certainly preferable to exterior additions. The regulations
warn of particular rehabilitative techniques which are generally not
appropriate for historic buildings and which are likely to lead to denial
of certification.[54] Physical treatments which may cause or accelerate
deterioration of buildings are to be avoided; these include improper
repointing techniques, improper exterior masonry cleaning methods,

[54] 36 C.F.R. § 67.7(b).

and the introduction of insulation into cavity walls of historic wood frame buildings where damage to historic fabric would result. Certification denial will also result from exterior additions that duplicate the form, material, style, and detailing of a structure to the extent that they compromise its original historic character.

2.3.2 Application of Standards

Based upon its experience in reviewing and evaluating rehabilitation projects, the NPS has identified a number of special problem areas which are most likely to result in certification denial if not properly handled.[55]

1. New heating, ventilating, and air-conditioning (HVAC) systems. Justification must be provided before installing window-mounted or through-the-wall HVAC units. This should include the effect of the new equipment and ductwork on the building fabric, a description of alternative systems considered and why the system in question was chosen, and a life cost analysis comparing the system proposed for use with two or more alternative systems. In general, owners are strongly encouraged to utilize systems which cause least damage to the historic fabric of the building.

2. Window replacement. Owners must justify the replacement of existing window sash with evidence, including photographs, of severe deterioration, and must provide data on the cost of repairing existing sash versus installing new sash. Where replacement is warranted and where windows are an integral part of the building's design and character, new sash should match the original in material, size, general pane configuration, and reflective qualities. Owners are encouraged to repair and retain existing sash where such sash add to the historic and architectural character of the building.

3. Removing interior plaster work and exposing masonry surfaces. Documentation of the existing condition of the interior walls, and the effect on existing woodwork, should be provided before undertaking work of this type. In general, owners are discouraged from exposing bare masonry surfaces unless this condition is supported by historical evidence.

4. Exterior masonry cleaning. Cleaning should be accomplished by the gentlest means possible, and an indication should be provided that other cleaning methods were considered but found less appropriate for

---

[55] *See* Rogers, "Reasons for the Denial of Certification for the 25 Percent Investment Tax Credit," 3 *Preservation L. Rept.* 2001 (1984); "Review of Rehabilitation Work under Section 2124 of the Tax Reform Act of 1976," Technical Preservation Services Division, Heritage Conservation and Recreation Service, U.S. Department of the Interior (1980).

the condition of the material to be cleaned. Owners are strongly encouraged to clean masonry only when necessary to halt deterioration or to remove graffiti and stains.

5. Exterior masonry repair. Documentation should indicate which areas require repair, and should provide evidence that repointing mortar will match the original in composition, proportions, color, and texture. Owners are encouraged to repoint only those portions of a building that require repair.

6. Storefront alterations. Before altering an existing storefront, information should be provided on when the existing storefront was constructed; what physical conditions are present; and if a historical treatment is planned, on what research or physical evidence the proposed new storefront designs are based. Owners are strongly discouraged from introducing a storefront or new design element on the ground floor which alters the character of the building and its relationship with the street, or which causes destruction of significant historic fabric.

7. Incompatible new use or addition. Even when the economic viability of a historic building requires alteration to accommodate a new use or expansion to accommodate existing uses, any changes made must comply with the Standards for Rehabilitation. Economic hardship may be grounds for alteration under local law, but it is not a sufficient condition for certification. The standards do not necessarily discourage additions, but they must be compatible with the building and the neighborhood. At the same time, however, additions should be clearly distinguishable from the original building and should not attempt to give the property an earlier appearance.

8. Incompatible treatments to building interior or rear or secondary elevations. As with alterations or additions, evaluation of interior, rear or secondary elevation treatments for certification purposes may differ from the scrutiny required under local law. Many local ordinances, for example, permit a commission to analyze only proposed treatments of building exteriors. Certification decisions, however, are based upon an entire project, and involve review of all impacts upon a building. It is important to emphasize that issuance by a local commission of a certificate of appropriateness for an alteration by no means guarantees NPS certification.

The regulations acknowledge that, in limited cases, "it may be necessary to dismantle and rebuild portions of a certified historic structure to stabilize and repair weakened structural members and systems."[56] Extreme intervention of this type will be considered as part of a certified rehabilitation if three conditions are met:

---

[56] 36 C.F.R. § 67.7(c).

1. The necessity for dismantling is justified in supporting documentation.
2. Significant architectural features and overall design are retained.
3. Adequate historic materials are retained to maintain the architectural and historic integrity of the overall structure.

As further guidance in applying the Standards for Rehabilitation, the NPS, through the Technical Preservation Services Division, periodically publishes a bulletin entitled *Interpreting the Secretary of the Interior's Standards for Historic Preservation Projects.* These publications briefly describe a particular application for certification of a rehabilitation project, state whether the request was approved or denied, and explain the rationale for the positive or negative determination. While the decisions discussed relate to the particular facts and circumstances of a given application, they do highlight the sorts of rehabilitative techniques which are likely to encounter disapproval from the NPS. A number of the major bulletins will be found in Appendix D.

2.3.3   Certification Procedure

The NPS will review any proposed, ongoing, or completed rehabilitation project for compliance with the Standards for Rehabilitation. Owners are encouraged to obtain a determination of compliance before undertaking work. Prior approval enables owners to proceed knowing that the work meets necessary requirements, and will also expedite certification of the completed project. Owners who undertake rehabilitation projects without securing approval in advance do so at their own risk. Not surprisingly, most certification denials have arisen in situations in which the SHPO or NPS has not been consulted until after work has been substantially completed.

Projects involving the use of federal funds such as grants from the Department of Housing and Urban Development or Urban Development Action Grants generally require certification prior to the final commitment of the federal funds.

If a proposed or ongoing project has been approved, the owner must promptly report to the NPS substantive changes in the work as described in the application.[57] This must be accomplished by written statement, with a copy to the appropriate state official. The NPS will notify the owner if the revised project continues to meet the standards, and the owner can thereby assure continued conformance.

---

[57] *Id.* § 67.6(d).

Whether or not prior approval is obtained, final certification of the rehabilitation will be issued only after the project has been completed. Requests for certification are made by means of Part 3 of the Historic Preservation Certification Application, discussed in Section 2.2.2, which is entitled, "Request for Certification of Completed Work." Requests must include the project completion date; a signed statement indicating that, in the owner's opinion, the completed project meets the Standards for Rehabilitation and is consistent with the work described in Part 2 of the application; a statement of costs attributed to the rehabilitation; and photographs adequate to document the completed work.

The NPS charges fees for processing rehabilitation certification requests.[58] For reviewing proposed or ongoing rehabilitation projects estimated to cost over $20,000, the fee is $250; no fee is charged for projects under that amount. Fees for reviewing completed projects depend upon the dollar amount of costs attributed solely to the rehabilitation in the Historic Preservation Certification Application, in accordance with the following table:

Size of rehabilitation	Fee
$ 20,000 to 99,000	$ 500
100,000 to 499,000	800
500,000 to 999,000	1500
1,000,000 or more	2500

If review had already been undertaken before a project was completed, the initial fee of $250 will be deducted from these fees.

Generally, each rehabilitation of a separate certified historic structure is considered a separate project for purposes of computing fees. If, however, the project involves structures judged to have been functionally related to serve an overall purpose, the fee for preliminary review will be $250 and the fee for final review will depend upon total project costs. If multiple projects under the same ownership (for example, row houses or loft buildings) are submitted for review at the same time, the maximum total fee is $2500. The fee for preliminary review will be $250 per building, and the fee for final review will be based upon total rehabilitation costs, up to a maximum of $2,500. If $2,500 has already been expended in preliminary fees, no further fee will be charged.

Upon receipt of a completed application, the NPS will determine if the project is consistent with the Standards for Rehabilitation, in accordance with the timetable described in Section 2.2.4. If a project does not meet the standards, a letter will be sent to the owner explaining the

[58] *Id.* § 67.12.

reasons for that determination, and where possible, advising him of the revisions necessary to meet the standards.[59] It should be noted again that prior approval of a project by federal, state, or local agencies does not ensure certification by NPS. The Standards for Rehabilitation take precedence over other regulations and codes in determining whether a rehabilitation project is consistent with the historic character of a building and/or the district in which it is located.[60] If certification is denied, that determination may be appealed as discussed in Section 2.2.4.

The large majority of projects presented to the NPS are certified.[61] Of a total of 7,009 certification decisions made between Fiscal Years (FY) 1977 and 1983, 683 initial denials were issued, amounting to 9.7-percent initial denial rate. In FY 1983 alone, 245 requests for certification were denied, representing 10 percent of the decisions rendered.

In FY 1983, 127 initial denials, or 52 percent, were appealed. This rate has risen in recent years; from FY 1977 through FY 1980, a average of only 26 percent of initial denials were appealed. The number of denials overturned on appeal has declined somewhat. From FY 1977 through FY 1980, 56 percent of initial denials were overturned; in FY 1983, 44 percent of appeals resulted in reversal.

According to the NPS, submission of information on appeal which was not available during initial certification review by the regional office accounted for over 90 percent of the reversals on appeal. This statistic highlights the importance of providing complete information in the certification application, to minimize the danger of initial denial and reduce ultimate processing time. Professional error in applying the Standards for Rehabilitation, by NPS estimates, accounted for less than 10 percent of initial denials. Those were most often the result of erroneous determinations of incompatibility of new additions and replacement window sash.

Receiving certification from the NPS does not entirely discharge a building owner's obligations. The NPS reserves the right to make inspections at any time up to five years after completion of the rehabilitation. Certification may be revoked, after 30 days' notice to the owner, if it is determined that the rehabilitation was not undertaken as presented in the application and supporting documentation, or that the owner, upon obtaining certification, undertook further unapproved alterations as part of the rehabilitation project which were inconsistent with the standards.

---

[59] *Id.* § 67.6(c), (f).

[60] *Id.* § 67.7(d).

[61] All statistics are taken from "Tax Incentives for Rehabilitating Historic Structures: Facts and Figures 1977–1983" (National Park Service 1984), available from the National Trust for Historic Preservation.

Three

Financing Rehabilitation: Private Syndication

Introduction

An initial and important step in historic rehabilitation, as in any real estate venture, is the choice of ownership entity or syndicate.[1] The decision is guided by an evaluation of which vehicle will best achieve the purposes of raising the equity capital and securing the financing needed to acquire and rehabilitate the building and, thereafter, assuring its ongoing operation. For several reasons, which will be discussed in this chapter, the limited partnership has been the preferred form of association.

In addition to organizational advantages, a real estate limited partnership can offer numerous tax benefits. A major attraction of real

[1] For a thorough discussion of this subject *see* T. Lynn and H. Goldberg, *Real Estate Limited Partnerships*, 2d ed. (1983). *See generally* Madison and Dwyer, *The Law of Real Estate Financing* (1981, Supp. 1984). Entity choices include tenancy in common, joint venture, general partnership, limited partnership, corporation, land trust, or real estate investment trust.

estate investment in general is that it is not subject to the tax law's provisions that, in other areas, limit the income sheltered from tax to the amount that the investor has risked. Under present law, properly managed real estate investments may produce deductions larger than the amount invested. One who invests as a limited partner can normally claim a share of both depreciation of the building and deductions for other business expenses of the partnership, such as interest on borrowed funds. The advantages of undertaking the rehabilitation of a historic or other old building through a partnership may be further enhanced by the availability of investment tax credits which may be shared among the individual investors.

Projects undertaken through limited partnerships will often go through the syndication process, discussed more fully in this chapter and in chapter 8. This involves the preparation of a memorandum or prospectus for potential investors, which discloses all material information relating to the project, as required by federal and state laws governing securities offerings. The memorandum or prospectus typically contains *pro forma* financial projections of income (and losses) over an extended period, both for the project as a whole and for each prospective limited partner. Portions of a typical *pro forma* projection, showing expected cash-flow and after-tax consequences of a project over a 17-year period, for both general and limited partners, are found in Appendix G.

3.1   LIMITED PARTNERSHIPS

Limited partnerships are a valuable financing tool in the rehabilitation process. Most simply, a limited partnership allows an investor to pool resources with other investors to rehabilitate a building in a manner that the individual investor may not have the resources to accomplish alone. In addition to providing a vehicle by which individual investors can extend their financial reach by pooling resources, the law provides incentives which may make this form of association more advantageous to the investors than others.

A limited partnership is a partnership created under state statute that allows an investor to contribute or lend money or property to a business or venture managed by at least one general partner, without subjecting the investor-limited partner's personal assets to the full risk of the enterprise. The Uniform Limited Partnership Act (ULPA)[2] defines a limited partnership as "a partnership formed by two or more persons . . . having as members one or more general partners and one or more limited partners. The limited partners as such shall not be bound by the obli-

---

[2] *6 Uniform Laws Annotated* 559 *et seq.* (West Publishing Co., 1969).

gations of the partnership." In addition to limiting an investor's liability, the limited partnership form provides an investor with tax benefits that other organizational forms do not.[3]

To receive this favorable treatment, a limited partner must remain passive in the management of the partnership's affairs. All management and control of the partnership's business and property must be left to the general partner(s). The investor who does not heed this requirement and seeks to participate in business decisions risks a finding that he is a general partner, not a limited partner, thereby exposing himself to full personal liability.[4] However, if this restriction is heeded, the limited partnership can combine the legal and business benefits available from the corporate form of organization with the tax advantages otherwise available only to individuals. While a corporation can offer similar limitations on liability, corporate shareholders do not enjoy the tax benefits that limited partners do: Profits are taxed at both the corporate and the individual level, and tax deductions and credits for which the corporation is eligible cannot be passed through to the investor.[5] A general partnership or individual ownership, on the other hand, may provide the same tax benefits as a limited partnership, but general partners and individual investors bear full personal liability for the obligations and liabilities incurred in real estate ventures. For these reasons, limited partnerships have been a popular vehicle for investors wishing to contribute capital for the purpose of rehabilitating old buildings.

3.1.1. Creation Governed by State Law

A limited partnership is created under state statute.[6] Every state (except Louisiana) and the District of Columbia has enacted a statute which governs the formation and conduct of a limited partnership. These statutes are based on either the original 1916 ULPA[7] or a version of the

---

[3] *See* Section 3.2.3.

[4] *See* U.L.P.A. § 7.

[5] Some of the advantages of limited partnerships can be achieved, in appropriate circumstances, by adopting the corporate form and electing to be treated as a Subchapter S corporation. This form will, for example, permit direct pass-through of profits and losses to shareholders. For recent liberalization of the Subchapter S requirements, *see* Subchapter S Revision Act of 1982, P.L. 97-354 (1982); these changes may make this a more attractive alternative in some investment situations. However, there may also be disadvantages to Subchapter S election. It may, for instance, subject an entity to statutory provisions which reduce the total amount of losses which can be passed through to investors when compared to limited partnerships. Fuller consideration of the Subchapter S provisions is beyond the scope of this book.

[6] There were no limited partnerships under common law.

[7] The original ULPA is contained in 6 *Uniform Laws Annotated* (West Publishing Co., 1969).

1976 revised ULPA, adopted by the National Conference of Commissioners on Uniform State Laws in 1976. Although specific state statutes may parallel the language of these model acts, the precise wording and scope of each state's statute may differ from the model acts, and the specific statute of the appropriate jurisdiction should always be consulted.

Generally, a limited partnership must file a certificate in a central recording office of a designated public official in the jurisdiction. The certificate must specify such things as a description of the partnership and its business, the names of the general and limited partners, the amount of cash or other property contributed by the limited partners, provisions governing dissolution of the partnership, requirements for assignment of partnership interests, and provisions governing distributions of partnership property. In some cases, the partnership agreement is filed in lieu of, or in combination with, the certificate. The agreement is often filed as a subsequent amendment to a short-form certificate once it has been prepared.

Completion of the certificate is crucial. Failure to conform to the requirements of the state's law subjects the limited partners to the risk of the loss of limited liability. However, substantial compliance in good faith with the certificate filing requirements is sufficient to form a limited partnership.[8]

3.1.2. Role of General Partner

A general partner in a limited partnership is subject to all of the liabilities of a partner in a general partnership.[9] This includes full liability for all debts and obligations incurred by the partnership, as well as for wrongful acts performed by other partners.[10] A general partner in a limited partnership also possesses most of the rights and powers a partner would exercise in a general partnership. Certain of these powers, however, must be specifically provided for in the partnership agreement or by the unanimous written consent or ratification of all the general partners.[11] Without such agreement, for example, a general partner may not confess judgment against the partnership; admit a person as a general or limited partner; or continue the business with partnership property on the death, retirement, or insanity of a general partner.

---

[8] U.L.P.A. § 2(2).

[9] *Id.* § 9.

[10] Uniform Partnership Act §§ 24–26; contained in 6 *Uniform Laws Annotated* (West Publishing Co., 1969).

[11] U.L.P.A. § 9.

General partners owe a fiduciary duty to the partnership and to the limited partners.[12] They may be liable to the limited partners if they breach the duty of fair dealing in their conduct of the partnership business. Moreover, general partners act as agents of the partnership, and commitments within their actual or apparent authority are binding on the entity as a whole.[13]

A limited partnership automatically dissolves upon the death, retirement, or insanity of a general partner, unless the certificate authorizes the remaining general partners to continue the business, or all of the members consent to their doing so, and the business is in fact continued.[14] In order to avoid adverse tax consequences[15] and repetition of the administrative formalities of formation, it is often advisable to provide for these contingencies in the partnership agreement.

3.1.3. Role of Limited Partner

Generally speaking, the limited partner has fewer rights and powers, because he has entrusted management of the enterprise to the general partners. The limited partner may, however, take steps to insure that his investment is being protected. He has a right to inspect the partnership books, to demand full information about the partnership activities, to receive a share of partnership profits or compensation as provided in the certificate, and to compel a dissolution in certain circumstances.[16] He may also be able to bring an action against the general partners for breach of their fiduciary duty.[17] Moreover, upon the occurrence of certain conditions precedent or after giving specified notice, a limited partner may demand the return of his original contribution.[18]

A limited partner may assign his partnership interest, but, absent amendment of the certificate, the assignee succeeds only to the economic rights in that interest.[19] The assignee is entitled to receive a share of the profits and other compensation, or to a return of the original contribution, but can exercise none of the other rights possessed by the limited partner. However, if the certificate is properly amended, the substitute limited partner acquires all rights and powers associated with the interest, as well as all restrictions and liabilities, except for those liabilities of which he was not aware or on notice when he became a

---

[12] See e.g., Korn v. Franchard Corp., 388 F. Supp. 1326 (S.D.N.Y. 1975).

[13] U.L.P.A. § 9.

[14] Id. § 10.

[15] See Lynn & Goldberg, supra note 1, at Chap. 8.

[16] U.L.P.A. § 10.

[17] See Lerman v. Tenney, 425 F.2d 236 (2d Cir. 1970); Riviera Congress Assocs. v. Yassky, 268 N.Y.S.2d 854 (1st Dep't), aff'd, 18 N.Y.2d 540, 277 N.Y.S.2d 386 (N.Y. 1966).

[18] U.L.P.A. § 16.

[19] Id. § 19.

limited partner. Assignment of a limited partnership interest and admission of the assignee as a limited partner often requires consent of some or all of the other partners, pursuant to the partnership agreement.

So long as a limited partner does not take part in the control of the business, he does not assume the liability of a general partner.[20] His financial liability to the partnership is limited to making up the difference between the amount of his capital stated in the certificate and the amount actually contributed.[21] (If he does not make up this difference, cash distributed to him may have to be repaid unless the certificate is amended.) The case law is unclear in identifying those powers whose exercise will abrogate limited liability.[22] Some state versions of the ULPA do specify particular powers which may be granted to limited partners without burdening them with general liability.[23] The applicable statutory and decisional law should be carefully consulted by the drafter of a limited partnership agreement to avoid potentially troublesome involvement of limited partners in the business affairs of the partnership.

3.2 TAX TREATMENT OF PARTNERSHIPS AND PARTNERS

A partnership is not a taxable entity under the Internal Revenue Code.[24] Instead, the tax consequences of the actions of the partnership flow through to the individual partners. The income, gain, loss, deductions, and credits of the partnership become the income, gain, loss, deductions, and credits of the individual partners. The ability to escape taxation at the entity level is often a primary reason for using the partnership form.

3.2.1. Recognition as a Partnership by the IRS

A limited partnership recognized under a state statute will not necessarily be recognized as such for federal income tax purposes.[25] This is because the IRS's definition of a limited partnershp varies from that of the states. Since one of the primary reasons why individuals choose to organize as a limited partnership is to take advantage of tax benefits, attention must be paid to the independent criteria of the federal tax laws.[26]

[20] *Id.* § 7.

[21] *Id.* § 17.

[22] *See e.g.,* Weil v. Diversified Properties, 319 F. Supp. 778 (D.D.C. 1970) and cases cited therein.

[23] *See* 6 *Uniform Laws* at 582–86 and supplement.

[24] I.R.C. § 701; however, the partnership is required to file an information income tax return form. I.R.C. § 6031.

[25] Treas. Reg. § 301.7701-1(c).

[26] *See generally* Lynn & Goldberg, *supra* note 1, at Chap. 2.

For tax purposes, the question is whether the business entity will be classified as one with the characteristics of a partnership or with the features of a taxable association or corporation.[27] This question often arises in the context of a request by the partnership for an advance ruling by the IRS on its tax status.[28] The answer is based on the result of a balancing test. The factors to be weighed are the presence or absence of characteristics that have been identified as typifying the corporate form. For purposes of this test, the six characteristics of a corporation that the Treasury regulations identify are as follows:

1. Associates
2. An objective to carry on business and divide gains therefrom
3. Continuity of life
4. Centralization of management
5. Liability for corporate debts limited to corporate property
6. Free transferability of interests[29]

Since the first two characteristics—associates and business undertaken for common purposes—are shared by both corporate entities and partnerships, the inquiry focuses on the latter four characteristics. To obtain recognition as a partnership for federal tax benefits, it is necessary to show that two of these four remaining corporate characteristics are not present and that the entity "has no other characteristics which are significant in determining its classification."[30]

The IRS is most likely to find that the two characteristics of corporate association not present are continuity of life and free transferability of interests, with the possible addition of limited liability. The case law[31] leads to the conclusion that when investors are organized under the terms of the ULPA, the resulting entity will be treated as a limited partnership for federal tax purposes.

A limited partnership agreement drafted according to a state statute which corresponds to the ULPA lacks continuity of life, because a general partner has the power to dissolve the partnership.[32] The agreement and certificate should additionally state that the entity's life dissolves

---

[27] A partnership, for federal tax purposes, "includes a syndicate, group, pool, joint venture, or other unincorporated organization through or by means of which any business, financial operation, or venture is carried on, and which is not . . . a corporation or a trust or estate." I.R.C. § 761(a).

[28] See Rev. Procs. 74-17, 1974-1 C.B. 438; 72-13, 1972-1 C.B. 735.

[29] Treas. Reg. § 301.7701-2(a)(1). See Morrisey v. Comm'r., 296 U.S. 344 (1935).

[30] Treas. Reg. § 301.7701-2(a)(3).

[31] See e.g., Philip G. Larson, 66 T.C. 159 (1976); Zuckman v. United States, 524 F.2d 729 (Ct. Cl. 1975). These cases rejected IRS attempts to interpret the regulations restrictively to find limited partnerships organized under the U.L.P.A. to be taxable associations.

[32] Treas. Reg. § 301.7701-2(b)(3).

upon the occurrence of a certain event, which may be the termination of the partners' relationship based on the death, insanity, retirement, resignation, bankruptcy, or expulsion of a general partner. The partnership may in fact continue based on the consent of the remaining general partners or members, provided the consent is given after the event which triggers dissolution.[33] Prior consent by a provision for automatic continuation or even a fixed date when the partnership will dissolve may not be sufficient to avoid the continuity of life characteristic and may be seen by the IRS as an indication that the entity is not to be treated as a partnership for tax purposes.

Limited partnership agreements drafted in accordance with the ULPA also will lack the corporate characteristic of free transferability of interests, which refers to the ability of members to substitute fully a person outside the organization for themselves.[34] If consent of the other members of the organization is required prior to this substitution or if something less than all of the powers of the interest can be conferred, then this does not constitute free transferability.

The corporate characteristics of limited liability are not present when some member of the entity is personally liable for the debts and obligations of the organization.[35] In a limited partnership, the general partner has personal liability. There are, however, circumstances in which the general partner is not deemed to be personally liable, and such circumstances could lead the IRS to challenge the characterization of the entity as a partnership. A general partner may not be considered personally liable if he does not hold substantial assets, exclusive of his interest in the partnership, which could be reached by a creditor, and if he is a mere "dummy" acting as an agent of the limited partners.[36] For individuals acting as general partners, the definition of "substantial assets" is not well defined in either the regulations or case law.

Centralization of management is another characteristic of corporate organization. According to the regulations, centralization of management exists when the limited partners own "substantially all the interests" in the limited partnership.[37] Avoiding centralization of management in drafting a limited partnership agreement is difficult for two reasons. First, the concept of substantially all the interests is not clearly defined in the regulation or case law. And second, many limited partnerships are formed to obtain investment capital for which an ownership interest in most of the entity's profit is exchanged. It would seem that this might well be characterized as substantially all the interests.

---

[33] *See* Rev. Rul. 74-320, 1974-2 C.B. 404, modified by Announc. 75-23, 1975-11 I.R.B. 87.

[34] Treas. Reg. § 301.7701-2(e)(1).

[35] *Id.* § 301.7701-2(d)(1).

[36] *Id.* § 301.7701-2(d)(2).

[37] *Id.* § 301.7701-2(c)(4).

Avoiding centralization of management is perhaps most easily accomplished by giving the limited partners a right to take part in the conduct of the business in certain respects. That, however, runs the risk of jeopardizing their limited liability.

There is a rule of thumb used by the IRS to trigger its assertion of centralization of management. Centralization of management may be deemed to exist if the general partners own less than twenty percent of the partnership's capital and profits.[38] Since over eighty percent is often owned by limited partners, especially in the early years, most limited partnerships investing in real estate will be found to possess the corporate characteristic of centralized management.

When the sole general partner is a corporation, the IRS applies a somewhat stricter test in determining whether to issue an advance ruling classifying an entity as a partnership for tax purposes. In addition to showing that no more than two of these four corporate characteristics are present, the partnership must meet the four additional requirements of Revenue Procedure 72-13.[39] First, the limited partners may not own, directly or indirectly, individually or in the aggregate, more than 20 percent of the stock of the corporate general partner or its affiliates. Second, the corporate general partner must meet certain net worth requirements. If total contributions to the limited partnerships are less than $2.5 million, the general partner's net worth must equal $250,000 or 15 percent of the total contributions, whichever is less. If total contributions exceed $2,500,000, the corporate general partner's net worth must equal the sum of all of the net worth requirements pertaining to all of the partnerships in which it has an interest. Third, the purchase of a limited partnership interest cannot entail the purchase of any type of security of the corporate general partner or its affiliates. Finally, the limited partnership must be created and operated in accordance with the applicable state governing statute.

Structuring a limited partnership to undertake a rehabilitation project requires a balancing of the business needs of the parties, the requirements of state partnership law, and the characteristics which the IRS will look to. Beyond the substantive requirements, a limited partnership wishing to obtain an advance ruling on its tax classification must meet three additional conditions (independent of those relating to partnerships with corporate general partners). Under Revenue Procedure 74-17,[40] the general partner or partners must have at least a 1-percent interest in each item of partnership income, gain, loss, deduction, or credit. Second, to obtain an advance ruling, the aggregate

---

[38] Staff of Joint Committee on Internal Revenue Taxation, *Tax Shelters: Use of Limited Partnerships, Etc.*, 94th Cong., 1st Sess., at 4 (Comm. Print, 1975).

[39] 1972-1 C.B. 735.

[40] 1974-1 C.B. 438.

deductions to be claimed by the partners as distributive shares of partnership losses for the first two years of operation cannot exceed the amount of equity contributions to the partnership. Since the "at risk" rule, discussed next, does not apply to real estate partnerships, equity contributions are often lower than would otherwise be necessary, and thus compliance with this condition may be troublesome for some real estate syndicates. Third, no creditor who extends nonrecourse financing to the limited partnership shall have acquired, as a result of making the loan, any interest in partnership profits, capital, or property other than that of a secured creditor. These rules apply solely to requests for advance rulings; they are not used as classification criteria.

3.2.2 Tax Benefits Available to Partners

A limited partnership is an attractive method of organization because of the tax benefits available to each partner. The tax benefits accrue because the partnership is not itself a taxable entity. All tax benefits pass through the partnership to the individual partners. The partners are then taxed on their distributive shares of these gains, losses, income, and credits.

Limited partners thus enjoy the tax benefits available for old building rehabilitation described earlier in this book. This includes the investment tax credit of up to 25 percent on rehabilitation expenses.[41] This can be based upon the fully leveraged cost of the property, rather than the basis of the partners' investment. In addition, the partnership can depreciate the cost of the building plus the expenses of rehabilitation, for another potentially sizable deduction.[42] The depreciation deduction may, of course, be taken only when the building is first placed in service.[43]

The Tax Reform Act of 1976 and the Revenue Act of 1978 retained the real estate exception to the at risk rule, which generally forecloses deductions in excess of the basis of that portion of the partnership property with respect to which a partner is personally liable.[44] While other forms of limited partnerships are now subject to the at risk limitation, real estate syndications can continue to benefit from the tax shelter effect of nonrecourse financing.

A partner can also claim a deduction based on partnership losses. The amount of the deduction is limited by the basis of the partner's interest in the partnership,[45] which is equal to the cost of the partner's

---

[41] *See* Section 1.2.1.

[42] *See* Section 1.3. The taxpayer must, however, reduce the basis of the rehabilitation cost if the credit is taken, by either 50% or 100% depending upon the credit. I.R.C. § 48(q)(1).

[43] Treas. Reg. § 1.167(a)-10(b).

[44] Lynn & Goldberg, *supra* note 1, at 62-63.

[45] I.R.C. § 704(d).

share in that entity. This cost is the amount of cash and the value of property contributed to the partnership, as well as any obligations or liabilities of the partnershp assumed by the partner, reduced by any return of capital or property.[46] Because basis may include a proportionate share of nonrecourse liabilities, it may exceed the amount which would be deemed at risk for an individual partner.

During the construction period, costs will be incurred that are not associated directly with actual construction. Among these are interest on construction loans, taxes, various financing, service and management fees, and other so-called soft costs.[47] Some of these expenses are deductible by the partner in the taxable year in which they are incurred. The general rules are that the expense must be ordinarily and necessarily incurred in the course of profit-seeking activities, that the useful life of the asset acquired not extend beyond the taxable year, and that the expense be in a reasonable amount and for a stated purpose.[48]

Some costs may not be immediately deducted, but must be added to the basis of the partnership's property and depreciated or amortized over the property's useful life.[49] These include costs of construction (which include builder's fees and profits, raw materials, and labor utilized in construction); furniture; fixtures; equipment; and architectural, landscape, design, and legal fees associated with the project. Under the Tax Reform Act of 1976, construction period interest and taxes generally must also be capitalized rather than immediately deducted.[50]

Often, most or all of the partnership tax benefits are allocated to the limited partners to attract potential investors. A limited partner's tax benefits from the partnership's activity are allocated according to the partner's distributive share. This share is generally set out in the partnership agreement,[51] which should specify the partners' respective shares of income, gain, credit, deduction, and loss. If the partnership agreement does not address any of these items, the partner's distributive share is determined by his interest in the partnership.

Under some circumstances, it is also possible to make "special allocations" of specific partnership items. This may be intended to make investment in the partnership especially advantageous for particular individuals, or to allocate tax benefit items to those partners for whom they will do the most good. Special allocations must be defensible, however, and can be subject to challenge and disallowance if they can-

[46] I.R.C. § § 705(a),722, 742; Treas. Reg. § § 1.722-1, 1.742-1.
[47] For a detailed discussion of the tax implications of these costs, see Lynn & Goldberg, supra note 1, at 117 et seq.
[48] Treas. Reg. § 1.162-3, 7.
[49] Treas. Reg. § 1.263(a)-1.
[50] I.R.C. § 189. Exceptions to this rule apply to expenses associated with, inter alia, low-income housing. § 189(d).
[51] I.R.C. § 704(a).

not be justified. Under the Tax Reform Act of 1976, special allocations are permissible only if they can be shown to have "substantial economic effect."[52] If the special allocation is disallowed, any affected items are reallocated in accordance with the partner's interest in the partnership, which is determined by taking into account all the facts and circumstances of a given situation.

3.2.3 Tax Shelter Registration

In an effort to crack down on tax shelter abuses, the Tax Reform Act of 1984 adopted new provisions mandating registration of certain tax shelters and listing of tax shelter investors. These requirements will be applicable to many real estate limited partnerships, and promoters must be careful to comply with them before any limited partnership interests are offered for sale.

Section 6111 of the Internal Revenue Code requires registration of tax shelters if two conditions are met. First, the investment must have been represented as offering a "tax shelter ratio" of better than two to one. The tax shelter ratio is the ratio, in a given year, which the aggregate deductions plus 200 percent of the credits potentially allowable to an investor bears to the total amount invested by the investor. Second, the investment must either be (1) required to be registered under federal or state securities law; (2) exempt from registration under a provision requiring the filing of a notice with a federal or state agency regulating securities; or (3) substantial—that is, involving five or more anticipated investors and an aggregate amount exceeding $250,000. Pursuant to temporary regulations issued by the IRS,[53] registration is accomplished by filing Form 8264, "Application for Registration of a Tax Shelter."

The new law also requires that any organizer or seller of a "potentially abusive tax shelter" must maintain a list of investors available to the IRS upon request.[54] A potentially abusive tax shelter is one required to register under Section 6111, or one which the Secretary of the Treasury determines by regulation to have the potential for tax avoidance or evasion. The information provided by these registration and filing requirements is intended to enable the IRS to monitor tax shelters more efficiently, and increased auditing of such shelters can be anticipated. The IRS has announced that it will issue prefiling notification letters to investors in certain potentially abusive tax shelters, thereby warning such investors that their returns will be audited if they claim tax benefits from the shelter.[55]

---

[52] I.R.C. § 704(b)(2).
[53] 49 Fed. Reg. 32712 (1984); Treas. Reg. § 301.6111-1T.
[54] I.R.C. § 6112.
[55] Rev. Proc. 83-78, 1983-2 C.B. 595.

3.3   FEDERAL SECURITIES LAWS AFFECTING LIMITED PARTNERSHIPS

Although neither the Securities Act of 1933 nor the Securities Exchange Act of 1934 mentions limited partnership interests by name,[56] it is well established that such interests constitute "securities" whose issuance, sale, and transfer is subject to federal and state regulation. The Supreme Court has defined "security" as "an investment in a common venture premised on a reasonable expectation of profits to be derived from the entrepreneurial or managerial efforts of others."[57] The typical limited partnership interest involves a sharing of risks and benefits of a profit-seeking enterprise without direct involvement in the control of the business, and thus clearly falls within this definition.[58]

In general, federal law requires registration of securities offerings, registration of broker-dealers, and compliance with disclosure and anti-fraud provisions. Section 5 of the Securities Act of 1933 requires registration prior to the sale of securities, unless the issuance in question falls within one of the several strictly defined categories of exemption. Registration is a time-consuming and expensive process, and thus qualifying for an exemption is quite advantageous. Offerings of interests in real estate limited partnerships undertaking preservation or rehabilitation work will often fall within one of the available exemptions, which are therefore discussed at some length in this chapter. It should be noted, however, that full disclosure of all material information relating to an offering is required whether a security must be registered or not, and that the anti-fraud provisions, discussed below, apply regardless of any registration requirement.[59]

3.3.1   Regulation D

In 1982 the Securities and Exchange Commission (SEC) adopted regulation D,[60] which set out the criteria for several categories of exemption

[56] The Securities Act of 1933 is codified at 15 U.S.C. § § 77a-77aa, and the Securities Exchange Act of 1934 at 15 U.S.C. § 78a-78jj. Citation will be made to the sections of the acts rather than to the U.S. Code. The securities acts variously define "security" as "certificate of interest or participation in any profit-sharing agreement," "investment contract," and "instrument commonly known as a 'security.'" Securities Act of 1933 § 2(1), Securities Exchange Act of 1934 § 3(a)(10).

[57] United Housing Foundation v. Forman, 421 U.S. 837, 852 (1975); see also SEC v. W. J. Howey Co., 328 U.S. 293, 301 (1946).

[58] See Goodman v. Epstein, 582 F.2d 388 (7th Cir. 1978), cert. denied, 440 U.S. 939 (1979). There may, however, be special situations in which a limited partnership interest does not qualify as a security, as, for example, if the limited partner contributes some managerial skills to the enterprise. See Frazier v. Manson, 651 F.2d 1078 (5th Cir. 1981).

[59] See Securities Act of 1933 §§ 12(2), 17; Securities Exchange Act of 1934 § 10(b); SEC Rule 10b-5, 17 C.F.R. § 240.10b-5.

[60] 17 C.F.R. §§ 230.501 – 230.506, effective April 15, 1982.

from the registration requirement of Section 5. Regulation D, which comprises SEC Rules 501–506, superseded preexisting exemption standards in several respects and introduced some new concepts and definitions which are critical in determining whether an issuance of securities qualifies for exemption. Regulation D is not exclusive, and attempted compliance with its provisions does not prohibit an issuer from claiming any other exemption that may be available. Nevertheless, complying with the relatively clear and straightforward terms of Regulation D will generally be easier than attempting to justify exemption on some other ground, and anyone planning or assisting with an offering should be thoroughly familiar with its provisions. The text of Regulation D is found in Appendix H.

3.3.2   General Provisions of Regulation D

A central concept in Regulation D is that of the "accredited investor." Exemption requirements, as explained in this chapter, are somewhat more liberal when accredited investors constitute some or all of the purchasers in a given offering. Under Rule 501(a), accredited investors include banks, insurance companies, investment companies, and private business development companies. Charitable organizations with assets exceeding $5 million also qualify. Individuals must meet one of several financial requirements to qualify as accredited investors. These requirements are satisfied by anyone whose net worth exceeds $1 million, or whose annual income exceeds $200,000. In addition, accredited investors include those who purchase at least $150,000 of the securities being offered, so long as the investment does not constitute more than 20 percent of the purchaser's net worth.

Under the doctrine of integration, the SEC may treat several ostensibly distinct offerings as one combined offering for purposes of determining eligibility for exemption from registration. If this were not so, issuers might be tempted to subdivide offerings into enough units to qualify for the small offering exemption, and the intent of the registration requirement would be thwarted. Conversely, if no limitations on integration existed, an offering solely to intrastate residents might be integrated by the SEC with one of broader scope, thus making the former ineligible for the intrastate offering exemption.

Rule 502(a) adopts a "safe harbor" rule, compliance with which will insure that integration does not occur. If there are no offers or sales of securities by or for the issuer of the same or a similar class of securities during the six-month periods before and after a Regulation D offering, there will be no integration of that offering with offers or sales outside that time period. If the safe harbor requirements are not met, integration will depend upon the facts and circumstances of the particular issuance. In making a determination, the SEC will consider a number of factors,[61] including whether the offerings are part of a single plan of financing, whether they are made at or about the same time, and whether they are made for the same general purpose.

Rule 502(b) specifies the type of information which must be furnished to potential investors under Regulation D. The specific disclosure requirements applicable to an offering depend upon its nature and the exemption for which it qualifies. No specific disclosure conditions are imposed upon small offerings under Rule 504, or to offerings solely to accredited investors under Rules 505 and 506. Real estate limited partnerships are not typically subject to the periodic reporting requirements of the Securities Exchange Act, and so will be required to comply with the set of Rule 502(b) information requirements applicable to nonreporting, rather than reporting, entities.

In general, the rule requires nonreporting issuers to furnish information "to the extent material to an understanding of the issuer, its business, and the securities being offered." More specifically, for offerings up to $5 million, the issuer must provide information of the kind required by Part I of SEC Form S–18, a prospectus for offerings not to exceed $5 million. For larger offerings, the issuer must furnish the kind of information required in Part I of a formal registration statement. Guide 5, published by the SEC,[62] provides guidance as to the categories of information which the staff looks for in reviewing offering statements. It is not official or binding, however, and not all of the information detailed in the guide need necessarily be included in the statement.

Rule 502(b) also requires an issuer to give each purchaser an opportunity to ask questions about the terms and conditions of the offering, and to supply additional information which can be acquired without unreasonable effort and which is necessary to verify the accuracy of information already disclosed.

3.3.3　Exemptions under Regulation D

When all of its general requirements are met, Regulation D provides three exemptions from registration for offerings by real estate limited partnerships. Under Rule 504, an issuance is exempt if the aggregate offering price does not exceed $500,000.[63] If the offering complies with this dollar limitation, there are no restrictions as to number or type of offerees or investors. Nor is there any specific information disclosure

---

[61] Securities Act Release No. 4434, 26 Fed. Reg. 11896 (December 6, 1961); Securities Act Release No. 1459, 27 Fed. Reg. 11316 (November 6, 1962).

[62] Securities Act Release No. 6384, 47 Fed. Reg. 11480 (March 3, 1982).

[63] This amount must be reduced by the "aggregate offering price for all securities sold within the twelve months before the start of and during the offering of securities under this Rule 504 in reliance on any exemption under section 3(b) of the Act or in violation of section 5(a) of the Act."

requirement under Rule 502(b). There may, however, be state disclosure standards, and federal antifraud rules discussed herein, will apply in any event.

To obtain exemption under Rule 505, the offeror must comply with two sets of restrictions. First, the total amount of the offering cannot exceed $5 million,[64] and as with the other Regulation D exemptions, the integration standards of Rule 502(a) apply. Second, the entity issuing the securities (here, limited partnershp interests) must reasonably believe that there are not more than 35 purchasers in the given offering. In calculating the number of purchasers, however, accredited investors may be excluded.

The first criterion for exemption under Rule 506, like Rule 505, is that the issuer reasonably believe there are no more than 35 purchasers in the offering, excluding accredited investors. The second criterion is somewhat more complex. The issuer must reasonably believe that each purchaser who is not an accredited investor "either alone or with his purchaser representative(s) has such knowledge and experience in financial and business matters that he is capable of evaluating the merits and risks of the prospective investment." The specific content of this "sophistication" requirement is not spelled out in the rule, and will probably involve a number of factors, including the complexity of the offering, the amount of information disclosed, and the level of knowledge of the purchaser. If the purchaser himself does not possess adequate sophistication, the rule will be satisfied if his representative, who assists him in evaluating the prospective investment, is sufficiently knowledgeable. The status of the representative must, however, be acknowledged by the purchaser in writing, and certain conflicts of interest between the issuer and representative must be either disclosed or avoided altogether.[65]

3.3.4   Other Exemptions

Regulation D was intended to simplify the conditions of eligibility for several of the statutory exemptions from registration under the Securities Act of 1933. Its procedures do not, however, represent the exclusive means of complying with the statute. An issuer who does not wish to satisfy all of the requirements of Regulation D, cannot do so, or attempts to do so and fails, may attempt to avail himself of other avenues to the statutory exemptions discussed below. In seeking such exemption, however, the burden of proof as to each requisite element is on the issuer.[66] Thus, compliance with Regulation D wherever possible is likely to be the preferable course.

---

[64] The same deductions must be made in calculating this amount as are required under Rule 504.

[65] "Purchaser representative" is defined in Rule 501(h).

[66] *See* Lively v. Hirschfeld, 440 F.2d 631, 632 (10th Cir. 1971).

Section 4(2) of the Securities Act exempts from registration "transactions by an issuer not involving any public offering." The Supreme Court has held that the construction of this exemption "should turn on whether the particular class of persons affected need the protection of the Act."[67] The SEC adopted Rule 506, which replaced Rule 146, as one means of compliance with the statutory "private placement" exemption. An issuer can, however, claim to fall within Section 4(2) even without complying with Rule 506.

Both Regulation D and the older Regulation A, which provides a somewhat more complicated exemption procedure, were adopted under the authority of Section 3(b) of the Securities Act of 1933, which authorized the SEC to create exemptions for small offerings. As just discussed, however, the SEC regulations are not all-encompassing, and failure to comply with their requirements does not automatically make the statutory exemption of Section 3(b) unavailable. An offeror may be able to demonstrate eligibility for exemption directly under the terms of the statutory provision. In addition, Section 4(6) of the Securities Act, enacted in 1980, offers an exemption related, although not identical, to that of Section 3(b). Basically, Section 4(6) exempts transactions in which the total offering price is less than $5 million, and in which there is no advertising or public solicitation. Each section may, in certain circumstances, offer an exemption where the specific requirements of the other, or of the subsidiary regulations, would not.

Section 3(a)(11) of the Securities Act exempts from registration offerings extended only to residents of the state in which the issuer does business. This exemption is somewhat less useful than it appears. An offer to even a single nonresident makes the exemption inapplicable, so an issuer should be certain that he will qualify for the exemption before placing reliance upon it. Rule 147[68] defines the significant terms in Section 3(a)(11) in a rather technical manner. Exemption under either the statute or the rule does not, of course, provide exemption from registration requirements under the state securities laws.

3.3.5 Fraud and Rule 10b–5

Even when an offering qualifies for one of the exemptions just discussed, the issuer must be careful not to run afoul of the anti-fraud provisions of the securities laws. In general, the issuer's duty is twofold: He must not misrepresent any material fact in the disclosure provided to potential purchasers of the security, nor may he withhold any material information. If he does either of these, he risks a claim for rescission of the purchase agreement or for damages under, *inter alia*, Rule 10b-5 of the SEC.[69]

[67] SEC v. Ralston-Purina Co., 346 U.S. 119, 125 (1953).
[68] 17 C.F.R. § 230.147.
[69] *Id.* § 240.10b-5.

Rule 10b-5 was issued to enforce Section 10(b) of the Securities and Exchange Act, which contains a general prohibition on the use of manipulative or deceptive devices in connection with the purchase or sale of any security. This broad antifraud provision is fleshed out by Rule 10b-5, which makes it unlawful, in connection with the purchase or sale of any security—

(a)   to employ any device, scheme or artifice to defraud;

(b)   to make any untrue statement of a material fact or to omit to state a material fact necessary in order to make the statements made, in the light of the circumstances under which they were made, not misleading; or

(c)   to engage in any act, practice, or course of business which operates or would operate as a fraud or deceit upon any person.

Because limited partnership interests generally qualify as securities, a general partner selling such interests is likely to be subject to this proscription.[70]

The breadth of Rule 10b-5's coverage can perhaps best be illustrated by listing the elements of a fraud claim under that rule:

1. *Purchase or sale.*   Under the so-called *Birnbaum* doctrine,[71] adopted by the Supreme Court in *Blue Chip Stamps v. Manor Drug Stores,*[72] only a purchaser or seller of securities can maintain an action under Rule 10b-5. An individual who asserts that he decided not to buy or sell as a result of the misrepresentation or omission of material information is barred from bringing a fraud claim under the rule.

2. *Manipulation or deception.*   The rule reaches a broad range of untrue statements, misleading statements, or omissions. The fraudulent statements can be made in any manner: in filings with the SEC, in disclosure statements to offerees, or in oral communications. In the partnership context, allegations of misrepresentation by general partners of the value and permanence of their contributions to the partnership, and of the transfer of other businesses to the partnership, have been held to support a 10b-5 claim by the limited partners.[73] The withholding of information about obstacles to accomplishing the object of the partnership, prior to the final payment of the limited partners' contributions, has also been held to be fraudulent under 10b-5. [74]

3. *Materiality.*   For purposes of Rule 10b-5, information misrepresented or omitted is considered material if a reasonable man, in deter-

[70] *See, e.g.*, Goodman v. Epstein, 582 F.2d 388 (7th Cir. 1978).

[71] Birnbaum v. Newport Steel Corp., 193 F.2d 461 (2d Cir. 1952).

[72] 421 U.S. 723 (1975).

[73] McGreghar Land Co. v. Meguiar, 521 F.2d 822 (9th Cir. 1975).

[74] Goodman v. Epstein, *supra*.

mining his choice of action as to the transaction in question, would attach importance to the information.[75] If the offeree's decision whether to buy, or the investor's decision whether to sell, would reasonably be affected by the information, it is material.

4. *Scienter.* Although the terms of neither the statute nor the rule seem to require it, the Supreme Court has imputed a requirement that the putative offender must have acted with scienter, or intent to deceive, manipulate, or defraud in order to be liable under 10b-5.[76] Under this view, mere negligent conduct does not rise to the level of fraud and hence is not actionable under this provision.

5. *Reliance or causation.* There must be a causal relationship between the fraudulent act and the sale or purchase in order to state a 10b-5 claim. Where the alleged act involves misrepresentation, the plaintiff must demonstrate that he relied on the misrepresentation.[77] Where the fraud involves omission, however "positive proof of reliance is not a prerequisite to recovery. All that is necessary is that the facts withheld be material in the sense that a reasonable investor might have considered them important in the making of this decision."[78]

The Securities Act contains a broad antifraud provision which forbids substantially the same conduct as Section 10(b) of the Securities Exchange Act. Section 17(a) of the Securities Act[79] makes it unlawful to use fraudulent devices in the offer or sale of any securities in interstate commerce. The two provisions have only "minimal differences."[80] Section 17(a) applies solely to the sale of securities, rather than purchase or sale, but it is typically in relation to sale that fraudulent activities are alleged. The major difference between the two relates to the availability of private enforcement. It is well established that private suits may be brought by aggrieved parties under Section 10(b).[81] By contrast, while federal courts have indicated that a private remedy is also available under Section 17(a),[82] the Supreme Court has not affirmed that reading of the law.[83]

---

[75] SEC v. Texas Gulf Sulphur Co., 401 F.2d 833, 849 (2d Cir. 1968), *on remand*, 312 F. Supp. 77 (S.D.N.Y. 1970), *aff'd*, 446 F.2d 1301 (2d Cir.), *cert. denied*, 404 U.S. 1005 (1971).

[76] *See* Ernst & Ernst v. Hochfelder, 425 U.S. 185 (1976).

[77] *See, e.g.,* Safecard Services, Inc. v. Dow Jones & Co., 537 F. Supp. 1137 (E.D. Va. 1982).

[78] Affiliated Ute Citizens v. United States, 406 U.S. 128, 153–54 (1972).

[79] 15 U.S.C 77q(a).

[80] Stephenson v. Calpine Conifers II, Ltd., 652 F.2d 808, 815 (9th Cir. 1981).

[81] Ernst & Ernst v. Hochfelder, *supra*, 425 U.S., at 196.

[82] *See e.g.,* Stephenson v. Calpine Conifers II, Ltd., *supra*); Kirshner v. United States, 603 F.2d 234 (2d Cir. 1978).

[83] Herman & MacLean v. Huddleston, 459 U.S. 375 n.2 (1983).

3.4 STATE SECURITIES REGULATION

The power of the states to regulate securities activities is specifically reserved by the federal securities laws.[84] The legal requirements imposed by a state are independent of the requirements under federal law. Thus, in addition to any federal requirements, a securities transaction in the form of a transaction involving limited partnership shares may be subject to the further requirements of those states where the interests are being offered or sold.[85]

Every state has enacted laws regulating securities. These "blue sky" laws[86] typically contain any or all of three central components: (1) registration requirements for securities to be sold in the state, (2) antifraud and disclosure requirements, and (3) registration requirements for brokers and dealers. Some states also have suitability requirements governing both investors and the merits of the offering. Although the types of restrictions are generally similar, the specific provisions of various blue sky laws vary significantly.

Many states require that securities be registered with a state's securities officer or commission prior to sale within the state. This frequently involves filing an application form with a filing fee, providing disclosure information through a prospectus or memorandum, and furnishing other material exhibits and additional information. This process is simplified in some states which utilize a method called coordination, by which a copy of the prospectus and information filed under the Securities Act of 1933 is submitted in full or partial satisfaction of the state requirement. If the offering is not registered with the SEC, then the issuance must meet the state's "qualification" requirements before any sales can be made. In this regard, the powers of some state securities commissions are broader than those of the SEC. While the SEC can only require full disclosure prior to an offering, some state securities administrators can actually refuse registration of offerings if they find them unfair to investors.

Sales of limited partnership interests may benefit from exemptions from registration that exist in many states. Generally, these involve offerings to a limited number of investors, but their precise terms vary significantly. Some count offerees, others purchasers, and the maximum number allowed varies from 3 to 50. Moreover, New York and New Jersey require compliance with laws relating specifically to real estate syndications, in addition to their general securities laws.

---

[84] Securities Act of 1933 § 18; Securities Exchange Act of 1934 § 28(a).

[85] *See generally* Lynn & Goldberg, *supra* note 1, at Chap. 10.

[86] The term *blue sky* comes from a U.S. Supreme Court case, Hall v. Geiger-Jones Co., 242 U.S. 539 (1917), which described the aim of these laws as the prevention of "speculative schemes which have no more basis than so many feet of blue sky."

In some states, exemptions are self-activating: Qualifying issuances do not have to comply with any procedural requirements. Other states require the filing of a notice or other information with the state securities commission. In some states, the rules of eligibility for exemption are clearcut; in others, availability of the exemption rests within the broad discretion of the commission. The securities laws of New York and California are representative of approaches taken by different states, and are discussed in some detail in the following sections.

3.4.1 New York Securities Law Governing Real Estate Limited Partnerships

Sales of limited partnership interests in real estate which are offered in New York must comply with the Real Estate Syndicate Act.[87] Any public offering of securities which "consist primarily of participation interests or investments in one or more real estate ventures" is subject to the act's requirements.[88]

No public offering of these interests may be made in or from New York until an offering statement or prospectus has been submitted to the Department of Law. The information that must be included in the submission is described in detail in Section 352–e(b) of the General Business Law, and includes a description of the property, the nature of the interests being offered, and information on the principals involved in the offering. Whether New York filing requirements apply depends upon whether the offering is made in New York, not whether the property is located there.[89]

The statute does not define the parameters of a "public" offering. In certain circumstances, however, the attorney general in his discretion will furnish a "no action" letter confirming that registration is not required. This is generally possible when an offering is made to no more than nine offerees who are wealthy and sophisticated and who have a preexisting relationship with one or more principals of the issuer or promoter.

In addition, an issuer may apply for an exemption from Section 352-e filing requirements under Section 352-g. That section permits an exemption, at the attorney general's discretion, for an offering to up to 40 individuals, or for an offering which has been either registered under, or exempted from, the federal securities laws. The first basis of exemption requires that the entire offering throughout the world be made to no more than 40 people. If the number of offerees cannot be controlled (and the application should indicate how the number will be

---

[87] Gen. Bus. Law § 352-e to 352-j.

[88] *Id.* § 352-e.

[89] *See* Ledgebrook Corp. v. Lefkowitz, 354 N.Y.S.2d 318 (Sup.Ct. 1974).

restricted), this exemption will not be available. The attorney general may, in addition, require some representation as to the net worth of the investors to establish exemption on this ground.

Alternatively, an issuer may seek exemption on the basis that the offering has been exempted from federal securities law registration. The exemptions may not be precisely coextensive, however. For example, New York does not accept the Regulation D provisions for distinguishing accredited from nonaccredited investors. While the attorney general does not impose a uniform standard in determining whether an offering is suitable for all potential investors without registration, he may require that some showing be made that all purchasers, not just accredited investors, meet an adequate suitability standard. Section 352-g will also provide exemption from offerings exempted under Section 4(2) of the Securities Act of 1933,[90] where Regulation D exemption cannot be obtained. Intrastate offerings exempted under Section 3(a)(11) of the Securities Act,[91] however, are not eligible for a New York exemption under Section 352-g.

If an issuer falls within the scope of Section 352-e, he will also be required to register as a securities "dealer" under Section 359-e. Exemption from the dealer registration requirements is available, however, if the issuer is involved only in a limited offering of real estate limited partnerships to no more than 40 individuals.[92] Application for both exemptions (from the public offering and securities dealer registration requirements) can be filed together with the Department of Law. It should be noted that, as a general matter, using a broker or dealer as an agent in selling partnership interests will not exclude an issuer from the definition of a dealer.[93]

3.4.2 California Securities Law Governing Real Estate Limited Partnerships

Securities transactions in California are governed by the Corporations Code.[94] The statute requires the issuers of securities in California to comply with qualification requirements, analogous to registration procedures under federal law and filing procedures under New York law.

There are two means of avoiding the burden of the California qualification requirements that may be available to offerors of limited part-

---

[90] See Section 3.3.4 above.

[91] See id.

[92] Gen. Bus. Law § 359-f(2.)(d).

[93] Id. § 359-e(1.)(a).

[94] Cal.Corp.Code §§ 25000 to 25706.

[95] Id. § 25102(f).

nership interests. The first is the limited offering exemption.[95] The criteria for this exemption largely track those of federal Regulation D, but do vary somewhat. The statute requires a maximum of 35 purchasers, with some of the same exclusions from the count as are permitted under SEC Rule 501: purchasers of at least $150,000 of securities in the offering, purchasers with net worth exceeding $1,000,000 or annual income exceeding $200,000, and certain persons affiliated with the offeror. Every purchaser must have either a preexisting personal or business relationship with the offeror or an individual associated with it, or personally or together with his professional advisers be capable of protecting his own interests in connection with the transaction. There is no limit on the number of offerees under this exemption, nor is any particular disclosure required.

The commissioner of corporations may also, in certain circumstances, permit an offering to be qualified by means of a limited offering qualification, which relieves the offeror of the duty of compliance with some or all of the requirements.[96] Eligibility for this treatment depends upon meeting suitability requirements for California investors (generally, net worth of $250,000 to $500,000 or more and substantial income) and providing investors with written disclosure of areas in which the offering does not meet the commissioner's usual standards. There will also be restrictions on transfer of partnership interests.

Four

Financing Rehabilitation: Public and Nonconventional Sources

Introduction

Having put together the appropriate organizational structure for undertaking a rehabilitation project, the next step is to obtain the necessary funding. In most cases, syndicates will first approach banks or other private lenders for construction period or longer-term mortgage financing. In addition, however, there are public and private sources of preservation and rehabilitation funding which should be explored in putting together the most advantageous funding package.

In recent years, public funds earmarked exclusively for rehabilitation have become quite scarce. However, a number of governmental funding sources exist which, while not primarily designed to support rehabilitation, can be used for that purpose. One example of such a program source is the federal Urban Development Action Grant program, discussed in greater detail in this chapter. While development often means new construction, it is also true that rehabilitation can be an important tool in community development and revitalization. Other housing and urban development programs similarly make funds available for rehabilitation purposes. By moving into these more general financing programs, the preservationist will encounter an application process that placed no special emphasis upon the goal of preservation. Under the UDAG program, for instance, urban development is the objective, and that end will be sought without a programmatic preference for either rehabilitation or new construction. Consequently, it will be necessary to develop more aggressive strategies to "sell" rehabilitation proposals to these funding sources.

Similarly, foundation and corporate donors may not give special priority to preservation projects, but may be willing to provide support for them. The primary objective of such funding sources may be civic improvement or cultural development, goals to which historic preservation can make a significant contribution. The challenge, again, is to demonstrate that rehabilitation can compete for funding in these categories.

Not only should more varied sources for funding be explored, but financing should be sought for limited portions of a project. For example, certain energy programs may provide assistance for insulating old houses. With less money available, no single source is likely to cover all financing needs, and owners should be familiar with the full breadth of available programs in order to assemble a financing package.

There are a number of factors which a developer should consider in attempting to build a broader financial base.

1. *Eligibility.* Each funding source has its own eligibility criteria. An applicant must determine whether the funding source considers his project eligible for financial support.

2. *Use of structure.* Some funding sources are available only if the building is put to a particular use. Thus, an applicant who intends to rehabilitate a building as, for example, a senior citizen community center, may find that the project qualifies for funding not because it is a rehabilitation project, but because it will benefit senior citizens.

3. *Project location.* Some funding programs seek to provide incentives aimed at improving the condition of particular areas. Thus, a program may make funds available only for use in urban neighborhoods, rural areas, or commercial districts.

4. *Size of undertaking.* The size of the undertaking is a significant consideration, since it affects both the type and amount of financial assistance that must be injected to get the project off the ground. In some cases, private financing will suffice to initiate a venture; in other circumstances, a large grant or loan guarantee will be required before other financing is forthcoming.

5. *Project time constraints and the funding bureaucracy.* The time which elapses between submitting an application and receiving funding approval varies among programs. The applicant attempting to assemble a funding package must coordinate agency schedules with his own funding requirements.

Since both sources and amounts of funding available change over time, current information on public and private funding sources is essential. Up-to-date information can be acquired from reference sources at many community and university libraries. Libraries of nonprofit organizations can be extremely helpful and many of these are accessible to the general public.[1]

4.1   PUBLIC SOURCES: THE FEDERAL GOVERNMENT

A wide array of financing assistance is potentially available at the federal level. It ranges from direct grants and loans to indirect forms of assistance such as tax incentives and loan guarantees. In addition to absorbing part of project costs, government financing assistance can lower interest rates or extend a loan's term, or share some of the project's risk by providing protection for private lenders. The types of assistance include the following:

Formula grant. Allocation of money to states or their subdivisions according to a distribution formula prescribed by law or administrative regulation, with or without restrictions on the purposes for which funds may be used.

Project grant. The funding of specific projects for a fixed period.

Direct payments for specified use. Financial aid provided directly to eligible beneficiaries by the federal government with restrictions on how the money is spent.

Guaranteed/insured loans. Loan guarantee or insurance programs to protect a lender and thereby encourage it to extend financing it might not otherwise provide. These programs insure the lender against default in whole or in part by the borrower. Many lending institutions which have not previously loaned money for rehabilitation projects view them as a larger risk than conventional construction programs, and indemnification may be essential to their participation.

Direct loan. Direct lending of federal moneys for a specific period of time, without interest or at below-market rates.

---

[1] For the most current information regarding the funding status and application procedures of all of the following federal funding programs, *see* U.S. Office of Management and Budget, *Catalogue of Domestic Federal Assistance Programs*, issued annually by the Government Printing Office.

4.1.1 Historic Preservation Programs

Federal assistance targeted for historic rehabilitation has been dramatically reduced in recent years. While many of the programs remain authorized and intact in name, appropriations have been slashed. Programs that once provided grants and loans to eligible projects now may insure loans only if the applicant can find financial backing privately. Other programs no longer participate in rehabilitation at all and are limited to property surveying and information dissemination.

The National Historic Preservation Fund grants program, authorized by the National Historic Preservation Act (NHPA),[2] continues to be funded in Fiscal Year (FY) 1985. Once a source of funds for acquisition or development of historic properties, however, historic preservation grants-in-aid are no longer available for that purpose. Since FY 1983, grant uses have been restricted to financing state staff salaries, equipment, materials, and travel necessary to expand and maintain the National Register. In FY 1983, $51,000,000 was appropriated; that has been reduced to $25,480,000 for FY 1985.[3]

The historic preservation grant program is administered by the NPS. States, territories, and the National Trust may apply to the Secretary of Interior for annual grants for anticipated projects. These entities may subgrant to public and private parties to accomplish program objectives. States must subgrant 10 percent of the funds to local governments that have met certain requirements.

Assistance is provided on a formula and matching grant basis. Federal grants can fund up to 50 percent of project costs, while the remaining 50 percent must come from state and/or private funds and/or allowable in-kind donations. However, if the funds are for state and local historic resource surveys, the federal share is increased 70 percent.[4]

To become eligible for these funds, a state must appoint a state historic preservation officer (SHPO) to oversee a program of federally specified preservation activities, including participation in the National Register nomination process. Within the federally mandated framework, however, each state makes its own decisions regarding which projects to fund.

4.1.2 Community Revitalization and Development Programs

The United States Department of Housing and Urban Development (HUD) administers a number of community development programs which may be used to assist rehabilitation activities. Although not in-

[2] 16 U.S.C. § 470a(d).
[3] 3 *Preservation L. Rept.* 1059 (1984).
[4] 16 U.S.C. § 470b(a)(3).

tended exclusively for preservation funding, HUD's block grant programs have provided more funding for these activities than have programs tailored specifically for preservation. These programs provide a variety of financing, including direct grants, loans, guaranteed loans, and mortgage insurance.

A group of the more significant programs are under the general supervision of HUD's Office of Community Planning and Development. However, it is important to note that these programs often have their own offices with separate administering officials.

4.1.3   Urban Development Action Grants

States, local governments and urban counties may obtain Urban Development Action Grant (UDAG) funding to help alleviate severe economic distress or physical and economic deterioration in particular areas.[5] A UDAG is awarded on a competitive basis as a categorical grant—that is, a grant to support projects of a statutorily defined nature. These units of nonfederal government then use the federal funds for loans to developers of eligible projects which will spur economic development in distressed areas. Among the project uses to which these grants can be applied is preservation and restoration of historic properties.

While authority for implementing the UDAG program is vested in HUD's Office of Community Planning and Development, many crucial decisions are made at the local level, including how aggressively the local government will compete to obtain UDAG funds from the federal government. The UDAG program is a large federal development assistance undertaking, both in dollars expended and number of projects funded, but funding is increasingly competitive as appropriations are reduced and more communities become aware of the prospective benefits of UDAG support. The purpose of the UDAG program is to provide sufficient public financial support for individual qualifying projects so as to spur private investment. In this sense, UDAGs are a catalyst. They cannot be the sole source of a project's support, but rather leverage their financing with other private or public funds committed to a project. UDAGs are designed to bridge the gap between the return on investment a private investor could expect in a profitable location and that anticipated in a deteriorated area.[6] Support from UDAG funds is available only when necessary to bring a project to fruition.

New York City, for example, lends UDAG funds to project developers at rates which are generally below the market rate, under the auspices of the New York City Public Development Corporation. UDAG funds

[5] 42 U.S.C. § 5318.
[6] C. Duerksen, ed., *A Handbook on Historic Preservation Law* at 324 (1983) (hereinafter cited as "Handbook").

have been used in New York City for acquisition of land and buildings, infrastructure improvements, construction, and renovation. The loans have been sizable; their average amount is $300,000.

UDAGs are awarded based upon a set of "distressed factors." Although criteria vary somewhat depending upon the size of the recipient community, eligibility of cities and counties for grants depends upon (1) the percentage of housing constructed before 1940, (2) per capita income change, (3) the percentage of poverty, and (4) population growth lag or decline.[7] In addition, to achieve eligibility a locality must have "demonstrated results in providing housing for low and moderate-income persons and in providing equal opportunity in housing and employment for low and moderate-income persons and members of minority groups."[8]

4.1.4 Community Development Block Grants

Community Development Block Grants (CDBGs)[9] are also administered by HUD's Office of Community Planning and Development. These grants are designed to assist communities in meeting development needs, particularly by providing housing to their citizens, enhancing both their neighborhood and business environments, and broadening economic opportunities for low and middle-income individuals. One of the objectives of the program is the restoration and preservation of buildings of special value for historic, architectural, or aesthetic reasons.[10]

Metropolitan cities, urban counties, certain units of local government, and suburban areas are eligible for these funds, although the requirements for cities of differing populations vary. Communities with a population exceeding 50,000 receive the grant as an entitlement; the amount of grant assistance is determined by a statutory formula. Other entitlement recipients are central cities of federally designated standard metropolitan statistical areas (SMSAs) and urban counties with a population of more than 200,000 excluding entitlement cities. Communities with a population of less than 50,000 are eligible on a competitive basis. Although the actual applicants are units of state, county, and local government, the benefits are intended to pass through to low and moderate-income residents (generally defined as families with less than 80 percent of the median family income for the area). Grantees must spend at least 51 percent of the funds they receive on activities to benefit these residents.

[7] 24 C.F.R. § 570.452.
[8] 42 U.S.C. § 5318(b).
[9] Statutory authorization for CDBGs is found in 42 U.S.C. §§ 5301-17.
[10] *Id.* at § 5301(c)(7).

There are many possible uses of CDBGs for rehabilitation projects. Communities develop their own programs and funding priorities consistent with federal standards and the program's primary aims of benefiting low and moderate-income persons, eliminating blight or slums (or preventing their formation), and meeting urgent community development needs. Localities exercise significant discretion. Some of the activities that communities may fund for rehabilitation purposes include acquisition of property, relocation, code enforcement, energy efficiency maintenance, rehabilitation counseling, direct expenditures for rehabilitation of residential and nonresidential structures, and the establishment of revolving funds for these purposes. Moreover, the regulations specifically provide that "CDBG funds may be used for the rehabilitation, preservation, and restoration of historic properties, whether publicly or privately owned." Historic properties are defined to include those listed or eligible for listing on the National Register, those listed on a state or local inventory of historic properties, or those designated as state or local landmarks or historic districts.[11]

Communities may use the funds to support their own rehabilitation projects, or they may contract with other local agencies or nonprofit organizations to carry out all or part of their neighborhood revitalization programs. Communities may also designate subgrantees, such as neighborhood-based nonprofit organizations, Section 301(d) Small Business Investment companies, or local development corporations, to engage in projects designed to revitalize a neighborhood or enhance community economic development. CDBG funds may also be used by cities to provide financial assistance to profit-making businesses to engage in these activities, so long as they meet program objectives.

Citizen participation regarding the use of CDBG funds is required. Local governments must schedule and conduct public hearings on proposed projects.[12]

CDBGs offer greater flexibility than any other community development program. Financial assistance may take the form of grants, loans, loan guarantees, interest subsidies on private loans, or principal-subsidized loans.

In New York City, one use of federal community development funds has been to provide low-interest home improvement loans. Lower and moderate-income families who are owner-occupants of one to four-unit homes are eligible to borrow up to $10,000 for such purposes as rehabilitation. If the home is in a historic district, the city's Department of Housing Preservation and Development will provide assistance in meeting the Landmarks Preservation Commission's standards for approval of the work. Application for the loans is made at participating

[11] *See* 24 C.F.R. § 570.202(d).
[12] 24 C.F.R. § 570.301(a)(2)(ii).

lending institutions which provide Federal Housing Administration Title I loans, discussed in the next section. Supplementing commercially available moneys with public funds allows a reduction of interest rates to well below the market rate.

4.1.5 Federal Housing Administration Insurance Programs

The Federal Housing Administration (FHA) administers a large number of guaranteed and insured loan programs. By sharing the risks involved in extending credit with private lenders, the government hopes to encourage a better flow of financing for targeted activities. FHA loan insurance programs require the benefited projects to meet HUD standards for rehabilitation. These standards are expressed either as minimum design criteria (including construction standards) or as minimum property standards.

Under the Title I Home Improvement Loan Insurance program,[13] HUD insures loans that applicants obtain through private lending institutions for improvements, repairs, and alterations of both residential and commercial buildings. Participating private lenders process applications and determine eligibility criteria. Loans may be extended up to $8,750 per unit, to a maximum of $43,750 in a multiunit building. Single-unit structures can receive a maximum of $15,000. The maximum term of the loan is 15 years.

The Historic Preservation Loan Program[14] insures loans for rehabilitation of properties either listed on the National Register or certified by the Secretary of the Interior as meeting National Register criteria. Administered by HUD, this program is a variant of Title I property insurance designed to stimulate lending in depressed areas.[15]

Rehabilitation Mortgage Insurance, also known as 203K loans,[16] represent another program that may be available as a source of rehabilitation funds. Through the FHA, HUD provides loan guarantees and insurance to protect lenders against loss on residential financing. These loans may be provided to individuals to finance both the acquisition and rehabilitation of one to four-family dwellings. In addition, the loans may be used to refinance outstanding debts from rehabilitation activities.

4.1.6 HUD Housing Programs as Rehabilitation Incentives

Housing assistance programs provide an incentive to rehabilitate building sites either through direct loans, loan guarantees, or rent subsidies. In theory, the latter provide an indirect incentive to rehabilitate

[13] 12 U.S.C. § 1703.

[14] *Id.*

[15] *See Handbook, supra* note 6, at 313–14.

[16] 12 U.S.C. § 1709(k).

because an owner, after rehabilitating, can (subject to local rent control or stabilization laws) charge lower and middle-income tenants the market rate for these units and the federal government will cover the difference between this market rate and a fixed percentage of the tenant's income. However, the bulk of federal rental subsidies has been provided through HUD's Section 8 housing assistance payments program.[17] In 1983 that program was amended to deny further funding for new construction or for substantially rehabilitated housing. The moderate rehabilitation program remains in effect, but the Section 8 program overall no longer provides much assistance for preservation and restoration of old and historic buildings. Two new programs were enacted in 1983, however, as part of the Housing and Urban/Rural Recovery Act,[18] and were designed to fill part of the void left by the restriction of Section 8 assistance. These new programs are discussed in the following sections.

Patterned after the UDAG program, Housing Development Grants (HODAGs)[19] are awarded by HUD to localities which, in turn, provide funds as grants or loans to sponsors of new construction or substantial rehabilitation projects, including historic preservation. Subsidies may be used for rental, cooperative, or mutual housing projects and can fund up to 50 percent of construction or rehabilitation costs. Funding is available to 1,470 cities nationwide, based upon the extent of housing built before 1940 occupied by lower-income households, overcrowding among those households, the level and duration of rental housing vacancies, and the need for new housing. Projects receiving assistance must remain as rental housing for at least 20 years.

Appropriations for this program are modest in comparison to those which had been available under Section 8. Two hundred million dollars were available in FY 1984, and $115 million in FY 1985. As of the first application deadline, August 14, 1984, $510 million in grant funds had been requested. While there are no guarantees that the program will continue to be funded in future years, it appears to have strong support in Congress and within HUD.

The Rental Rehabilitation Program was also enacted in 1983.[20] Unlike HODAG, this program supports rehabilitation of rental properties which are initially occupied by low-income households but which may not continue to be. Aided units need not be retained as low-income housing for any prescribed period of time, and rents on the units are permitted to rise to free market levels. Protection for lower-income tenants is provided through rent subsidies rather than through rent controls. Projects must, however, be located in low-income neighborhoods.

[17] 42 U.S.C. § 1437f, repealed by Pub. L. 98-181 (1983).
[18] Pub. L. No. 98-181, 97 Stat. 1153 (1983).
[19] 42 U.S.C. §§ 1437, 3535(d).
[20] *Id.*

Grants may cover as much as 50 percent of rehabilitation expenses, up to a maximum of $5,000 per unit (adjustable upwards in some areas). Restrictions are rather stringent: Funds may be used only to correct substandard conditions, make essential improvements, or repair major systems in danger of failure. Rent subsidies may be provided to eligible tenants, to a maximum of the difference between 30 percent of the tenant's income and either market level rents or the actual rent on the occupied unit, but no more than a specified ceiling. Funds are awarded by formula, not competitively, to states and eligible localities.

Under the Urban Homesteading program,[21] individuals can obtain properties at low cost from local governments by agreeing to rehabilitate them to code standards within 3 years. They must also agree to live in the rehabilitated building for a minimum of 5 years. Title is conveyed after the 5-year period has expired. Properties that are used for the homesteading program generally are transferred to the local government from the federal government through HUD, the Veterans Administration, the Federal Housing Administration, or the Farmers Home Administration, which acquire the properties through foreclosure in their mortgage programs.

4.1.7 Farmers Home Administration Programs

Low and moderate-income households in rural areas are eligible under Farmers Home Administration programs for grants and low-interest loans under the following programs.

When conventional financing for purchasing, constructing, repairing, or rehabilitating a home in a rural area cannot be obtained, low-interest long-term loans (up to 33 years) may be available to low and middle-income individuals under the Section 502 program.[22] A town is deemed rural if it has a population of no more than 10,000, is rural in character, and is not closely associated with an urban area. Residents of a town may also be eligible if the town's population is between 10,000 and 20,000 persons and it is outside an SMSA. Applicants living in suburbs are eligible.

The maximum loan varies depending on the state of the housing market in the particular area. The rate of interest is adjusted to reflect the applicant's economic circumstances.

An owner-occupant of a home in a rural area who does not have the financial resources to qualify for a loan under the Section 502 Rural Housing program may be eligible for a Section 504 loan, grant, or combined loan and grant to repair his home to make it a decent, safe, and sanitary place to live.[23] The loans and grants under this program

[21] 12 U.S.C. § 1706e.

[22] 42 U.S.C. §§ 1471, 1472.

[23] 42 U.S.C. § 1474.

are, however, fairly small. Eligible individuals may receive a loan no greater than $7,500. If the eligible individual is more than 62 years old, he may receive a grant of up to $5,000.

Under the Rural Rental Housing Loan program, loans may be awarded or insured for the purchase, construction, or rehabilitation of rental housing in rural areas for low and moderate-income individuals, families, or senior citizens.[24] The type of project usually funded is multiunit or cooperative housing. The loans may be for up to $2 million with a maximum term of 50 years. Loans are available to nonprofit corporations and consumer cooperatives, while individuals, nonprofit and profit-making corporations, cooperatives, limited partnerships and nonfederal government agencies are eligible for loan insurance.

On a smaller scale, the Farmers Home Administration extends direct long-term loans (up to 40 years) to construct, enlarge, expand, or rehabilitate existing community facilities.[25] Facilities must be located in rural areas or in towns with populations of no more than 20,000. The public use facility must be one which provides an essential service, such as a library, hospital, fire station, transportation center, or community building providing social, cultural, or recreational benefits. Eligible applicants for the loans include nonprofit organizations and units of local governments, including special purpose districts such as school districts.

4.1.8 Access to Credit

Many preservation projects require access to credit. Unfortunately, credit has not always been forthcoming for neighborhood revitalization, often because banks have been reluctant to make loans for projects in rundown neighborhoods.

For this reason federal laws were enacted to monitor the availability of funds for projects in these credit-starved neighborhoods. The federal role in this regard is carried out through the Community Reinvestment Act (CRA).[26] Under the CRA, a lender must help to meet the credit needs of all the neighborhoods in its community, including moderate and low-income neighborhoods. Agencies which exercise supervisory responsibilities over financial institutions, such as the Federal Reserve Board, the Federal Deposit Insurance Corporation, the Comptroller of the Currency, and the Federal Home Loan Bank Board must consider a financial institution's credit and investment practices in the community when the bank or other lending institution applies for a new branch, relocation of a home or branch office, bank corporate reorganization (such as merger or consolidation with another financial institution) or charter. If the financial institution is found to be "redlining"—that is,

[24] 42 U.S.C. § 1485.
[25] 7 C.F.R. § 1942, Subpart H.
[26] 12 U.S.C. §§ 2901–05.

uniformly denying credit to applicants in particular neighborhoods—the agency can deny the lender's application.

Where neighborhoods have been "disinvested" by lending institutions, these laws can provide a valuable tool to end this discriminatory practice. In Richmond, Virginia, the threat of using Community Reinvestment Act sanctions helped bring bank executives to the negotiating table. The result was a commitment to improve credit policies in neighborhoods where investments had not occurred. The bank promised to lend 20 percent of the dollar amount of mortgages and rehab loans for the entire city to disinvested neighborhoods, to advertise the availability of this credit to residents of the targeted areas, and to sell $5 million worth of tax-exempt bonds annually through state and city housing authorities, with proceeds to go to prospective homebuyers in the most distressed areas.

4.1.9 Financing for Commercial Projects

The Small Business Administration provides direct and guaranteed/insured loans to low-income persons or small businesses which cannot acquire business financing through normal lending channels.[27] This program does not have as its primary intent the rehabilitation of historic and old buildings, yet many of these buildings are suitable for small business use and are located in areas of high unemployment. Thus, these funds could be used to serve rehabilitation purposes. Although federal funding for this program continues, there has been an effort to shift its emphasis away from providing any direct loans and toward providing assistance exclusively in the form of loan guarantees or insurance.

Once an important source of rehabilitation funds, the development programs of the Economic Development Administration that could be used for rehabilitation activities have now been all but eliminated. Several years ago the Department of Commerce's economic development and public works programs provided a broad package of financial assistance to aid rehabilitation projects. The programs sought projects that produced jobs and contributed to a community's economic development, and were used across the country to revitalize communities, and to acquire and renovate all types of old buildings, including factories, warehouses, courthouses, and hotels. A description of these programs follows, but as in other areas it must be noted that appropriations generally have been severely slashed or are nonexistent.

Public Works and Development Facilities grants and loans[28] are directed toward economic revitalization and toward improving em-

---

[27] 15 U.S.C. § 636.
[28] 42 U.S.C. §§ 3131–44.

ployment opportunities. To achieve the aim of long-term economic expansion, projects which will attract business or will recycle old industrial and commercial facilities are favored. Rehabilitation of public service facilities and downtown revitalization projects are such eligible activities. States, counties, and towns, and public and private nonprofit organizations may apply for assistance for these projects.

Special Economic Development and Adjustment Assistance grants, known as Title IX grants,[29] are available to assist state and local governments in reversing declining economic conditions in their areas. Regions of the country facing long-term economic deterioration are aided by grants to plan activities, including recycling of land and structures, which will assist in the creation of jobs. Preservation and rehabilitation projects are quite compatible with the program's aims. Funds may be used to develop plans to revitalize aging downtown districts or residential neighborhoods. Grants may be used to establish a revolving fund for these purposes, as well as for specific preservation activities.

States, cities, and counties, public and private nonprofit corporations, and native American groups are eligible. The funding level of the grants is based on the community's economic needs.

4.1.10 Special Purpose Funding

Those seeking financial aid should consider all possible relationships between the work that has to be accomplished and the wide range of subjects that government programs touch. As previously noted, in most instances the major purpose of these governmental programs is something other than rehabilitation or development, but a rehabilitation project can be structured to take advantage of the funding which the programs provide. For example, the Department of Transporation (DOT) may not seem to be a likely source of preservation funds, but if a project pertains to old public transportation facilities or structures, assistance from DOT may be available.

Consider the Amtrak Improvement Act of 1974.[30] In this program, administered by DOT, federal funds are provided for adaptive reuse of old railroad stations that are listed or eligible for listing on the National Register. Many communities have such stations, which are often rundown and abandoned. Frequently they are located in the heart of the town's downtown area, so that rehabilitation of the railroad depot can be an important step in revitalizing the entire downtown. The Amtrak Improvement Act does require that after renovation the station be used for a transportation function, such as a mass transit station for bus, subway, or light railroad. Nevertheless, whether the station is placed

[29] 42 U.S.C. §§ 3241–45.
[30] 49 U.S.C. § 1653(i).

back into service to the community as a transportation center or station with help from this act or as a private commercial establishment with the support of other programs, preservation objectives have been furthered.

Another illustration of a program which could be exploited is one established by the Surface Transportation Act of 1978.[31] This act authorizes the identification of bridges of historical significance by DOT, with the assistance of the State Historic Preservation Office, and provides grants for rehabilitation or replacement of these bridges. Most of the grants so far have been used to fund replacement rather than rehabilitation, but the program need not be so limited.

Other funding opportunities exist as well. It may be helpful to analyze a rehabilitation project in terms of the many smaller discrete tasks of which it is composed. Thus, although the Department of Energy will not make funds available to underwrite full rehabilitation projects, it will assist with the portion of rehabilitation that involves windows and insulation.[32] Other portions of the rehabilitation project may involve the jurisdiction of other departments and agencies. The National Endowment for the Arts supports activities in the fields of architecture, urban design and planning, landscape architecture and other related areas.[33] Under the Design Arts Program, for example, activities such as adaptive use studies and district revitalization planning are eligible for grants.

Similarly, the National Endowment for the Humanities, whose purpose is to further research, education and projects in the humanities, assists programs related to history.[34] Preservation projects that further public understanding of history, whether of a geographic area or a conceptual area such as architecture, may find funding support here.

4.2   PUBLIC FINANCING: STATE AND LOCAL AID

As with regulatory protection for historic sites, many important programs relating to preservation funding are carried out at the nonfederal level. This is largely due to the federal government's delegation of many of the decisions and much of the administration of its programs to nonfederal levels of government, which has caused a number of communities to create local funding sources.

The types of funding available at the state and local level are wide-

---

[31] 23 U.S.C. § 144.

[32] *See, e.g.,* 42 U.S.C. § 8231.

[33] Statutory authority for these grants is found at 20 U.S.C. § 954.

[34] Statutory authority for these grants is found at 20 U.S.C. § 956.

ranging. Particularly popular is the revolving fund (discussed more fully later in this chapter), because of its ability to regenerate itself for further preservation activities through repayment of the loans it awards. Many states, local governments, and nonprofit corporations have developed revolving funds to obtain maximum benefit from their scarce resources.

Other strategies include direct loan programs or loan guarantees or subsidies, as well as seed grants to get projects started. Indirect financing through income and property tax relief, as discussed in chapter 5, has also worked successfully and is available in a number of areas. Even those areas which have found it impossible to make funds available directly because of budgetary constraints, can and do provide in-kind aid through, for example, expert architectural, survey, historical, and design assistance and advice.

4.2.1 Funding Sources

State and local governments wishing to support preservation activities must first determine how to raise necessary funding. Many communities which have not already allocated tax revenues for preservation and rehabilitation financing may, at present, find it politically difficult to raise taxes or implement new taxes or special assessments for preservation activity. The ability to utilize tax revenues will, of course, vary depending on the community and on the degree of public awareness of the benefits that can accrue to the community from preservation programs.

Another source of revenue for preservation activities is the sale of municipal bonds. Such bonds are attractive to investors because the interest is tax-free, thus allowing the municipality to sell them at an interest rate lower than that for taxable bonds. Several types of bonds are typically used: revenue bonds, to be paid from the earnings of the projects which they support; general obligation bonds, to be paid back from general tax revenues; and special assessment bonds, to be paid back from anticipated specific revenue sources.

The Internal Revenue Code contains tax-exemption provisions for two types of bonds which may be particularly adaptable to the purchase and rehabilitation of old buildings. Industrial development bonds (IDBs)[35] may be issued by units of local government for new construction or rehabilitation of rental housing or commercial projects (with a ceiling of $10 million on issuances for the latter purpose). Rehabilitation expenditures must exceed 15 percent of the cost of acquiring the buildings. The 1984 Tax Reform Act imposed new limits on issuance of IDBs.[36]

[35] 26 U.S.C. § 103(b).
[36] 26 U.S.C. § 103, as amended.

A state may annually issue bonds with a face value of no more than the greater of $150 per capita, or $200 million in total. The $150 limitation is to be reduced to $100 in 1986. The annual ceiling may be carried over for up to 3 years for specified projects, including certain projects affecting residential rental property. Moreover, IDBs issued to finance particular types of multi-family residential rental property are entirely exempted from the volume limitations. Unless states specify otherwise, 50 percent of the volume ceiling will be allocated to the state and 50 percent to local jurisdictions.

Qualified mortgage subsidy bonds[37] may be sold to provide financing for first-time homebuyers. Proceeds from the bonds may be used as mortgage financing for purchasers of single-family homes, condominium or cooperative units, or two-to-four-family houses in particular situations. They can also be used to support residential rehabilitation projects. Both types of financing are subject to numerous technical restriction in the Internal Revenue Code, which govern their eligibility for federal tax exemption.

In 1979 Alabama enacted legislation allowing communities to set up local historic preservation authorities to issue tax-exempt industrial revenue bonds.[38] The revenue generated by the sale of these bonds can be utilized to acquire and renovate buildings listed in the National Register. These properties are then leased to developers who pay the debt service on the bonds. When the bonds are paid off, the developer gets title for a nominal fee and an agreement to maintain and insure the properties. The program has been successful: In Montgomery alone, 20 buildings had been renovated by early 1983.

4.2.2 Funding Entities

Many states have created quasi-public corporate entities to finance and undertake development projects by issuing tax-exempt bonds or through other mechanisms. While the projects undertaken by these organizations are rarely targeted at historic preservation directly, economic development programs in older urban areas often serve that goal.

One model of such an entity is the New York State Urban Development Corporation (UDC).[39] The UDC was established to provide a more effective means of dealing with problems of physical deterioration, economic stagnation, unemployment, shortage of housing, and lack of civic facilities in the state. To combat these problems, it is authorized to plan and implement projects to supply housing for low and moderate-income

[37] 26 U.S.C. § 103A.

[38] Code of Alabama §§ 41-10-141(11), 142, 147.

[39] The statutory authority for the UDC is found in N.Y. Unconsolidated Laws §§ 6251–6285.

families, to assist in the industrial and commercial development of areas of high unemployment or economic blight, and to provide needed educational, cultural, community, or other civic facilities. The UDC may finance projects in support of these purposes by issuing tax-exempt bonds and notes as required.

Many of the UDC's projects have involved rehabilitation of old and historic property. In one instance, UDC funding was instrumental in the renovation of the nineteenth-century Customs House in lower Manhattan. A current large-scale UDC project, the revitalization of the Forty-second Street corridor and Times Square area of Manhattan, is described in detail in chapter 8.

4.3 NONCONVENTIONAL SOURCES FOR FINANCING REHABILITATION

Even in the best of times, financing a preservation project can be difficult. Conventional sources for construction financing are often wary of what they perceive as the high risk associated with the rehabilitation of older buildings, which may be compounded when the buildings are located in older, more neglected areas which are less obviously capable of yielding a satisfactory return on investment. Accordingly, preservationists and rehabilitators may be well advised to look at some or all of the following as potential revenue sources.

4.3.1 Revolving Funds

When financing in the form of public funding or bank loans is difficult to obtain, a revolving fund can provide an alternative source of capital. This method of financing takes its name from the characteristics that make it distinctive and give it value and flexibility as an alternative financing source. Generally, moneys from a fund are used to purchase threatened historic property, or are loaned to individuals or groups to engage in preservation projects. Through acquisition and subsequent resale with protective covenants, the property may be preserved and the fund replenished. Sometimes the property is rehabilitated prior to resale, while on other occasions agreement may be reached to sell a building as is for subsequent rehabilitation by the new owner. The money that is received from the resale of the property renews the fund, and its availability for subsequent reinvestment provide a continuing source of preservation financing assistance in a community.

The characteristic of recycling funds makes this financing technique particularly suitable for certain projects, such as multibuilding or large multiphase developments. In such projects, the limited capital available to complete one phase can generate enough capital to finance the next phase, and so on until the project is completed.

4.3.2 Operation and Administration

Revolving funds have been administered by both public and private entities. While there may be advantages and disadvantages to each, private control has generally been found preferable.[40] Public administration of revolving funds is often hampered by unpredictable budgetary strings, and maintaining a sufficient level of appropriations has proven difficult. Private entities have greater flexibility and can adapt more quickly to changing circumstances without the encumbrances of state or local laws and regulations.

Although other legal entities can be used to administer private revolving funds, they are most often organized as nonprofit, tax-exempt corporations. This requires both incorporation under the appropriate provisions of state law and certification by the IRS of tax-exempt status.[41] Such status provides dual tax benefits: It relieves the organization itself of tax liability, and permits donors to deduct their contributions for income tax purposes. Moreover, nonprofit status may be necessary to qualify for certain financial assistance. This legal form is most appropriate when the money used to purchase and resell the buildings will be solicited from private donors and charitable foundations. On the other hand, when a major portion of the fund will be composed of public general revenue or block grant funds, it may be necessary to organize a public or quasipublic organization.

A variety of preservation goals may be served by revolving funds, and the objectives of a particular fund should be clearly defined. Revolving funds have been used to save single buildings, limited areas, and entire neighborhoods. Both residential and commercial buildings can be restored, depending upon the needs of the local community.

Next, the fund should formulate an operational plan, with immediate and long-term objectives and procedures to achieve them. Responsibilities should be assigned and coordinated, and the talents of those involved in organizing the revolving fund should be assessed to determine if outside assistance will be required. The purpose of the fund will dictate the responsibilities of its administrators. Functions that will need to be performed include finance (auditing of funds spent and received), construction (planning, architectural design, supervision, and execution of restoration activities), marketing (promotion and sale of restored properties), purchasing, fund-raising, and community and government liaison.

---

[40] HCRS Historic Preservation Fund Grants: Potential Source for Local and Statewide Revolving Funds (HCRS Supplement 11593, October 1978).

[41] For additional information on tax-exempt status, *see* B. Hopkins, *The Law of Tax Exempt Organizations* (1983).

The distribution of authority in the Lafayette Square, St. Louis, Restoration Committee is instructive.[42] The purchase and sale of property must be approved by a 13-member board of directors. The power to grant approval of expenditures for ongoing rehabilitation activities and the responsibility for implementation is held by the vice president for redevelopment. The general membership must approve proposed loans from the revolving fund. When a purchased house has been rehabilitated and is ready for resale, the housing committee chairperson is authorized to sign the contract of sale, after its review and approval by both the redevelopment vice president and the full board.

4.3.3 Funding

A revolving fund must generate a corpus of working capital sufficient to finance basic activities and to insure that the fund never becomes so depleted that its programs must be suspended or terminated. Because sales, rents, and loan repayments will often recoup an amount less than that invested in a property, an effective fund management will require continuing fundraising efforts and the development of financial management strategies to maintain the fund. It should be remembered, however, that the purpose of the fund is not preservation of a large bank balance, but preservation of historic resources. In some cases, a worthwhile project may not return to the fund all that had been invested. To achieve the fund's aims, its administrators must be sensitive to the fact that not every project can break even, and that requiring every project to do so may jeopardize a fund's ultimate effectiveness.

Initial seed money for a fund may be obtained in the form of grants, gifts, or loans. Government programs, private organizations, foundations, corporations, membership dues, or individual contributions, and even in some cases tax revenues[43] are all potential sources of funds. In particular, a number of government programs described earlier can be used to support a revolving fund, including the Urban Development Action Grants, Community Development Block Grants, and Special Economic Development and Adjustment Assistance (Title IX).

Reduced levels of appropriations for these government programs, however, mean that revolving fund administrators, like sponsors of other preservation activities, will increasingly have to look elsewhere for financial support. Because of the localized nature of most revolving fund efforts, fund-raisers should give particular consideration to foun-

---

[42] *See* Coffey, "Revolving Funds for Neighborhood Preservation: Lafayette Square, St. Louis" at 5, 12 (National Trust for Historic Preservation, 1977).

[43] In 1975 New Orleans voters levied a special tax on real property within a targeted area to establish a fund to allocate $200,000 for historic preservation activity over a 2-year period.

dations, corporations and individuals with an interest in the community to be benefited. These potential sources are discussed later in this chapter.

4.3.4   Preservation Strategies

Depending on the circumstances, the revolving fund may purchase historic property for resale or, on occasion, for rental. The fund may rehabilitate a property itself prior to resale or rental, or may require the purchaser or tenant to commit to rehabilitating the building for reuse. In other instances, it may wish to acquire options to be used to hold property off the market. A revolving fund may also serve as a lender when other sources of loan funds are unavailable. In this capacity it may provide direct loans, loan guarantees, or participate in the lending of money with other financial institutions.

Resale of unrestored property should be accompanied by restrictions in the deed, and by an agreement for rehabilitation or restoration. The Galveston, Texas, Historical Foundation, for example, uses these two basic legal instruments to protect the property it resells after acquiring it with revolving fund money.[44] Its deed restrictions require that no demolition, exterior changes, or new construction occur without foundation approval, and that the owner maintain the building exterior and its structural soundness. These restrictions bind all subsequent owners of the property. In the written contract, the purchaser agrees to restore the exterior in an appropriate manner and to invest an agreed-upon sum of money to renovate the interior for active use.

A fund should be cognizant of the psychological impact of preservation activities, and use it to full financial and other advantage. Purchases, for instance, should be made quietly lest property prices soar with the disclosure that an area has been targeted for preservation activity. On the other hand, early sales may be made publicly and even at significant loss, to ignite a flurry of preservation activity and to create a positive and exciting atmosphere which may spur further investment. After the anchor projects are set in motion, however, a fund must emphasize recovery of costs to remain financially sound.

4.3.5   Private Foundations

Foundations are a potentially fruitful source of preservation funding. There are, of course, thousands of foundations in the United States, so the initial task is to identify those that will look favorably on a request for funds for preservation purposes. Foundation interests can be determined by looking at current funding patterns. Many foundations

[44]Brink, "Commercial Area Revolving Funds for Preservation" at 8 (National Trust for Historic Preservation, 1976).

limit their giving to particular subject areas and geographic locations. Based on these considerations, the would-be recipient should develop a list of foundations with a funding history in his area.

The thousands of foundations generally fall into one of five types:

1. *General purpose foundations.* These are the smallest in number but have the greatest assets and make grants on a broad basis (though usually in accordance with set priorities) to health, education, social welfare, and international activities, among others.

2. *Special purpose foundations.* These give in a specific area of interest and/or restrict their giving to a particular geographic area.

3. *Family foundations.* These are by far the largest in number but are generally much smaller in assets, and are usually concerned with the personal philanthropic interests of the family.

4. *Corporate foundations.* These are the instruments through which corporations make their contributions. They are, in theory, largely independent of the corporations, although in practice they are generally governed by the corporation's officers and directors.

5. *Community foundations.* These entities, committed to the improvement of particular communities, may be especially helpful in supporting local preservation activities. There are more than 250 community foundations in the United States, with total assets of $2.5 billion. They exist in most large cities, but their resources vary considerably. Some small community foundations have no more than a few thousand dollars. At the other extreme are groups such as the San Francisco Foundation with assets of more than $350 million.

Community foundations are composed of large numbers of individual trusts managed by one "parent" foundation. The donor specifies what, if any, restrictions are to be placed on the trust. The foundation provides professional management of its assets, and a governing board of directors and a professional staff administer its funds.

Community foundations are public charities and thereby offer donors tax benefits not available through other types of foundations. Because their boards and staff are involved in funding other civic activities and projects, they can provide more than money. They can help to negotiate aid from local interests, may serve as a catalyst for other funding sources, and can provide valuable contacts and technical assistance.

Foundation grants can be particularly helpful as seed money for revolving funds. In 1971 a $15,000 grant from the Putnam Foundation started the Historic Preservation Loan Fund of the New Hampshire Charitable Fund. This modest financial beginning has been the catalyst for the fund's participation in the financing of a number of projects. In 1974, for instance, it loaned $40,000 for the adaptive reuse of the 1823 Belknap Mill to convert it into an arts center and office, gallery, and meeting space for the community. Among the many other preservation projects that this fund has participated in was the preservation of the home of Robert Frost.

The job of locating those foundations that may be willing to part with their money is made easier by a number of guides to foundations. A vital resource in this regard is the Foundation Center, a nonprofit educational organization dedicated to providing information to assist those searching for grants.[45] The *Foundation Directory, Grants Index,* and *News* are three of many publications, reports, and other works published by the Foundation Center or contained in its libraries in a number of cities across the country. These works are helpful in identifying which foundations are giving grants, whom they are supporting, and for what purposes.

Another particularly valuable tool of the Foundation Center is the COMSEARCH PRINT OUT, a computer-produced guide to foundation giving. Organized by state, the list gives the foundation name and location, the amount it contributed and the recipient, a description of the project, and a source to contact for further information. Moreover, foundation activities are broken down into subject categories. Microfiche file #50 for 1983, for example, is entitled, "Architecture, Historical Preservation/Historical Society." Its scope includes 319 grants with a total value of $12,284,493, made *inter alia* to historical societies, architectural associations, cultural organizations, and governmental and community development organizations for capital support, general operating support, renovations and restorations, and conversion of historic structures into facilities for present-day use. In addition to the Foundation Center publications, the fund-raiser can learn more about funding sources from a foundation's annual report and from IRS information returns, which are available for all private foundations.

4.3.6 Corporate Donations

Private corporations may also be willing to support preservation activities. Their donations are often funneled through corporate foundations set up for the purpose of making charitable contributions. As a general rule, corporations prefer to give their money to activities in communities where they have a presence, such as a headquarters, factory, or regional office. A company with facilities in a particular location has a stake in the development of that community, both for its own economic viability and as a pleasant place for its employees to work.

For example, the Brown-Forman Distillers Corporation donated $100,000 to two neighborhood organizations in Louisville, Kentucky, to rehabilitate property in the residential area near its plant. Similarly, the Atlantic Richfield Company (ARCO) contributed $300,000 to the

---

[45] The Foundation Center's central library is located at 888 Seventh Avenue, New York, NY 10019. To learn if one of its area libraries is near you, call 800–424–9836.

National Trust to fund inner-city rehabilitation projects in the company's seven regional and headquarter cities—Houston, Denver, Louisville, Philadelphia, Chicago, Dallas, and Los Angeles. Standard Oil of Ohio (SOHIO) has made similar contributions. Another example is corporate support for the revolving fund that the Galveston Historical Foundation has operated since 1973 to restore a historic downtown area known as the Strand. That revolving fund was, in fact, started with grants from the Moody Foundation ($200,000) and from the Kempner Fund ($15,000), both run by companies located in Galveston.

The Foundation Directory, the same basic text as is used in researching foundation grants, is also a good resource for corporate donations. This book can provide answers to questions that arise after a group seeking funding has identified corporations to solicit. For example, it will specify the individuals who make decisions on grant requests, and the guidelines on which those decisions are based. Corporations differ on whether these decisions are made at the headquarters or locally, and knowing whom to approach is useful information for the funding applicant.

4.3.7 National Nonprofit Organizations

Perhaps the most significant sources of nonpublic support for preservation are the programs of the National Trust for Historic Preservation.

The National Trust operates seven grant and loan programs.[46] Among them is the Preservation Services Fund, which provides small matching grants to enable local preservation organizations to hire expert consultants to determine adaptive reuse plans for old buildings, to upgrade historic districts, to develop energy conservation plans for historic buildings, to furnish technical advice on a range of preservation activities, and to develop preservation education programs. In FY 1982, organizations in 33 states and the District of Columbia were awarded 77 grants totaling approximately $135,000.

A second National Trust program, the National Main Street Center, has as its goal the revitalization of small-town business districts. This program combines building preservation and business revitalization. Currently, approximately 130 Main Street projects are underway in small towns throughout the country, under the auspices of this Trust program. It has, in fact, been so successful that the Trust has expanded eligibility to include central business districts and neighborhood shopping districts in larger cities, and is planning demonstration programs in such areas.

The National Trust has also established the Inner City Ventures Fund to provide grants and loans to support conversion of old buildings to new uses. Since 1981 this fund has provided financial assistance totaling $1.4 million to 22 organizations, which have in turn used these funds

---

[46] These programs are administered under authority granted by 16 U.S.C. § 470a–470h.

to generate an additional $25 million from other sources for rehabilitation projects. In Portland, Oregon, for example, the 37-unit single-room occupancy Butle Hotel, in the historic downtown district known as Old Town Portland, was acquired and renovated on a total budget of $667,000. Of this, $40,000 was a grant and another $40,000 was a loan from the Inner City Ventures Fund. Remaining funds were derived from a grant and loan from the city, a foundation grant, and a seller loan.

Another National Trust program is the National Preservation Loan Fund, which provides nonprofit and quasi-public organizations with low-interest loans to help them establish and expand their own revolving loan funds. In addition, the Trust administers the Historic Properties Preservation Fund, which is intended as an incentive to stimulate fundraising for the preservation, stabilization, or restoration of particularly significant historic properties which are open to the public.

In all, the National Trust committed $2.3 million in grants and loans to preservation projects in 1984. Its financial resources come from governmental appropriations, private contributions, and recycling of the Trust's own reservoir of funds. By its own estimate, Trust programs in 1984 spurred the investment of more than $22 million in 146 preservation projects.[47]

4.3.8 Private Fundraising and Individual Donors

Finally, preservationists should not overlook the possibility of raising funds through direct contributions from individual donors. By far the largest share of money given to nonprofit organizations in general comes from individual donations, and there is no reason why this source cannot be tapped to support rehabilitation of historic properties.

To encourage donations, fund-raisers should always seek to give contributors a sense of personal involvement in the projects undertaken with their support. One ingenious illustration is provided by New York's Polonia Restoration Company, which sold stone pieces from the walls of the Brooklyn Bridge to raise funds for its restoration. The stones came from a part of a wall being excavated for a new entryway, and thus their removal did not affect the integrity of the bridge.

In seeking private contributions, as well as support from corporations and foundations, the legal status of the preservation organization may be important. Contributors generally prefer to give to groups if their donations will qualify for a charitable deduction for federal and state income tax purposes. For this reason, as just noted, it is advantageous for the recipient organization, if it qualifies, to obtain recognition of its tax-exempt status from the IRS.

[47] *Preservation News*, March 1985 at S-2.

Five

State and Local Incentives for Rehabilitation

Introduction

Complementing the federal tax incentives discussed in chapter 1 is a broad array of state and local programs designed to encourage rehabilitation of historic buildings. Because state and local regulation affects historic structures in more direct and diverse ways than does federal regulation, the range of preservation programs at those levels is far broader. The approaches generally adopted, like the federal program, rarely involve affirmative financial or other support for preservation activities. Rather, rehabilitated buildings are given special status and treatment under tax and regulatory programs which might otherwise make rehabilitation undesirable or uneconomic.

This chapter discusses those steps taken by state and local government entities to encourage historic preservation in three basic areas—property taxation, building codes, and zoning regulation. Local property taxes often create a major disincentive to the continued use of old buildings, and many states and municipalities have enacted special provisions to ease that burden. Reconciling the health and safety requirements of local building codes with preservation of the aesthetic and architectural integrity of historic structures is often difficult, but a number of constructive legislative efforts have been made to integrate these sometimes-conflicting sets of social objectives. Finally, local zoning laws may encourage the demolition of historic buildings in favor of more lucrative uses of property, and many cities have experimented with transfer of development rights programs to ameliorate the problem.

5.1 PROPERTY TAXES

The property tax system, a major source of revenue at the local level, can pose a serious threat to the continued existence of income-producing historic structures. Property taxes are typically assessed on both the value of the underlying land on which a building sits and the value of any improvements (including buildings) which are made on the property. Historic structures are often situated either in downtown areas or in desirable outlying neighborhoods. In either event, land values are likely to be comparatively high, and a substantial return from current uses of a building may be necessary to offset property tax liabilities. Yet historic buildings, by virtue of scale and structure, may not be amenable to high-return uses, and the temptation to replace them with buildings generating greater income may be strong. In addition, rehabilitation of historic buildings increases their value, leading to higher assessed values and tax bills, without necessarily producing a greater income stream.

State and local governments have implemented a wide variety of tax incentives to help save historic structures. Basically, the incentives can be classified either as direct relief from property taxes or as measures that are designed to reduce assessments on historic properties. These programs can have an important effect on the economics of a rehabilitation by both increasing and accelerating projected cash flow.

The tax provisions adopted by different states vary both in the nature and magnitude of the tax benefits they provide.[1] This section will examine a number of state and local tax incentive programs, and highlight the range of provisions that have been adopted. No attempt is made to provide a complete categorization of all the state and local tax incentives that have been implemented; rather, the chapter describes both the typical measures and some innovative approaches that have been developed in a number of jurisdictions.

----

[1] *See* Powers, "Tax Incentives for Historic Preservation: A Survey, Case Studies and Analysis," 12 *The Urban Lawyer* 103 (1980), for an extremely useful typology of state property tax incentives. *See also* D. Listokin, *Landmarks Preservation and the Property Tax* (1982).

5.1.1 Exemptions from Real Property Taxation

While a large number of states grant total exemption from property taxes to historic structures, exemptions are generally allowed only in very restricted circumstances.

Thus, some states exempt historic properties from taxation but limit the exemption to specific geographic areas. Puerto Rico has enacted legislation that grants exemption to property owners of rehabilitated historic buildings located in the Old San Juan Historic District.[2] An applicant who desires the exemption submits to the Secretary of the Treasury a total property tax exemption plan which details the rehabilitation and restoration work to be performed. If the Secretary approves the plans, the site is granted either a 10-year (for complete renovation) or a 5-year (for partial renovation) exemption from taxation. Once the initial exemption expires, the Secretary can extend the exemption provided that the Institute of Puerto Rican Culture certifies that (1) the property has not been substantial altered in its original design; (2) the property deserves to be preserved as part of Puerto Rico's cultural heritage for its historic or architectonic value; and (3) upon conclusion of the work pursuant to the requirements of the Institute, the property will be in equal or better condition than that presented when its first total restoration was carried out. In addition, any proceeds received from leasing a designated building are exempt from income tax.[3]

Tennessee exempts the value of improvements or restoration of a historic structure[4] from property taxes so long as the owner agrees to restore the property under the guidelines that are set out in the statute. Essentially, the restoration must either be part of an official government program or the plan must be approved by a historic properties review board. The exemption lasts for 10 years in the case of a partial restoration and 15 years in the case of a full restoration. At the end of the applicable period the structure is assessed at its full market value. However, if the structure is significantly altered or demolished during the

---

[2] P.R. Laws Ann. Tot.13, § 551 (1977).

[3] *Id.* A similar tax incentive was enacted as part of the Louisiana State Constitution. Art. XIV § 22A (until its repeal in 1974 by Art. VII § 18(C)) created the Vieux Carre Commission and gave it the responsibility of preserving the architecture of the New Orleans Old French Quarter. The New Orleans City Council was authorized to exempt from municipal and parochial taxation buildings which the commission designated as having historic or architectural value.

[4] "Historic structure" is defined as structures on the Tennessee or National Register or approved by a historic properties review board. The board is to designate as historic (1) any structure over 175 years old; (2) any structure over 125 years old unless it is established as nonhistoric; and (3) any structure over 75 years old subject to an individual review. Tenn Code Ann. § 67-519 (Supp. 1983).

exemption period, the property owner is immediately liable for the difference between the taxes paid and the tax that would have been due if the structure was assessed at its full market value.[5]

The Tennessee statute is subject to two limitations. First, the law only applies to counties with a population exceeding 200,000.[6] Second, the provisions of the statute only come into effect when either the county or a municipality within the county ratifies them by a majority vote.[7]

Another method that has been employed to fully exempt certain properties from taxation is contained in New York City's Historic Landmarks Act.[8] An owner of a historic structure can file an application with the Landmarks Preservation Commission to demolish or alter the structure if it is not capable of earning a reasonable return. In response, the commission has two alternatives. First, it can approve the property owner's plan. Second, it can develop a plan which allows a property owner to obtain a reasonable return by granting a full or partial tax exemption and/or remission. If the Board of Estimate approves the tax exemption and/or remission, the commission can then deny a certificate of appropriateness for alteration or demolition, thereby preserving a building the return on which would otherwise be so low that the commission would be required to grant such a certificate.[9]

Hawaii grants a property tax exemption to property owners who dedicate their land to public uses.[10] To apply, a landowner petitions the director of taxation. The application sets out the exact area of land that is to be dedicated, states how the land shall be used, and agrees that the land shall be used, improved, and maintained for the sole purpose of furthering the dedicated use. The director must approve the plan and grant the tax exemption, if he finds that the use of the dedicated land will have a benefit to the public that is at least as great as the value of the real property taxes for the land.

Once the plan is approved, the owner of the property is granted a tax exemption and forfeits the right to change the use of the land for a period of 10 years. The exemption and the accompanying limitation on the use of the property are automatically renewable indefinitely, subject to the right of either the owner or the director to cancel the exemption by giving 5 years' notice any time after the fifth year of the dedication. Failure to observe the restrictions on the use of the land

[5] Tenn. Code Ann. § 67-519 (Supp. 1983).

[6] *Id.* § 67-520.

[7] *Id.* § 67-521.

[8] The enabling legislation for this provision is N.Y. Gen. Mun. Law § 96-a.

[9] New York City Admin. Code § 207-8.0. Note that this procedure is instituted by the Landmark Preservation Commission and not by the owner of the historic property. The provision contains a procedure that will save the building from demolition even if the owner rejects the plan. *See* Section 7.4.5.

[10] Haw. Rev. Stat. § 246-34(a)(b)(1976).

subjects the owner to cancellation of the exemption. The owner is then liable for the difference between the taxes paid and the taxes that would have been due if there had been no exemption, plus an interest payment of 5 percent per year from the dates that the payments would have been due.[11]

Finally, a number of states provide for tax exemption for historic sites that are owned by historic associations that operate as not-for-profit corporations. For example, New York provides an exemption from taxation for any incorporated historical society that holds a historic site for nonbusiness purposes provided that the society does not hold more than six sites in any one locality.[12] New Jersey exempts from taxation any building which has been certified as a historic site by the Commission of Conservation and Economic Development and that is owned by a nonprofit corporation.[13] Ohio exempts historic buildings and sites from taxation when the property is not held for profit and when the site or building is dedicated to public use.[14]

5.1.2   Credit or Deduction against Real Property Taxes

Another way to provide an incentive for historic preservation is to grant owners of historic property either a tax credit or a tax deduction when they undertake preservation activities.

Maryland has passed two statutes that provide tax incentives for historic preservation.[15] The first statute enables counties to grant tax credits against local property taxes to help property owners offset the expense of the restoration or preservation of their historic structures. The statute applies to historic structures located in designated historic districts. It allows counties and municipalities to provide property owners with a credit against the local property tax of up to ten percent of the cost of renovating or restoring qualifying structures. Additionally, it allows counties and municipalities to provide a tax credit of up to five percent of the cost of the construction of architecturally compatible new structures in a designated area. Any tax credit earned in any given year can be carried forward for up to five subsequent tax years. A county or municipality can provide further conditions which must be satisfied in order for a property owner to qualify for a tax credit.[16]

[11] *Id.* § 246-34(a)(d).

[12] New York Not-for-Profit Corp. Law § 1408.

[13] N.J. Stat. Ann. § 54.4-3.52 (Supp. 1983).

[14] Ohio Rev. Code Ann. § 5709.18. *See also* Conn. Gen. Stat. § 12-81(7) (1983) (providing property tax exemption for property owned by Connecticut corporations which are organized exclusively for scientific, educational, literary, historical, or charitable purposes so long as the property is being used for these purposes).

[15] Md. Ann. Code art. 81 §§ 12G, 281A (1980).

[16] *Id.* § 12G.

Maryland's second statute provides for a state income tax deduction for amounts expended to restore or rehabilitate qualifying structures, along much the same lines as the federal program.[17] The owner of a "certified nondepreciable historic structure" can deduct from his state income taxes the amount that is expended on "certified rehabilitation." Certified nondepreciable historic structures are nondepreciable properties (as defined by Section 167 of the Internal Revenue Code) that are: (1) listed in the National Register, (2) located in a registered historic district and certified by the Secretary of the Interior as being of historic significance, or (3) located in a state or locally designated historic district which is certified by the Secretary of the Interior to be preserving and rehabilitating historically significant buildings in that district. Certified rehabilitation is rehabilitation that is consistent with the historic properties of the district. This determination is made by reference to regulations that are promulgated by the state comptroller.

The taxpayer qualifies for the deduction by filing an application with the state comptroller. The deduction can equal the portion of the property's basis that is attributable to the amounts expended for certified rehabilitation and can be calculated over a 60-month period. The taxpayer can elect to begin the 60-month deduction period either in the month following the month when the rehabilitation work is complete or in the taxable year following the taxable year in which the rehabilitation is complete.[18]

New Mexico also provides for a property tax credit for rehabilitation of qualifying historic structures. Eligible properties are those listed on the official New Mexico register and available for educational purposes in conformance with conditions set forth by the cultural property review committee.[19] Restoration, preservation, and maintenance expenditures which are approved by the cultural property review committee can qualify as a credit against local, city, county, and state taxes that are assessed on the property. Any credit earned in a given year can be carried forward for up to 10 subsequent years.[20]

5.1.3 Abatement through a Lower Tax Rate

A number of states have passed statutes which promote the preservation of historic property by taxing the property at a lower rate than other property.

---

[17] *Id.* § 281A.

[18] *Id.* § 281A(f).

[19] The cultural property review committee is composed of the state archaeologist, the state historian, and five other professionally recognized historians, anthropologists, architects, or historians who are appointed by the governor. N.M. Stat. Ann. § 18-6-4 (1980).

[20] N.M. Stat. Ann. § 18-6-13 (1980).

North Carolina provides for a 50-percent abatement of the property taxes of structures or sites that are designated as historic. Properties can be designated as historic by the passage of a local ordinance; once designated the property is taxed at 50 percent of its true value.[21] The difference between the taxes paid and the taxes that would have been paid without the special classification becomes a lien on the property. The lien is carried in the records of the taxing unit as deferred taxes. If the property loses its designation (except through fire or natural disaster), the additional taxes for the fiscal year that begins in the calendar year in which the disqualification occurs, the additional taxes for the three preceding fiscal years, and the interest that is due on unpaid taxes, all immediately become due.[22]

In Arizona, a property owner may apply to the county assessor to have the property classified as historic for the purposes of taxation.[23] Historic property is defined as real property that is listed in the National Register, is open to the public for at least 12 days per year, and meets the minimum standards of maintenance that are established by the Arizona State Parks Board.[24] The county assessor refers the application to the state historic preservation officer (SHPO), who may view the premises and approve or disapprove the application. If the preservation officer finds that the property is historic, he cannot disapprove the application because of the potential loss of revenue that may result. If the application is disapproved, the property owner can appeal to the state superior court.[25]

If approved, the property is classified as historic property for a 15-year term. In the last year of the term, the property owner may apply for a renewed classification for an additional 15-year term. Classified property is taxed at the rate of 5 percent of full cash value. This compares with a rate of 25 percent of full value for most commercial and industrial uses, 16 percent of full cash value for most land used for agricultural purposes, 18 percent of full cash value for most leased or rented residential property, and 10 percent of full cash value for most private use residential property.[26]

If the historic property becomes disqualified (except by transfer to an exempt organization, by fire or by act of God) a penalty is levied. The penalty is the lesser of 50 percent of the amount that the property taxes were reduced during the current term in which the property has been classified as historic, or 50 percent of the market value of the

[21] N.C. Gen. Stat. § 105-278(a) (Supp. 1981).

[22] Id. § 105-278(b).

[23] Ariz. Rev. Stat. § 42-139.01 (1984).

[24] Id. § 42-139.

[25] Id. § 42-139.01.

[26] Id. § 42-136, § 42-139.01, § 42-227.

property. An additional penalty of 15 percent of the base penalty is levied if the property owner fails to notify the assessor of the disqualification.[27]

5.1.4 Abatement for Rehabilitation or Renovation

A number of states provide a property tax abatement to help preserve historic property which is renovated or rehabilitated.

South Dakota provides a five-year "moratorium on the taxation of increased valuation due to restoration or rehabilitation of qualifying historic property."[28] To qualify, property must either (1) be on the state historic register and have received federal restoration grant-in-aid assistance, (2) have been substantially renovated with financing from the historic preservation loan fund, or (3) have been renovated with private funds with the approval of the board of cultural preservation. In addition, the property owner must sign a covenant which requires that he perform maintenance which preserves the restored portions of the property.

Illinois has a statute which was not enacted explicitly to help preserve historic property, but which can be utilized by owners of historic property.[29] Under the Illinois statute, "homestead property" is defined as real property containing residential buildings consisting of less than 55 dwelling units. Up to $15,000 per dwelling unit in improvements in homestead property is exempt from increases in taxation for up to four years after the improvements are completed. To qualify, a taxpayer must submit an application to the assessor which shows that the increase in value in the property is solely attributable to new improvements on an existing structure that is at least 30 years old, that the improvement does not increase the square footage of the structure, and that the improvement betters the total condition or energy efficiency of the structure.[30]

Colorado's statute[31] is similar to Illinois'. Properties that qualify for the abatement are residential buildings with three or fewer dwelling units and commercial buildings which are part of a development or redevelopment project area. Any increase in the actual value of the property commenced on or after July 1, 1976, due to the rehabilitation

---

[27] *Id.* § 42-139.02.

[28] S.D. Codified Laws §§ 1-19A-20, 1-19A-21 (1980).

[29] Ill. Ann. Stat. ch. 120 § 500.23.4 (Supp. 1983).

[30] *See also* N.Y. Real Prop. Tax Law §§ 488-a, 489, enabling large cities to provide exemption from taxation on the increase in assessed valuation of property due to specified improvements.

[31] Col. Rev. Stat. § 39-5-105 (1982).

or modernization of the properties, is exempt from taxation for the five years after the rehabilitation or modernization[32] is completed. However, the abatement ceases if the building changes ownership except by descent or inheritance.

5.1.5   Assessment Based on Current Use of Property

Generally, property is taxed based on its highest and best use. Obviously, this taxing scheme penalizes owners of historic property because their property often cannot be utilized to its maximum capability. Recognizing this problem, a number of states have enacted statutes which reduce taxes on historic property by valuing the property based on its current use.

California has provisions applying to "qualified historic property." Qualified historic property is private property that is not exempt from taxation and which is visually accessible to the public. In addition, the property must be (1) a registered California historic landmark,[33] (2) a property listed on the National Register, or (3) a property listed on a city or county historic register or inventory. For categories (1) and (2), the property must also be either (a) the "first, last, only or most significant historic property of its type," (b) "associated with an individual or group having a profound influence on the history of California," or (c) "a prototype of, or an outstanding example of a period, style, architecture movement, or construction or . . . one of the more notable works, or the best surviving work in a region of a pioneer architect, designer, or master builder."[34] A city or county, acting on the application of an owner of a qualified historic property, may create a historic zone. The city or county may then contract with the owner of the qualified historic property located within the historic zone. The contract must restrict the use of the property to maintenance of its historically significant character, and must last for a minimum of 20 years.[35] The city or county legislature can cancel the contract after holding a public

---

[32] For residential property, rehabilitation or modernization "includes repainting, reroofing, or rewiring; replacement of baths, fixtures, or heating systems; kitchen modernization; addition of air conditioning, fireplaces, patios, fences, or lawn sprinklers; and the finishing of a basement. Neither term includes room additions; the conversion of patios, porches, or garages into living areas; the addition of outbuildings; or the change of the use of a structure." Col. Rev. Stat. § 39-5-105(2)(a) (1982). For commercial properties rehabilitation or modernization includes any major renovation or rehabilitation. *Id.* at § 39-5-105(3)(a).

[33] For landmarks up until Register No. 769, the landmark site must conform to the existing criteria of the California Historic Landmarks Advisory Committee or the State Historic Resources Commission. Cal. Pub. Res. Code § 5031(a) (Supp. 1983).

[34] Cal. Pub. Res. Code § 5031.

[35] Cal. Gov't Code §§ 50280-81 (1983).

hearing, if the owner breaches one of the contract's provisions.[36] Finally, the city, county, or any other landowner can bring a court action to enforce the contract.[37]

"Qualified historic property" subject to a historical property contract described in the preceding paragraph is "restricted historical property" which is valued by a complicated capitalization of income method. This contrasts to the normal method of valuing property by evaluation of sales data from comparable properties.[38]

The District of Columbia has enacted a far simpler statute.[39] It applies to land and improvements which have been designated as a historic building by the Joint Committee on Landmarks of the National Capital. This property is assessed on the basis of the current use of the building, if this is less than the building's highest and best use. The District of Columbia can also require property owners to enter into a contract of at least 20 years' duration, that contains reasonable assurances that the property will be used and maintained to encourage the building's preservation. If the contract is violated, the city can collect the back taxes that would have been due if the property had not qualified under this provision, plus interest.

Some states provide for current use assessment for open space lands which include historic sites. The state of Washington, for example, defines open space land to include "any land area, the preservation of which in its present use would . . . preserve historic sites."[40] An owner of a historic site must apply to the county legislative authority, which decides, based on broad general criteria, whether or not to approve the application for classification as open space land. Approval can be conditioned on the owner's meeting specified conditions. If the application is approved, the value of the land for property tax purposes is determined by the current use of the land and improvements; other potential uses of the property will not be considered. Once approved, a site retains its special status until the owner notifies the county assessor of his intent to withdraw the property from the classification. However, a classified site cannot be applied to another use for at least 10 years. Failure to use the site for the purpose specified in the application results in a penalty equal to 120 percent of the tax that would have been due if the property had not been specially classified.[41]

---

[36] *Id.* §§ 50284, 50285. If canceled the owner must pay a penalty equal to 12½ percent of the full value of the property when the contract was canceled. *Id.* § 50286.

[37] *Id.* § 50287.

[38] Cal. Rev. Tax Code §§ 439–439.4 (Supp. 1984). Note that § 439.2 provides for different capitalization rates for owner-occupied single-family dwellings and for all other buildings.

[39] D.C. Code §§ 47-842 through 47-844 (Supp. 1978).

[40] Rev. Code of Wash. § 84.34.020(1)(b) (Supp. 1984).

[41] *Id.* § 84.34.020-.80.

Nevada also provides for current use assessment for open spaces that include historic sites.[42] Open-space use includes the utilization of land that will preserve historic sites that have been designated by the division of preservation and archeology of the State Department of Conservation and Natural Resources. An owner of real property can apply to the county assessor for an open-space use assessment. The application is referred to the board of county commissioners (and additionally, if applicable, to the governing body of the city). If the application is approved, the assessed value of the property is determined by calculating 35 percent of the value of the property's open-space use.

5.1.6 Assessment Based on Restricted Property Use

Historic property is often subject to restrictions as to its use. Many states have enacted statutes that attempt to take these restrictions into account by reducing the assessed value of historic properties that are subject to encumbrances.

In Virginia, either individual sites or entire districts can be designated as historic by the Virginia Landmarks Commission.[43] The commission can then obtain a contract with the landowner to limit the use of the landmark in order to perpetuate and preserve its historic character. In addition, the commissioner notifies the tax assessor, who is to consider the designation to be prima facie evidence that the value of the property has been reduced. When the assessor values the property, he must take into consideration the facts that commercial, industrial, and certain other uses of the properties within the district are restricted.

West Virginia has a similar procedure. Any county or municipality may create a historic landmark commission which has the power to designate historic landmarks and historic districts.[44] The commission may contract with the historic property owner to limit the property's use so as to perpetuate and preserve the features that led to its landmark designation. Once a historic district has been established, the commission must notify the tax assessor and specify the historic sites and the restrictions that the commission has obtained with regard to the sites. The assessor shall then take these factors into consideration when assessing the property.

In Connecticut, a municipality may acquire by any method (including purchase, condemnation, gift, devise, or lease) an interest in an area designated as open-space land which includes land used for historic and scenic preservation.[45] This interest can include outright purchase, ease-

[42] Nev. Rev. Stat. §§ 361A.040-.090, .190-.250 (1983).

[43] Va. Code §§ 10-138, 10-139, 10-142 (1978). *See also* § 10-140.

[44] W. Va. Code, title 8, Article 26A (1984). A similar tax incentive plan has been adopted by South Dakota. *See* S.D. Compiled Laws Ann. § 1-178-2025 (1980).

[45] Conn. Stat. Ann. § 7-131(b), 131(c) (1972).

ment, or restrictive covenant. The owner of an encumbered property may apply to the assessor, who shall value the property (during the next assessment) to reflect the existence of the encumbrance. An owner who is unhappy with the assessor's evaluation can appeal the determination to the board of tax review and from there to the superior court.

Kentucky also defines open-space land to include "land as an area which is provided or preserved for . . . historic or scenic purposes."[46] Its statute provides for acquisition of any interest in open-space land by local legislative bodies, through any means except eminent domain. Where a local legislature has obtained an encumbrance, the assessment of open-space property must reflect the change in the market value of the land that results from the encumbrance.

In Illinois, municipal authorities have the power to provide for landmark designation of land and structures which have a historical interest or value.[47] In connection with this power, the authority may promulgate regulations which limit the construction, alteration, demolition, or use of the designated properties. Alternatively, the authority can acquire an interest in the land to achieve the same objectives. Any depreciation of designated property that occurs because of encumbrances or restrictions imposed by the corporate authorities shall be deducted from the valuation of the property for property tax purposes.

Finally, New York enacted enabling legislation in 1980 which grants municipalities broad powers to regulate historic sites, districts, or buildings in the interest of preserving their historical, cultural, or aesthetic value. Where necessary to achieve these purposes, a locality may acquire by any appropriate means a "fee or any other lesser interest, development right, easement, covenant or other contractual right . . . to historical or cultural property within its jurisdiction."[48] The effect of the acquisition of property rights under these provisions "shall be taken into account" in valuing the property for real estate tax purposes.

5.1.7   Frozen Assessments

Finally, one state—Oregon—has enacted a statute which has proven quite effective in fostering rehabilitation. Oregon law provides a tax incentive for historic preservation by freezing the value of qualifying historic property for the purpose of the property tax assessment.[49] Historic property is defined as any property that is currently listed in the National Register of Historic Places, that is open to the public for sight-

[46] Ky. Rev. Stat. §§ 65.410, 65.440, 65.460 (1980).

[47] Ill. Ann. Stats. ch. 24 §§ 11-48.2-2, 11-48.2-6 (Supp. 1983).

[48] N.Y. Gen. Mun. Law § 119-dd.

[49] Or. Rev. Stat. §§ 358.475-.565 (1981).

seeing for at least one day for each calendar year, and that is maintained in accordance with the standards established by the state historic preservation offices.[50]

In order to be granted the special assessment, the owner of a historic property must submit an application to the county assessor and agree that the SHPO and the members of the state advisory committee on historic preservation can personally inspect the property. Upon receipt, the county assessor forwards the application to the state historic preservation officer. He reviews the application, with the assistance of the state advisory committee on historic preservation. If the property is properly classified as a historic property, the application cannot be denied solely on the grounds that the approval will result in a loss of revenue.[51]

If the SHPO approves the application he notifies both the applicant and the county assessor. The county assessor then classifies the property as historic property. For the 15 consecutive assessment years following the calendar year in which the application was made, he values the property at the true cash value of the property at the time that the application was approved. If the preservation officer denies the application, the property owner can appeal to the circuit court in the county where the property is located.

Historic property can be disqualified by written notice from the property owner to the assessor, sale or transfer of the property to a tax-exempt organization, or discovery (instituted by the county assessor and carried out by the preservation officer) that the property no longer qualifies as historic property. If the property becomes disqualified, a penalty is levied which is equal to the amount that the tax would have increased if the property had not been designated as historic multiplied by the number of years that the assessment was frozen, plus an additional 15 percent (unless the property owner personally notifies the assessor that the property no longer qualifies). The penalty is waived if the property is sold to a tax-exempt organization or if the property is destroyed by fire or act of God.[52]

5.2   LOCAL BUILDING CODES

The rehabilitation of older structures is often complicated by the application of state and local building codes. These codes have traditionally been written with new buildings in mind, and their requirements may be unduly onerous when applied to historic buildings undergoing

[50] *Id.* § 358.480.
[51] *Id.* § 358.485-.490.
[52] *Id.* § 358.495-.525.

restoration. Moreover, prescriptive provisions of building codes designed to protect health and safety may necessitate measures which jeopardize character-defining features of historic structures.

A number of different problems can be raised by the application of modern building codes to historic rehabilitation work. The expense of meeting code requirements can sometimes tip the balance away from renovating an existing building and toward construction of a new one. If an older building undergoing rehabilitation is subject to code provisions applicable to new buildings, it may be simpler and cheaper to demolish the existing structure and build a new one instead.

Code requirements may also disrupt or destroy a historic building's texture, architectural details, or even structure. Materials mandated for use in new construction may be incompatible with historic materials. External modifications required for fire egress or handicapped access may impair the appearance or historical integrity of a building facade, as may internal renovations to meet electrical, mechanical, ventilation, or fire detection standards.

In addition, entire building assemblies used in the construction of older buildings but no longer extant may have been dropped from building codes. This may make it impossible for local officials to determine whether outmoded materials or techniques continue to meet safety requirements, or what steps need to be taken to bring them up to acceptable levels. Instead, wholesale replacement may be ordered. Code provisions frequently do permit officials to sanction use of alternative technologies or methods which will achieve the same results as those prescribed.[53] However, a lack of technical basis for determining interchangeability, or simply a resistance to innovation, may make recourse to such provisions difficult to obtain in practice.

An owner contemplating rehabilitation should first determine whether the local building code in his jurisdiction will become applicable as a result of the anticipated work, and if so, to what extent. If code requirements will apply and are likely to be difficult to meet, owners should explore whether exceptions or variances are obtainable. Finally, special provisions of local codes, or entirely separate state or local provisions, may apply to historic structures, partially or totally in lieu of general building standards. Sample code provisions are cited here to illustrate typical approaches, but developers must, of course, refer to local codes applicable to particular projects.

[53] For example, Section 103.2 of the Building Officials and Code Administrators International (BOCA) Basic/National Existing Structures Code (1984) provides the following:

> *Modifications.* When there are practical difficulties involved in carrying out structural or mechanical provisions of this code or of any approved rule, the code official may vary or modify such provision upon application of the owner or the owner's representative, provided that the spirit and intent of the law shall be observed and public welfare and safety assured.

Local codes may prove especially problematic for developers who anticipate receiving a 25 percent investment tax credit for rehabilitation work. Code provisions may require alterations which cannot be accomplished in conformity with the Secretary of the Interior's Standards for Rehabilitation, which in turn must be complied with to qualify for the 25 percent credit. A project sponsor who plans to seek such a credit at the completion of the project, and who intends to structure financing on that basis, should review building code requirements at an early stage to determine whether it will be possible to undertake a certified rehabilitation within building code strictures.

5.2.1   Applicability to Existing Structures

Typically, building code provisions become applicable to existing structures when they are substantially altered or when they undergo a change in use. An illustration of this general approach is the New York State Uniform Fire Prevention and Building Code (UFPBC), which applies in every municipality in the state that has not enacted an equivalent or more stringent code.[54] The UFPBC becomes applicable when the occupancy or use classification of an existing building changes (as for example, from single-family to multifamily, or from residential to comercial).[55] The code also applies if an existing building (other than a mobile home) is physically relocated. UFPBC provisions apply to any alterations or additions made to a building; repairs, which are defined to include less extensive changes "for the purpose of maintenance, preservation or restoration," are not affected. Finally, UFPBC coverage is triggered for an entire building when "the cost of any additions or alterations made within any six-month period exceed 50 percent of the cost of replacement of the building at the beginning of that six-month period."[56]

A new approach has been taken by the Building Officials and Code Administrators International (BOCA), an organization which produces the Basic Building Code, one of the three widely used model codes in the United States.[57] In 1984 BOCA adopted a new Article 25 to its model building code, entitled, "Repair, Alteration, Addition, and Change of Use of Existing Buildings." The new section allows individual code require-

---

[54] N.Y. Exec. Law §§ 373, 379.

[55] New York State Uniform Fire Prevention and Building Code § 1231.1 (1984), *reprinted in* 19 N.Y.C.R.R. § 475 *et seq.*

[56] *Id.* § 1231.3b.

[57] The Basic Building Code, soon to be retitled the National Building Code, is in general use in the North and Midwest. The other model building codes are the Standard Building Code, used in the South, and the Uniform Building Code, used in the West. Although these model codes are not official documents, they frequently serve as the technical basis for codes promulgated by states and localities.

ments to be waived, so long as the overall degree of safety is equivalent to that under the provisions of the code which apply to new construction. In addition, the new provisions use a numerical rating system to provide an objective evaluation process for determining when minor code deficiencies can be compensated for through alternative methods which exceed other code requirements.

Also in 1984 BOCA published the first edition of the Basic/National Existing Structures Code, designed to apply to existing buildings in lieu of new construction codes. The Existing Structures Code provides an independent set of requirements, and specifies that alterations or repairs "may be made to any structure without requiring the existing structure to comply with all the requirements of the code for new construction provided such work conforms to that required by this code."[58] The code is based upon the "Code Enforcement Guidelines for Rehabilitation of Existing Structures" issued by the Department of Housing and Urban Development in 1980, and it remains to be seen how widely accepted this approach will be.

If a local housing code is determined to apply to an existing structure, or to become applicable as a result of a rehabilitation project, it may be possible to appeal the determination of a local building official that onerous work must be undertaken to meet code standards. The BOCA Existing Structures Code, for example, creates an appeals board, and permits appeal by "[a]ny person affected by any notice which has been issued in connection with the enforcement of any provision of this code."[59]

New York law, as part of the implementation of the UFPBC, requires the secretary of state to establish

> a procedure whereby any provision or requirement of the uniform code may be varied or modified in cases where strict compliance with such provision or requirement would entail practical difficulties or unnecessary hardship or would otherwise be unwarranted. Such procedure shall be designed to insure that any such variance or modification shall not substantially affect adversely provisions for health, safety and security, and that equally safe and proper alternatives may be prescribed. Requests for a variance shall be resolved within sixty days of the date of application unless a longer period is required for good cause shown. . . .[60]

5.2.2 Historic Building Provisions

Special building code provisions may exist at the state or local level whose applicability is restricted to historic structures. In some cases, these may be incorporated within new construction codes of general

[58] BOCA Basic/National Existing Structures Code § 102.1 (1984).
[59] *Id.* § 111.1.
[60] N.Y. Exec. Law § 381(1.)(f).

applicability, and may restrict or withdraw entirely the application of a code to historic buildings. The UFPBC, for instance, provides that

> buildings which are officially designated as historic buildings because of historical or architectural importance shall be permitted to be repaired for the purpose of historical preservation or restoration without conforming to the requirements of the Code provided that the existing use is continued and the repairs are acceptable to and deemed safe by the local authority having jurisdiction, except that requirements for facilities for the physically handicapped shall remain applicable.[61]

The benefits of this provision are somewhat more limited than they might at first appear. "Historic buildings" are defined to include those specifically designated as historically significant at the state or local level, which are listed in the National Register, or which are determined by the Secretary of the Interior to be eligible for listing. Thus, for example, buildings eligible for certified rehabilitation as contributing to the significance of a historic district would not be deemed historic for building code purposes. In addition, the provision does not apply if the use of the building is to be changed, often the case in adaptive reuse of older buildings. Moreover, the historic buildings exception only applies to repairs, not to more extensive alterations.

California has taken a different approach, adopting an entire State Historical Building Code (SHBC).[62] The SHBC sets out alternative building regulations which may be applied when a "qualified historical building or structure" is to undergo repairs, alterations, or change in occupancy which would normally bring it within the purview of the local building code of general applicability. "Qualified historical buildings or structures" are defined as those "deemed of importance to the history, architecture, or culture of an area by an appropriate local, state or federal government jurisdiction."[63]

Among other provisions, the SHBC permits the alteration or repair of historic buildings using original materials, providing no life-threatening hazards are thereby created.[64] It also sets out standards which can be met when it would be impossible for a historic building to conform to general building code requirements without impairing the historic fabric of the building. With respect to handicapped access, for example, the SHBC permits the local enforcement authority to use alternatives to strict compliance with access provisions so long as "reasonably equivalent" access is assured.[65] In "extreme conditions" exemption from the access provisions may be possible, but only if the historical

---

[61] New York State Uniform Fire Prevention and Building Code § 1233.1 (1984).
[62] Cal. Admin. Code tit. 24 Part 8.
[63] *Id.* § 8-302.
[64] *Id.* § 8-1006.
[65] *Id.* § 8-1303.

fabric or aspect of the building would be destroyed by meeting access requirements, and equivalent services for the handicapped are offered in a nonexempt location.

Until July 1, 1985 provisions of the SHBC were permissive, not mandatory, and cities or counties in California could choose to apply its provisions as a whole, or to utilize any combination of regular and alternative building code provisions.[66] After that date, however, the alternate provisions of the SHBC became mandatory on state agencies and local building departments.

5.3 TRANSFERABLE DEVELOPMENT RIGHTS

Transferable Development Rights (TDRs) are a land use technique through which the unused development potential of one piece of property is severed from that property and transferred to an area or site deemed capable of accommodating the increased density. TDRs assist in the preservation of a landmark or other low-density use (such as a theater) by mitigating the economic burden imposed on the landmark's owner.

A property's excess development "rights" consist of the difference between the actual size of the landmark and the larger building size allowed under the zoning law. The unutilized development rights can, under varying forms of restrictions, be purchased by a developer who will be permitted to construct a building on another site that is larger than the zoning regulations governing that site would otherwise allow.

TDR provisions attempt to mitigate economic burdens imposed on the landmark owner who may be unable to exercise the full bundle of development opportunities available under generally applicable zoning rules. There can be financial and preservation benefits for the municipality as well. TDRs prevent a loss of tax revenue to the city because the purchasers of the development rights pay taxes based on the increased value of their property attributable to its enhanced development potential. In addition, TDRs are a preservation technique that requires minimal expenditure of public funds. And, because they prevent the current use value of the landmark property from being lost to the owner, they may help to counter opposition to designating a site as a landmark.

The favorable eye cast by the Supreme Court in *Penn Central Transportation Co. v. New York City*[67] on the opportunity to transfer development rights above Grand Central Terminal suggests that a locality might be wise to include TDRs or a comparable mechanism in its preservation ordinance. The majority view, and the better reading of the

[66] Cal. Health and Safety Code § 18954.
[67] 438 U.S. 104 (1978).

Penn Central decision, holds that conferring landmark status and thereby restricting alteration or demolition of a structure is a legitimate exercise of the police power and does not in itself constitute a compensable taking of property rights.[68] Thus, in most cases, a TDR program should not be deemed essential to assure the constitutional validity of a well-drafted landmarks program. Nevertheless, particularly in localities whose landmark provisions are advisory only and do not bar alteration or demolition of historic structures, TDRs may, as a practical matter, be critical to a successful preservation program.

Moreover, in those communities where the use of TDRs is appropriate, the technique has often gained widespread community support. In Denver, for example, a city zoning provision expanding the transfer and sale of unused development rights from adjoining lots to other locations passed the city council unanimously after receiving broad support from local property owners, civic organizations, and city agencies.[69]

TDRs may not, however, be suitable for all communities. The value of a TDR program is dependent on the availability of a market for the transferable development rights. In some cities no purchasers may be interested in those rights, making a TDR plan at best a symbolic gesture.

Objections to TDRs have been raised on fairness grounds. Some claim that those who work or live in the area that receives the transfer must bear the cost of the more dense environment, while those who work or live adjacent to the landmark enjoy the benefits. This is a concern which should be addressed in determining the size of the transfer district.

TDR is a flexible tool, and municipalities can adopt different strategies for its use according to the community's needs. These differences regulate which properties are eligible, where the development rights can be transferred, how much can be transferred, and when transfer can be made. Each city must decide the boundaries of the transferor and receiving zones. Zoning requirements also differ, and this will affect the operation of a TDR program. The following discussion illustrates different approaches taken by several municipalities which have adopted TDR plans.

5.3.1 Eligible Property

The first step in drafting a TDR provision is to define the category of property which the program is designed to protect. A number of definitions, of varying degrees of liberality, can be applied: Federally

---

[68] *But see* Fred F. French Investing Co., Inc. v. City of New York, 39 N.Y.2d 587, 385 N.Y.S.2d 5, *cert. denied,* 429 U.S. 990 (1976), holding a taking of property unconstitutional even where the opportunity to transfer development rights is available.

[69] Denver Zoning Ordinance § 59-54(3)(m) (adopted 1982).

or locally designated landmarks, Register-listed or eligible properties, or buildings contributing to the significance of historic districts are illustrative of possible classifications. The chosen definition must balance the specific goal of protecting historic properties against the more general zoning goal of controlling the type and density of land use within a municipality or particular target areas (including historic districts), an objective which could be undermined by an overly broad TDR program.

The New York City TDR provision applies generally to "landmark buildings," which are defined to include "any structure designated as a landmark by the Landmarks Preservation Commission and the Board of Estimate" pursuant to the procedures established by the city's preservation ordinance.[70] The New York City Zoning Resolution also has a special provision relating to the Special Theatre Subdistrict, created to preserve New York's well-known Broadway theaters. In that subdistrict, the TDR program applies not only to landmark buildings but to buildings whose interiors are designated as landmarks as well.[71] This could bring many of the theaters, which are currently being considered for interior (but not exterior) landmark status, within the protection of the program.

The applicability of the Dallas TDR ordinance to a given property also depends upon the area in which the property is located. In the central business district, a building's development rights may be transferred if it is a designated historic landmark. In the city's West End Historic District, however, the program applies to any building which "is a contributing structure listed in the National Register of Historic Places."[72]

The San Francisco program has the broadest and most elaborate provisions on TDR eligibility.[73] A parcel may qualify as an "Eligible Transfer Lot" from which rights may be transferred if it—

1. Contains a "Significant Building" (one which is at least forty years old, is judged to be of individual importance, and is rated either excellent in architectural design or very good in both architectural design and relationship to the environment under the San Francisco landmarks program).
2. Contains a "Contributory Building" (which must also be at least 40 years old, be judged of individual importance, and achieve certain ratings for architectural design and relationship to the environment depending upon whether the building is located within or outside a locally designed Conservation District).

---

[70] N.Y.C. Zoning Resolution § 74-79.
[71] *Id.* § 81-747.
[72] Dallas City Code § 51.4.501(d)(2)(B) (amended 1982).
[73] San Francisco City Planning Code § 128(a).

3. Contains a building which is not designated as Significant or Contributory, but has undergone a "Compatible Rehabilitation" (that is, has been substantially altered to conform to the scale and character of Significant and Contributory buildings in its district) or is a "Compatible Replacement Building" (that is, a newly constructed building compatible with others in the district).

4. Contains an approved urban park, garden, or plaza.

5. Contains a building which has been restored to its original distinguishing qualities or character after an unlawful alteration or demolition.

6. Is a parcel to which development rights have been transferred but not used.

5.3.2 Transfer Zone

Perhaps the most significant issue in determining the viability of a TDR program is the size of the transfer zone, the area to which development rights may be transferred. The size of the transfer zone is important because it must include areas suitable for denser development and large enough to provide a sufficient number of potential receiver sites to absorb the transferred rights. Those downtown areas which would typically provide the strongest market for sellers of TDRs are also likely to be the areas in greatest danger of overcrowding. Again, a municipality may be forced into a delicate balancing of zoning and preservation objectives.

Seattle's ordinance, one of the most restrictive, permits transfer only between adjacent properties and properties located across an abutting alley.[74] New York City is slightly more liberal; its TDR provisions also limit transfers to adjacent lots, but define the term *adjacent* to include contiguous properties, properties across the street from each other, or properties fronting onto the same intersection.[75] Somewhat broader transfer is permitted in particular commercial districts. In those districts, an adjacent lot may be one across the street from a property which is part of a series of properties which includes the lot on which the landmark building is located, all of which are under common ownership. A similar provision applies in the Special Theatre Subdistrict.[76]

Some cities do not impose such rigid proximity restriction on TDRs, on the theory that the purposes of the program are better served by allowing transfer to a broader, though still homogeneous, area. Denver, for example, allows unused development rights to be transferred anywhere within the same zoning district.[77] Recipient properties under the

[74] Seattle Municipal Code § 24.46.110A.

[75] N.Y.C. Zoning Resolution § 74-79.

[76] *Id.* § 81-747.

[77] Denver Zoning Ordinance § 59-54(3)(m)(1).

Dallas ordinance can be situated anywhere within the city's core area.[78] San Francisco permits TDRs to be transferred not only to recipient properties but also to intermediary individuals or organizations who may in turn "hold them for subsequent transfer to other persons, firms, entities or to a Development Lot or Lots" where they may be used.[79]

5.3.3 Rights which Can Be Transferred

To prevent TDR transactions from wreaking havoc with carefully formulated zoning regulations, most programs impose some limitations on the amount of development rights which an individual transferee site can receive. Because of these limitations, it may be impossible to transfer the full bundle of unused development rights from a landmark site to a single receiver site. This handicap can, to a large degree, be ameliorated by permitting TDRs from a single landmark site to be sub-divided and transferred to more than one receiver site.

Under New York's TDR program, for example, the size of a building (measured in floor area) which is allowable on an adjacent lot to which development rights are transferred cannot be increased by more than 20 percent, except in certain sections of midtown Manhattan.[80] Unused development rights can, however, be transferred from the site of a landmark building to any number of lots, so long as all transferee lots satisfy the proximity requirements. Once completed, the transfer irrevocably reduces the amount of floor space which can be developed on the site occupied by the landmark. The reduction continues to apply to the lot even if the landmark is demolished, has its landmark designation withdrawn, or is enlarged or redeveloped.

The Dallas ordinance specifies that the maximum floor area ratio (the ratio between the allowable square footage of floor space of a building on a site and the square footage of the site itself) of a receiving site cannot be increased by more than 4 to 1.[81] A minimum of 20,000 square feet may be transferred under the terms of the ordinance. In Denver, the permissible enlargement depends upon the location of the transferee site. In the central business district, development rights can be transferred up to 25 percent of the "basic maximum gross floor area" allowed on the transferee site; in the lower downtown district, under certain conditions, a structure may be enlarged by up to 50 percent of the "supplementary maximum gross floor area."[82] The Denver ordinance permits a maximum of four transfers from a given zoning lot.

---

[78] Dallas City Code § 51.4.501(d)(4).

[79] San Francisco City Planning Code § 128(g).

[80] N.Y.C. Zoning Resolution § 74-792(2.)(d).

[81] Dallas City Code § 51.4.501(d)(5).

[82] Denver Zoning Ordinance § 59-34(3)(m).

5.3.4   Conditions of Transfer

Finally, TDR programs may require that certain conditions be satisfied before development rights can be transferred. Most commonly, these pertain to restoration of the transferor landmark. Denver, for instance, bars any transfer until the landmark has been renovated.[83] Dallas has a more elaborate provision, which requires that the landmark have been renovated within the past five years and that the total value of improvements exceed 50 percent of the assessed value of the building prior to restoration.[84] Moreover, in making that calculation, only restoration performed pursuant to an appropriate building permit may be counted.

The New York Zoning Resolution does not expressly require that renovation have occurred prior to a TDR transaction. An application to transfer rights, however, must include a "program for the continuing maintenance of the landmark."[85] In addition, before approving a transfer, the City Planning Commission must make the following findings:

1. The permitted transfer will not unduly increase the bulk of a new development, or the density of population or intensity of use in any city block to the detriment of occupants of nearby buildings.
2. Any disadvantages to the neighborhood from reduced access to light and air will be more than offset by the advantages of preserving the landmark.
3. The program for continuing maintenance will result in the preservation of the landmark.[86]

In the case of government-owned landmark sites, the commission must also condition the transfer on the owner's providing a "major improvement of the public pedestrian circulation or transportation system in the area." In the Special Theatre Subdistrict, when rights are to be transferred to a noncontiguous site, the commission must additionally find that any intervening lots are and will continue to be used as legitimate theaters or for other "theatre supportive uses" such as rehearsal space, recording facilities, or theater costume rental facilities.[87] The purpose of this provision is to preserve the integrity and homogeneity of the subdistrict as a focal point for the performing arts.

An even more far-reaching proposal for New York City has been offered by the city's Theatre Advisory Council, established to recom-

---

[83] *Id.*
[84] Dallas City Code § 51.4.501(d)(2)(C).
[85] N.Y.C. Zoning Resolution § 74-791.
[86] *Id.* § 74-792(5.).
[87] *Id.* § 81-747.

mend modifications to local zoning and other regulations to help preserve the Broadway theaters.[88] The council has proposed that existing
contiguity requirements no longer apply to TDR transfers from the
theaters and that unused development rights from parcels containing
Broadway theaters be transferable to a much larger receiving zone than
is now available under the Special Theater Subdistrict. In exchange for
this liberalization, however, an owner of a theater seeking to exercise
these transfer rights would be required to agree to two conditions. First,
the owner would be obliged to pay over a portion of the receipts from
the sale of unused rights to a newly created New York City Theatre
Trust Fund. The Trust Fund would use these funds, among other purposes, to purchase any Broadway theater threatened with demolition
or discontinued use as a legitimate theater, or to provide funding to
forestall demolition or change in use in other ways. Second, the owner
would be required to enter into a covenant, binding upon all subsequent
owners of the property, barring demolition of the theater and requiring
its continued use as a legitimate theater. Through this approach, use of
TDRs would be contingent upon conditions which would contribute to
the unique cultural character of the theater district.[89]

[88] Theatre Advisory Council, "To Preserve the Broadway Theatre: Report to the Planning
Commission of the City of New York on Theatre Preservation" (June 1984).

[89] For a comparison of this approach with an alternative that was considered by the Theater
Advisory Council, see Schneider, "Broadway's Newest Hit: Incentive Zoning for Preserving
Legitimate Theatres," 3 Cardozo Arts & Entertainment L. J. 377 (1984).

Six

Federal and State Statutory Protection for Historic Buildings

Introduction

A number of federal laws are designed to encourage the preservation of the country's historic resources. Chief among these are the National Historic Preservation Act of 1966 (NHPA), which has as its central focus the protection of historic resources, and the National Environmental Policy Act (NEPA), which encompasses such resources within the broadly defined environmental resources it is intended to protect. In addition, many states have enacted analogs to these federal laws which are intended to bolster the protection available to historic structures.

Unlike the local landmark laws discussed in chapter 7, which circumscribe the actions of private owners of historic properties, the statutes discussed in this chapter are not enacted in the exercise of the police power, which enables local governments to regulate the actions of their citizens in the interest of public health, safety, or general welfare. Rather, these statutes are addressed to the decision-making procedures of federal and state governments and their instrumentalities. Both the Section 106 and the NEPA processes described below (as well as their state analogs) are designed to ensure, not that individual structures are preserved, but that the impacts of proposed federal (or state) actions on historic resources are fully considered before potentially adverse action is taken. By requiring federal and state officials to be fully informed of the historic (and, more generally, the environmental) impacts of their actions, and to consider these impacts before acting, these statutes seek to foster their goal of governmental action which enhances, rather than destroys, the nation's historic resources.

6.1   THE NATIONAL HISTORIC PRESERVATION ACT

The NHPA[1] reflects the central policy that "maximum encouragement" should be given to the preservation[2] of historic property.[3] From its investigation prior to enactment of NHPA, Congress concluded that "historic properties significant to the Nation's heritage are being lost or substantially altered, often inadvertently, with increasing frequency."[4] The principal limitation of the NHPA's predecessor, the Historic Sites, Buildings, and Antiquities Act of 1935,[5] was that it could only protect historic resources of national significance.[6] It provided no protection to historic resources of state or local significance, a serious limitation since most of the historic sites in the country are of local or state rather than national importance. As discussed in chapter 2,[7] the NHPA authorizes the Secretary of Interior to expand the country's inventory of historically significant properties, the National Register of Historic Places, to include structures and sites important on the state and local, as well as national, level.

[1] 16 U.S.C. § 470 *et seq.*

[2] NHPA defines "preservation" or "historic preservation" as including identification, evaluation, recordation, documentation, curation, acquisition, protection, management, rehabilitation, restoration, stabilization, maintenance and reconstruction, or any combination of the foregoing activities. 16 U.S.C. § 470w(8).

[3] NHPA defines "historic property" or "historic resource" to mean any prehistoric or historic district, site, building, structure, or object included in, or eligible for inclusion on, the National Register; such term includes artifacts, records, and remains which are related to such a district, site, building, structure or object." 16 U.S.C. 470w(5).

[4] 16 U.S.C. § 470(b)(3).

[5] 16 U.S.C. §§ 461–67.

[6] See "Preservation of Historic American Sites, Buildings, Objects, and Antiquities of National Significance: Hearings on H.R. 6670 and H.R. 6734 before the Subcommittee on Public Lands of the House Committee on Interior and Insular Affairs," 74th Cong., 1st Sess., 4–8 (1975) (Testimony of Secretary Ickes).

[7] *See* Section 2.1.

The NHPA complemented its expanded effort to catalog resources in the National Register with provisions to protect these resources from federal actions that threaten their integrity. It requires that when there is a federal undertaking[8] which may directly or adversely affect any property listed on the National Register or eligible for listing on the Register or any National Historic Landmark, the head of the responsible federal agency "shall, to the maximum extent possible, undertake such planning and actions as may be necessary to minimize harm" to such property.[9] If the federal action will adversely impact a property listed on the National Register or eligible for listing, the agency must afford the Advisory Council on Historic Preservation (Advisory Council), created under the act,[10] a reasonable opportunity to comment. The agency must take these comments into account in its decisionmaking process.

The 1976 and 1980 amendments to the NHPA incorporated many of the requirements of Executive Order 11593,[11] which had in turn expanded the duties of federal agencies under the original 1966 Act. As amended, the NHPA imposes responsibilities on the heads of all federal agencies for the preservation of historic properties which are owned or controlled by those agencies.[12] Moreover, every agency is required to establish a program to locate and nominate any properties it owns or controls which appear to qualify for inclusion in the National Register. The amendments also apply the Act's protections to Register-eligible, as well as listed, properties[13]. The rationale for the latter provision is that the act implements the public policy of protecting all historic structures worthy of preservation. National Register listing has direct effects upon the owner of a property, including eligibility for tax benefits and in some circumstances, imposition of certain restrictions on use. For this reason, as explained in chapter 2,[14] an owner may veto such listing.

[8] An "undertaking" consists of a federal or federally assisted, or federally licensed "action, activity or program, or the approval, sanction, assistance, or support of any non-Federal action, activity or program." 36 C.F.R. § 800.2(c).

[9] 16 U.S.C. §§ 470f, 470h-2(f).

[10] 16 U.S.C. § 470i. The Advisory Council on Historic Preservation was established as an independent federal agency. It is charged with advising the president and Congress on matters relating to historic preservation. The members of the Council, who are appointed by the president, consist of the Secretaries of Interior, Agriculture, and four other agencies of the United States whose activities affect historic preservation; the president of the National Conference of State Historic Preservation officers; the chairman of the National Trust for Historic Preservation; the architect of the Capitol; one governor; one mayor; four preservation experts from the fields of architecture, history, archaeology, and other appropriate disciplines; and four members of the general public, one of whom chairs the Advisory Council.

[11] 36 Fed. Reg. 8921 (1971).

[12] 16 U.S.C. § 470h-2.

[13] 16 U.S.C. § 470f.

[14] *See* Section 2.1.4.

The fact that the owner does so, however, should not, and under the 1980 amendments does not, mean that such unlisted property is un-protected from the impacts of government action.

While the NHPA is the central federal law enacted to reflect the national policy of encouraging preservation, there are important limits on its ability to protect historic resources. In particular, the NHPA provides no certainty that a historic property will in fact be preserved. If a federal agency complies with the NHPA's procedures in arriving at its decision, the act requires no more, even if the result is the loss of an important historical structure.[15]

The protection afforded by the NHPA is also limited because, as just noted, the act applies only to actions involving the federal government. Actions by state or local government, private corporations, or individuals, for example, can threaten a historic property without triggering the application of the NHPA. The NHPA does not by itself prohibit these entities from taking any actions, including demolition, that they believe desirable with respect to either listed or eligible properties.

The program's effectiveness is further curtailed by the federal government's failure to appropriate sufficient funds. For example, grants to encourage rehabilitation activities, once an important component of the NHPA, are no longer funded. And the Advisory Council, which oversees the implementation of the NHPA, has seen its budget cut significantly, and, more recently, has had its authority challenged by other agencies.

6.1.1 Section 106 Review

A property that is listed or determined to be eligible for Register listing[16] is afforded certain protections under Section 106 of the NHPA.[17] Section 106 requires the head of a federal agency that proposes to take

---

[15] U.S. v. 162.20 Acres of Land, More or Less, 639 F.2d 299, 302 (5th Cir.), *cert. denied,* 454 U.S. 828 (1981).

[16] This includes properties eligible but not listed on the Register because of owner objection.

[17] 16 U.S.C. § 470f. Hereafter the statutory provision will be referred to by its popular name: § 106 review. The section provides that

the head of any Federal agency having direct or indirect jurisdiction over a proposed Federal or federally assisted undertaking in any State and the head of any Federal department or independent agency having authority to license any undertaking shall, prior to the approval of the expenditure of any Federal funds on the undertaking or prior to the issuance of any license, as the case may be, take into account the effect of the undertaking on any district, site, building, structure, or object that is included in or eligible for inclusion in the National Register. The head of any such Federal agency shall afford the Advisory Council on Historic Preservation established under sections 470i to 470v of this title a reasonable opportunity to comment with regard to such undertaking.

an action that would affect a listed or eligible property to notify the Advisory Council of that action and solicit its comments. Moreover, before taking any action, each federal agency is responsible for identifying all properties on the National Register or eligible for listing that are located within the "area of the undertaking's potential environmental impact" and that may be affected by the undertaking.[18] Minimum steps which must be taken to identify these properties are set forth in the regulations.[19] Telephone calls to the Advisory Council and to the State Historical Commission to ascertain whether a property is listed on the Register are insufficient to satisfy the agency's affirmative duty.[20] An agency cannot "passively rely on other agencies to satisfy its responsibilities under NHPA."[21]

It is noteworthy that the affirmative duty to identify historic properties is not dependent on a prior official determination that a property is eligible for listing. The language of the regulations has been interpreted to require an independent survey to locate and identify properties that may be eligible. In *Hough v. Marsh*,[22] the court found that "eligible property" is that which actually meets the National Register criteria, not that which has been determined to meet the criteria.

When the agency and the state historic preservation officer (SHPO) agree that a property is not eligible for the National Register, the requirements of the NHPA are satisfied and the agency action may proceed. However, if they do not, and a "question" exists as to the property's eligibility,[23] the regulations require that the Secretary of the Interior be requested to make a final judgment. If the agency and the SHPO reach the same conclusion pertaining to a property's eligibility, but employ differing reasoning, no dispute is said to exist and their determination of eligibility is final.[24]

If it is determined that the proposed federal action may affect property listed on or eligible for listing on the National Register, the Section 106 procedure must be undertaken. Generally speaking, the protection afforded by Section 106 is procedural and it does not prohibit conduct by the federal government, regardless of how damaging it will be to the historic property. The Advisory Council's comments, for example, must be "taken into account," and the agency must integrate them into its decision-making process.[25] However, the comments are advisory

---

[18] 36 C.F.R. § 800.4(a).

[19] *See id.* at (1), (2), and (3).

[20] Hough v. Marsh, 557 F. Supp. 74, 87-88 (D. Mass. 1982).

[21] *Id.* at 88.

[22] *Id.*

[23] 36 C.F.R. § 800.4(a)(3).

[24] Wilson v. Block, 708 F.2d 735, 756 (D.C. Cir.), *cert. denied*, 104 S.Ct. 371 (1983).

[25] 36 C.F.R. § 60.2(a).

only. No action can be required or prohibited by the Advisory Council. Once having complied with procedural requirements, the federal agency may adopt any course of action it believes is appropriate.

Nevertheless, the agency may not take lightly the responsibilities that Section 106 does create. Perfunctory consideration is not sufficient to meet the requirements of the NHPA. The language of the act is "mandatory and the scope is broad."[26] The procedure for complying with Section 106 is to be initiated as early as possible before an agency makes a final decision concerning an undertaking and, in any event, prior to taking any action that would foreclose alternatives or the Advisory Council's ability to comment.

6.1.2   Federal Agencies and Undertakings Affected

Section 106 review applies only to projects in which the federal government is involved. State, county, or local activity that threatens historic or architectural resources does not trigger the process, nor will the activity of private entities.

It appears that the term "federal agency" will be broadly construed in determing whether the Sectin 106 process is to be undertaken. In *Committee to Save the Fox Bldg. v. Birmingham Branch of Federal Reserve Bank of Atlanta,*[27] for example, the court found that the regional Federal Reserve bank which was planning to demolish a historic building was a "federal agency" under the NHPA.

In analyzing whether an action proposed by a federal agency is subject to review under Section 106, it must next be determined whether the action constitutes a federal or federally assisted "undertaking."

Federal regulations define "undertaking" broadly:

> "Undertaking" means any Federal, federally assisted or federally licensed action, activity, or program or the approval, sanction, assistance, or support of any non-Federal action, activity or program. Undertakings include new and continuing projects and program activities (or elements of such activities not previously considered under Section 106 or Executive Order 11593) that are: (1) directly undertaken by Federal agencies; (2) supported in whole or in part through Federal contracts, grants, subsidies, loans, loan guarantees, or other forms of direct and indirect funding assistance; (3) carried out pursuant to a Federal lease, permit, license, certificate, approval, or some other form of entitlement or permission; or (4) proposed by a Federal agency for Congressional authorization or appropriation. . . .[28]

Courts have also taken a broad view of the parameters of an undertaking. Undertakings have been deemed to include direct federal gov-

[26] United States v. 162.20 Acres of Land, More or Less, *supra*, 639 F.2d at 302.

[27] 497 F. Supp. 504 (N.D. Ala. 1980).

[28] 36 C.F.R. § 800.2(c).

ernment acts, such as the proposed demolition of a National Register courthouse by the U.S. Department of Housing and Urban Development (HUD) as part of an urban renewal project,[29] federal construction programs,[30] and military operations.[31] But government involvement need not be so direct. The federal government has also been deemed to be sufficiently involved with a project when it has granted its imprimatur by various means such as funding,[32] permitting,[33] or providing approval for a project when that approval was a condition of its initiation.[34] The determination as to whether a federal undertaking is present is based on the exercise of federal discretionary authority, regardless of who the ultimate actor is or where the action occurs. Thus, a federal undertaking may involve actions by nonfederal governmental units and private parties, when these actions are authorized, delegated, or supported by the federal government. And Section 106 review may be triggered even when the action occurs on land which is not federally owned or controlled. On the other hand, merely acquiring title to historically significant property through filing a declaration of condemnation has been declared a neutral act vis-a-vis the NHPA and not to be subject to Section 106 review.[35]

6.1.3   Effects of a Federal Undertaking

Once it has been determined that a federal undertaking is involved, the next threshold question is whether that undertaking will have an "effect" on the Register-listed or eligible property sufficient to require Section 106 review procedures. In general, the issue turns on whether the undertaking causes or may cause any change in the quality of those characteristics which qualify the property for the National Register. The regulations cite integrity of location, design, setting, materials, workmanship, feeling, or association of the property as being such significant qualities.[36] Because what may be significant in the case of a historic battlefield may not be significant with regard to a mansion (that is, location will certainly be crucial for the battlefield, but may or may

[29] See Save the Courthouse Comm. v. Lynn, 408 F. Supp. 1323, 1339 (S.D.N.Y. 1975).

[30] Commonwealth of Puerto Rico v. Muskie, 507 F. Supp. 1035, 1061 (D.P.R.), *vacated on other grounds sub nom.* Marquez-Colon v. Reagan, 668 F.2d 611 (1st Cir. 1981).

[31] Romero-Bercelo v. Brown, 643 F.2d 835, 858–60 (1st Cir. 1981), *rev'd on other grounds,* 456 U.S. 305 (1982).

[32] National Center for Preservation Law v. Landrieu, 496 F. Supp. 716 (D.S.C.), *aff'd,* 635 F.2d 324 (4th Cir. 1980); Ely v. Velde, 497 F.2d 252 (4th Cir. 1974).

[33] Coalition for Responsible Regional Dev. v. Coleman, 555 F.2d 398 (4th Cir. 1977); Weintraub v. Rural Electrification Admin. 457 F. Supp 78, 92–3 (M.D. Pa. 1978).

[34] See Weintraub v. Provident Nat'l. Bank, No. 78–1577 (E.D. Pa. May 11, 1978).

[35] United States v. 162.20 Acres of Land, *supra,* 639 F.2d at 304–05.

[36] 36 C.F.R. § 800.3(a).

not be for the mansion), "effect" must be evaluated in the context of the historic, architectural, archaeological, or cultural significance possessed by the particular property.

The effect which federal action may have on registered or eligible property may be either beneficial or harmful. Agency officials must determine whether either will result from the proposed undertaking.[37] The determination extends to effects that are indirect[38] as well as direct, and the distinction between direct and indirect effects is basically one of proximity. A direct effect occurs at the same time and place as the undertaking. An indirect effect may be less proximate—later in time or farther removed in distance. Examples of indirect effects include changes in the pattern of land use, population density, or growth rate that may affect historic properties. If the effect which may be caused is removed in time or distance to a degree that it is not reasonably foreseeable, then it is not even an indirect effect and Section 106 review is not required.[39]

Note that the Section 106 review process is required if there may be *any* effect; it does not adopt the higher standard of "significant effect" required by NEPA. This distinction is drawn more finely in the discussion on NEPA below.

6.1.4 The Consultation Process

If the federal officer, in consultation with the SHPO, concludes that the proposed action will have no effect on Register-listed or eligible property, the agency is required only to document the findings and to make them public. No further action is required under Section 106.[40] No review of this determination occurs unless an objection to the findings is registered with the executive director of the Advisory Council. If a timely objection is made, the executive director will review the case and issue his findings within fifteen days.[41] If there is a finding of no effect, Section 106 requires nothing more and the federal action may be undertaken.

If in the first instance, or as a result of an appeal to the executive director, it is found that there will be an effect, the next step is the determination of whether that effect will be "adverse." The agency, in consultation with the SHPO, must measure the effects of the action against regulating standards. Among the effects that are deemed to be adverse are the following:

---

[37] 36 C.F.R. § 800.4(b).

[38] *But see* Cobble Hill Ass'n. v. Adams, 470 F. Supp. 1077, 1091 (E.D. N.Y. 1979).

[39] 36 C.F.R. § 800.3(a).

[40] *Id.* § 800.4(b)(1); *see* Wilson v. Block, *supra*, 708 F.2d at 755.

[41] *Id.* § 800.4(b)(1).

1. Destruction or alteration of all or part of a property.
2. Isolation from, or alteration of, the property's surrounding environment.
3. Introduction of visual, audible, or atmosphere elements that are out of character with the property or alter its setting.
4. Neglect of a property resulting in its deterioration or destruction.
5. Transfer or sale of a property without adequate conditions or restrictions regarding preservation, maintenance, or use.[42]

This list is not exhaustive, and it may be argued that any action resulting in a deleterious change to the property constitutes an adverse effect. Note, however, that these criteria are to be applied with reference only to those characteristics of the property that contributed to its listing on the National Register.[43] Using this yardstick, if the agency finds that no adverse effect from the action will result, it must adequately document its determination and forward this to the Advisory Council for review. The documentation must include a statement explaining why each of the criteria of adverse effect in the regulations is inapplicable.[44]

It is rare for the Advisory Council not to ratify the agency's decision and to allow the action to proceed. If it does not, however, the Advisory Council's concerns as expressed by its executive director may sometimes be satisfied and its concurrence gained by the agency's written acceptance of specific conditions to ameliorate those aspects of the action which are undesirable. If the agency agrees to the conditions, they are incorporated into the agency's determination of effect and the Advisory Council's objection will be withdrawn. At this point, the agency's Section 106 responsibilities are considered satisfied and the action may proceed. In the event that conditional concurrence cannot be obtained, the agency official must provide supplemental documentation in the form of a Preliminary Case Report to the Advisory Council, notify the SHPO of this step, and extend to the Advisory Council the opportunity to comment on the proposed action.[45]

When the agency official finds that the proposed federal undertaking will result in an adverse effect or when, as just described, the official finds that there will be no adverse effect but this conclusion is disputed by the Advisory Council director upon review, then the consultation process is triggered.[46] The goal of this process is to avoid or minimize

---

[42] *Id.* § 800.3(b).

[43] *Id.* § 800.4(b).

[44] *Id.* § 800.13(a)(4).

[45] *Id.* § 800.4(d). The preliminary case report is prepared in accordance with § 800.13(b). It must be made readily available to the public. *Id.* § 800.6(b)(1).

[46] *Id.* § 800.6(b).

the proposed action's adverse effect. After the report is submitted, the consulting parties (the agency official, the SHPO, and the executive director of the Advisory Council) consider any feasible and prudent alternatives which could lessen or avoid entirely the adverse effects of the undertaking. Where more than one federal agency is involved, their consultation responsibilities may be coordinated through one lead agency. Meetings to solicit or disseminate information to or from the public are discretionary, but must be called at the request of any of the consulting parties.

Pending the outcome of this consultation process, the agency may not proceed with or sanction any action that would produce an adverse effect, or make any irreversible commitment that could result in an adverse effect or that would foreclose alternatives or modifications to the proposed undertaking that could avoid or minimize the adverse effects.[47]

6.1.5 Memorandum of Agreement

The product of these consultaltions is the memorandum of agreement (MOA). The MOA fulfills the Section 106 comment requirement. It reflects the understanding achieved between the consulting parties of a feasible and prudent alternative to the proposed federal undertaking which will mitigate or avoid its adverse effect.[48] If, however, the consulting parties agree that no such alternative exists and that the proposed undertaking is in the public interest, then the MOA will reflect the acceptance of the adverse effect. It may further describe any recording, salvage, or other procedures to be followed by the agency to minimize the adverse effects prior to its proceeding with its actions.

If no agreement is reached, the issue may be considered by the full Advisory Council or, alternatively, by a five-member panel.[49] Although assignment of matters to the Advisory Council or panels rarely happens, the decision to undertake those steps is principally the chairman's, upon the recommendation of the director. If the chairman declines to schedule a full council meeting, a summary statement of the federal undertaking is provided to each member. Three council members may object to this procedure, however, in which case the matter is scheduled.

Once the Council or panel meeting is scheduled, each of the consulting parties and the Secretary of Interior must submit written reports prior to its convening. Submission of oral testimony from other parties and the public may also be received, but to testify at the meeting, parties must provide advance notice of their interest to the executive director.[50]

---

[47] *Id.* § 800.4(e); *see* Nat'l. Trust for Historic Preservation v. U.S. Army Corps of Engineers, 552 F. Supp. 784 (S.D. Ohio 1982).

[48] 36 C.F.R. § 800.6(b)(5).

[49] *Id.* at (d)(2).

[50] *Id.* at (d)(4).

The executive director of the Advisory Council is responsible for preparing the MOA. However, it is based on a proposal by the relevant agency stating the actions agreed to by the consulting parties to avoid or mitigate adverse effects. If the adverse effects have been accepted, the proposal will reflect that. The executive director determines that the proposal reflects the agreement of the consulting parties, and within ten days forwards it to the chairman of the Advisory Council as an MOA for ratification. If the chairman decline to ratify, the MOA is returned to the agency for revision.

The final execution of an MOA is followed by a 30-day review period.[51] During this period the chairman of the Advisory Council may ratify the agreement, place it before the Council for consideration, or simply allow the time to elapse without taking action. Upon expiration of the 30 day period without action, the MOA is considered ratified.

The agency must then comply with the terms of the MOA.[52] If it fails to do so or modifies its actions, it must again seek the Council's comments based on its new action. And as before, while the renewed commenting process proceeds the agency must suspend actions that would lead to an adverse effect or that would foreclose the Council's consideration of modifications or alternatives to the proposed undertaking that could avoid or mitigate the adverse effect. The regulations allow amendment to the MOA if the original terms cannot be met or if a signatory believes a change is necessary.[53]

Several courts have reviewed the legal enforceability of the MOA. While it is not a settled question, it appears that a non governmental party can force compliance with the terms of the MOA.[54]

In sum, the NHPA provides for elaborate procedural steps that may delay an adverse action, but includes no substantive protections that can be used to halt a project. Even this limited role has been subject to challenge. The Office of Legal Counsel in the Justice Department has argued that the Advisory Council is exercising "substantive regulatory powers" over federal agencies which are more extensive than the power to comment on federal undertakings which is authorized by the act.[55]

It should be noted that the requirements outlined here are derived from the Advisory Council's regulations implementing Section 106. Those regulations, however, note that federal agencies may enact "counterpart regulations," which would provide an alternative route to sat-

---

[51] *Id.* at (c)(2).

[52] *Id.* at (c)(3).

[53] *Id.* at (c)(4).

[54] Courts have implicitly suggested that an MOA, in appropriate circumstances, would be enforceable. *See* United States v. 162.20 Acres of Land, 733 F.2d 377 (5th Cir. 1984); Don't Tear It Down, Inc. v. Pennsylvania Ave. Dev. Corp., 642 F.2d 527 (D.C. Cir. 1980).

[55] Memorandum for Michael J. Horowitz, OMB, and John M. Fowler, ACHP, October 28, 1983.

isfying the duties placed upon agencies by Section 106. When a counterpart regulation has been adopted, under the Advisory Council provisions, it "may, as appropriate, supersede the requirements" adopted by the Council.[56] When an agency has not adopted its own procedures for implementing Section 106, however, it remains bound by those of the Advisory Council.[57]

6.1.6 Phased Development

There is come controversy as to whether Section 106 requirements must be complied with through every stage of a continuing federal undertaking. The better view holds that as long as the agency retains discretionary authority over a project, Section 106 must be obeyed, and the Advisory Council retains review jurisdiction over continuing federal actions.[58] This problem often arises in connection with projects involving highway construction[59] and urban development.[60] WATCH V. Harris[61] held that Section 106 requirements "apply until the agency has finally approved the expenditure of funds at each stage of the undertaking," but the court there acknowledged that a different conclusion had been reached in other cases. More recently, the Third Circuit concluded that the "NHPA is applicable to an ongoing project at any stage where a Federal agency has authority to approve or disapprove Federal funding and to provide meaningful review of both historic preservation and community development goals. . . ."[62] Moreover, an undertaking cannot be severed into stages without each stage of development being subjected to Section 106 review. Were this not true, Section 106 could be effectively circumvented every time a project was extensive in time or geography.

The case law is also somewhat contradictory as to an agency's obligation to undertake a comprehensive review of all effects at the inception of a phased project. In *Save the Courthouse Committee v. Lynn,*[63] a case in which a citizens' group successfully sued to bar demolition of a courthouse complex until the Section 106 review had been completed, the court held that the regulations required agency officials to take steps "at a time when it was still possible to effect changes in the undertaking

---

[56] 36 C.F.R. § 800.11.

[57] *See* Save the Courthouse Comm. v. Lynn, *supra,* 408 F. Supp. at 1338.

[58] Morris County Trust for Historic Preservation v. Pierce, 714 F.2d 271, 279–81 (3d Cir. 1983); WATCH v. Harris, 603 F.2d 310 (2d Cir.), *cert. denied,* 444 U.S. 995 (1979).

[59] *See* Thompson v. Fugale, 347 F. Supp. 120 (E.D. Va. 1972).

[60] *See* WATCH v. Harris, 603 F.2d 310 (2d Cir.), *cert. denied,* 444 U.S. 995 (1979).

[61] *Id.* at 319.

[62] Morris County Trust for Historic Preservation v. Pierce, 714 F.2d 271, 280 (3d Cir. 1983).

[63] 408 F.Supp. 1323 (S.D.N.Y. 1975).

in order to circumvent an adverse impact." However, in *Vieux Carre Property Owners, Residents and Associates, Inc. v. Pierce*,[64] the court held that in the phased development of a city's use of Urban Development Action Grant (UDAG) funds, the city properly limited its historic preservation review to the phase of development for which the grant was specifically made rather than including speculative future phases.

Sometimes an agency will meet its responsibility to identify historic properties at the inception of a project, but become aware of a previously unidentified eligible property after construction has started. When this happens, Section 106 can be satisfied by complying with the requirements of the Archeological and Historic Preservation Act, which requires the Secretary of the Interior to protect historical and archaeological data from the potentially damaging effects of federal activities.[65] The regulations, however, reserve to the Secretary the right to require the agency to request the comments of the Advisory Council if he deems such consultation to be warranted based on the significance of the property, the effect of the original action, and any proposed mitigation actions.

Section 214[66] of the NHPA grants the Advisory Council the authority to exempt entire federal programs or undertakings from any or all of the act's requirements. The exemption must be determined to be consistent with the purposes of the NHPA and must take into consideration the magnitude of the exempted undertaking or program and the likelihood of the impairment of historic properties.

6.1.7   Enforcement of Section 106 Requirements

Litigation under Section 106 usually entails the claim that the agency did not initiate the review process when it should have or that after beginning the process it did not follow proper procedures—that it did not seek comments from the Advisory Council or did not take its comments into account. The regulations explicitly make a responsible official's failure to act reviewable as agency action by courts or administrative tribunals under the Administrative Procedures Act and under other applicable law.[67] Executive Order 11593 provides a second prong of attack on agency actions, by allowing plaintiffs to claim violation of both the statute and the order in appropriate cases.[68] If the agency has

---

[64] 719 F.2d 1272 (5th Cir. 1983).

[65] 16 U.S.C. § 469a-1. *See* 36 C.F.R. § 800.7(a).

[66] 16 U.S.C. § 470v.

[67] 40 C.F.R. § 1508.18.

[68] Aluli v. Brown, 437 F. Supp. 602 (D. Hawaii 1977), *rev'd on other grounds*, 602 F.2d 876 (9th Cir. 1979); Save the Courthouse Comm. v. Lynn, 408 F. Supp. 1323 (S.D.N.Y. 1975).

not given the Advisory Council an opportunity to comment, then an action for injunctive relief preventing the federal agency from taking further action is the appropriate remedy.

If the Section 106 process has resulted in an MOA that has not been complied with, then an action sounding in contract and seeking specific performance, a declaratory judgment or mandamus may be appropriate. As just mentioned, however,[69] whether a private party can force the agency to comply with the terms of the MOA is not a settled legal question.

6.1.8   Section 106 Litigants

Actions to enforce Section 106 may be brought by private citizens[70] or by groups.[71] The Advisory Council has the authority to institute suits to challenge compliance, but an aversion to public airing of interagency disputes greatly diminishes the likelihood of its doing so. Instead, the Advisory Council is more often brought into a suit by the citizen-plaintiff. Other groups or individuals are also potential defendants: The acting federal agency and the SHPO are the most obvious. If federal powers are delegated or benefits conferred to nonfederal levels of government or private individuals, these parties should also be considered as possible defendants in the action.[72]

The traditional standing test of injury-in-fact,[73] requiring a "personal stake in the controversy" demonstrated by the suffering of a "distinct and palpable injury,"[74] is utilized by the courts to determine who has standing to bring an action under the NHPA.[75] A citizen need not be the owner of the historic property or allege economic harm to show injury-in-fact allowing him to bring suit.[76] Generally, any resident or owner of property proximate to the property in question will have standing. Some courts have expanded this standard so that the test is not merely one of proximity. Under this view, residents of a town are con-

[69] See Section 6.1.5.

[70] See Aluli v. Brown, supra, 437 F. Supp. at 609; contra, Carson v. Alvord, 487 F. Supp. 1049 (N.D. Ga. 1980).

[71] See e.g., Neighborhood Dev. Corp. v. Advisory Council on Historic Preservation, 632 F. 2d 21 (6th Cir. 1980); Save the Courthouse Comm. v. Lynn, 408 F. Supp. 1323 (S.D.N.Y. 1975).

[72] See, e.g., Biderman v. Morton, 497 F.2d 1141, 1147 (2d Cir. 1974); Save the Courthouse Comm. v. Lynn, supra, 408 F.Supp. at 1344; but see Woonsocket Historical Soc. v. Woonsocket, 387 A.2d 530 (R.I. 1978).

[73] Sierra Club v. Morton, 405 U.S. 727 (1972).

[74] Warth v. Seldin, 422 U.S. 490, 498–99, 502 (1975).

[75] See Neighborhood Dev. Corp. v. Advisory Council on Historic Preservation, 632 F.2d 21 (6th Cir. 1980).

[76] See id. at 24.

sidered "users" of the property and beneficiaries of the environment that its unchanged character produces. Legal interest in preserving the property thereby accrues to each of the residents.[77]

In a Sixth Circuit case, the court held that a plaintiff deprived of enjoyment of an aesthetic resource has standing even if the plaintiff is not a resident of the neighborhood in which the building is located.[78] The court stated that "[t]he deprivation of the use of an aesthetic resource is not merely an abstract injury. . . . By alleging 'use' of the building's aesthetic and architectural value, plaintiffs met the *Sierra Club* standard [of injury-in-fact]."[79] The court went on to assert, "We do not believe that injury-in-fact is suffered only by residents of the neighborhood in which the historically and architecturally significant buildings are located." Nor, it stated, is "standing . . . to be denied because the alleged injury is commonly shared."

A "nonprofit membership organization devoted to protection of the built environment, with a specific interest in preservation of buildings of architectural and historic value in the District of Columbia" was granted standing in another case.[80] So, too, was a historical association with some of its members residing in the vicinity of a National Register building.[81] In a third case, the plaintiff, a concerned group of citizens organized as an unincorporated association whose members claimed interest in architecturally and historically significant buildings in that city was allowed to bring suit, because "a showing by plaintiff of the imminent demolition of a building listed on the National Register of Historic Places is sufficient to establish plaintiff's standing to sue.[82] That was held by the court to be a sufficient showing of prospective injury.

SHPOs also have standing to sue to enjoin demolition of historic buildings.[83]

Another possible basis of standing would be as a claimed third-party beneficiary of a MOA. Such status may, however, be difficult to establish.[84]

[77] Edwards v. First Bank of Dundee, 393 F. Supp. 680, 682 (N.D. Ill. 1975), *rev'd on other grounds*, 534 F.2d 1242 (7th Cir. 1976); River v. Richmond Metropolitan Auth. 359 F. Supp. 611,625 (E.D. Va.), *aff'd*, 481 F.2d 1280 (4th Cir. 1977).

[78] Neighborhood Dev. Corp. v. Advisory Council on Historic Preservation, 632 F.2d 21,24 (6th Cir. 1980).

[79] *Id.* at 23–24.

[80] *See* Don't Tear It Down, Inc. v. Pennsylvania Ave. Dev. Corp., 642 F.2d 527, 531 (D.C. Cir. 1980).

[81] *See* Wicker Park Hist. Dist. Preservation Fund v. Pierce, 565 F. Supp. 1066 (N.D. Ill. 1982).

[82] Committee to Save the Fox Bldg. v. Birmingham Branch of Federal Reserve Bank of Atlanta, 497 F. Supp. 504, 509 (N.D. Ala. 1980).

[83] Weintraub v. Rural Electrification Admin., 457 F. Supp. 78,88 (M.D. Pa. 1978).

[84] *See* Citizens Comm. for Envtl. Protection v. U.S. Coast Guard, 456 F. Supp. 101, 115–16 (D.N.J. 1978).

6.1.9　Procedural Issues in Section 106 Litigation

Federal subject matter jurisdiction in Section 106 suits may be claimed under the Administrative Procedure Act (APA).[85] The Supreme Court has held that under the provisions of the APA, a federal district court will generally have jurisdiction to review a federal agency action.[86] The exception to this is where a defendant can show by clear and convincing evidence that Congress intended to restrict judicial review under the specific substantive statute governing the agency's action.[87] However, a number of courts have held or implicitly found that a federal district court has the power to decide disputes relating to agency action or inaction with respect to the requirements of the NHPA.[88]

A federal district court also has subject matter jurisdiction under the general federal question jurisdiction statute, 28 U.S.C. § 1331(a). The grant of federal question jurisdiction for issues arising under the NHPA has been called "incontrovertible."[89]

It may be necesary to exhaust administrative remedies prior to seeking judicial review of controversies relating either to National Register listings or to proposed federal undertakings. A federal court in New York, for instance, dismissed a complaint seeking removal of a National Register listing because plaintiffs failed to exhaust the administrative procedure established by the regulations.[90]

Decisions of the Advisory Council are likely to be difficult to overturn in court. As the agency created to administer Section 106, the Advisory Council's determinations "are owed great deference unless those determinations are clearly in error."[91] This is consistent with the general principle that an agency action made in accordance with prescribed procedures will not be overturned under the Administrative Procedures Act unless it is found to be arbitrary, capricious, an abuse of discretion, or otherwise not in accordance with law.[92]

The 1980 amendments to the NHPA authorize the awarding of litigatin costs—attorney's fees, expert witness fees, and other related costs—to any person who brings an action to enforce the NHPA and

---

[85] 5 U.S.C. §§ 701–706 (1982).

[86] Abbott Laboratories v. Gardner, 387 U.S. 136, 140 (1967).

[87] Id.; Citizens Comm. for Hudson Valley v. Volpe, 425 F.2d 97, 101 (2d Cir.), cert. denied, 400 U.S. 949 (1970).

[88] Save the Courthouse Comm. v. Lynn, 408 F. Supp 1323, 1330–31 (S.D.N.Y. 1975), and cases cited therein.

[89] Id. at 1331.

[90] White v. Shull, 520 F. Supp. 11 (S.D.N.Y. 1981).

[91] National Trust for Historic Preservation v. U.S. Army Corp of Engineers, 552 F. Supp. 784, 791 (S.D. Ohio 1982).

[92] Aertsen v. Landrieu, 488 F. Supp. 314, 318 (D.Mass), aff'd, 637 F.2d 12 (1st Cir. 1980), quoting 5 U.S.C. § 706(2)(A).

"substantially prevails."[93] The court may award the costs of litigation in an amount it deems reasonable. Legislative history suggests a congressional purpose of encouraging actions which might not otherwise be brought because of the expense of litigation.[94] One court, discussing the legislative history of this provision of the NHPA, noted that "Congress has framed the amendment in very broad terms . . . without any limitations as to which defendants are liable for fees."[95]

6.1.10 Other Federal Agency Responsibilities under the NHPA

The NHPA creates other responsibilities for federal agencies in addition to compliance with Section 106. Section 206 of the 1980 amendments[96] makes federal agencies responsible for preserving historic buildings under their ownership or control. There is, moreover, a general responsibility to engage in historic preservation activities to the extent consistent with program objectives.[97]

6.2 OTHER FEDERAL PRESERVATION STATUTES

Resources of national importance may be designated as national historic landmarks.[98] Such designation results in both listing on the National Register and inclusion in the National Historic Landmarks Program, and hence in protection under both Sections 106 and 110(f) of the NHPA.[99] Section 110(f), in language similar to that of Secton 106, requires federal agencies to minimize harm to national landmarks from federal undertakings.

Similar in intent to Section 106 is the statutory requirement commonly known as Section 4(f) of the Department of Transportation (DOT) Act of 1966, although it is now recodified elsewhere.[100] This provision

---

[93] 16 U.S.C. § 470w-4.

[94] H.R. Rep. No. 1457, 96th Cong., 2d Sess., *reprinted in* 1980 U.S. Code Cong. and Admin. News at 6378.

[95] WATCH v. Harris, 535 F. Supp. 9, 13 (D. Conn. 1981).

[96] 16 U.S.C. § 470h-2(a)(1).

[97] 16 U.S.C. § 470h-2(d).

[98] 16 U.S.C. § 462(b). National historic landmarks are properties "which possess exceptional value as commemorating or illustrating the history of the United States." 36 C.F.R. § 65.1(b)(1). They must "pertain to the development of the Nation as a whole rather than to a particular State or locality." 36 C.F.R. § 65.2.

[99] 16 U.S.C. § 470h-2(f). Properties listed in the National Historic Landmarks Program were incorporated into the National Register in 1966. *See* 36 C.F.R. § 65.1.

[100] 23 U.S.C. § 138. Section 4(f), which had been codified at 49 U.S.C. § 1653(f), was repealed in 1983. Federal "use" of historic sites which triggers the application of this section has been broadly defined by the courts. *See, e.g.,* Monroe Cty. Conservation Council v. Adams, 566 F.2d 419, 424 (2d Cir. 1977), *cert. denied,* 435 U.S. 1006 (1978). The regulations implementing this provision are found at 23 C.F.R. Part 771.

bars any federal transportation program from using land from a historic site of national, state, or local significance unless "(1) there is no feasible and prudent alternative to the use of such land, and (2) such program includes all possible planning to minimize harm to such . . . historic site resulting from such use." The protection provided by this provision differs from that available under Section 106 in two respects. First, the regulations state that Section 4(f) applies to historic properties determined to be of national, state, or local significance, while Section 106 encompasses properties listed on or eligible for listing on the National Register. Second, the protection afforded by Section 4(f) is more than procedural. If a feasible and prudent alternative to a project involving the use of a historic site for highway purposes exists, the Secretary of Transportation is prohibited from approving or funding the project.

Other federal laws affect the use of historic properties which are federally owned. These restrict both the demolition and, in certain circumstances, the transfer of such properties to nonfederal entities, and also require the General Services Administration and other federal agencies to acquire and use historic properties to the greatest extent possible in carrying out their programs.[101]

6.3   THE NATIONAL ENVIRONMENTAL POLICY ACT

In general, NEPA[102] requires the federal government to consider the impact on the environment of proposed federal actions. Section 102 of NEPA provides that, before taking "major Federal actions significantly affecting the quality of the human environment," all federal agencies must, "to the fullest extent possible" prepare a "detailed statement" (known as an environmental impact statement or EIS) which analyzes the environmental impact of the proposed action.[103] The statute also makes it the "continuing responsibility of the Federal Government to use all practicable means, consistent with other essential considerations of national policy," to preserve and enhance the environment, including the "historic, cultural, and natural aspects of our national heritage."[104]

Thus, NEPA permits broad review of the effect of federal activities on both the environment in general and on historic resources in particular. The agency must consider direct, indirect, or cumulative impacts of its proposed action.[105] As part of this process, the agency must consider alternatives to the proposed action, focusing on those that would avoid or mitigate the harm.

[101] See 40 U.S.C. §§ 304a-2, 484(k)(3)(A), 601a; 16 U.S.C. § 470h-2(a)(1).
[102] 42 U.S.C. §§ 4321-4347.
[103] Id. § 4332.
[104] Id. § 4331(b).
[105] 40 C.F.R. § 1508.8.

The protection provided under NEPA is similar to that of the NHPA in that both establish procedural requirements. Both statutes impose a duty on the agency to consider the adverse impact, but not necessarily to avoid it.[106] By requiring such consideration, the statute aims to produce governmental action informed by intelligent review of environmental concerns, not to dictate a particular result in any given case.

6.3.1 Major Federal Action

Several factors must be analyzed to determine whether a proposed project involves a "major federal action."

Whether an action is major is determined by its effect.[107] An act such as a demolition may itself be a large, involved, significant project, but that is not the crucial consideration and that alone does not make the action major. The focus of the inquiry is on how great an impact that project will have on the surrounding physical environment.

A difficult problem arises when a federal permit or license affecting only a small portion of a project is a "but for" condition without which the entire project cannot proceed. In deciding whether the federal action involved is major, the permitting agency must determine whether to consider the entire project or only that portion subject to the permit. Several courts have indicated that the proper resolution of this question depends upon three factors: (1) the degree of discretion exercised by the federal agency over the federal portion of the project, (2) whether the federal government has given any direct financial aid to the project, and (3) whether overall federal involvement is "sufficient to turn essentially private action into federal action."[108] Generally, only if an agency has legal control, as opposed to factual control, of an entire project is project-wide environmental review likely to be required.[109] At the same time, when a pending project and other proposed projects are "so interdependent that it would be unwise or irrational to complete one without the others," the impact of all such projects must be considered in the EIS.[110]

A major action is federal when it is "potentially subject to Federal control and responsibility."[111] If a federal entity holds discretionary

---

[106] Preservation Coalition, Inc. v. Pierce, 667 F.2d 851,859 (9th Cir. 1982).

[107] 40 C.F.R. § 1508.18. The term "major" reinforces but does not have a meaning independent of "significantly." 40 C.F.R. § 1508.18. Actions include a responsible official's failure to act. *Contra*, State of Alaska v. Andrus, 429 F. Supp. 958 (D. Ala. 1977), *aff'd*, 591 F.2d 537 (9th Cir. 1979).

[108] Winnebago Tribe of Nebraska v. Ray, 621 F.2d 269 (8th Cir.), *cert. denied*, 449 U.S. 836 (1980), and cases cited therein.

[109] *Id.*

[110] Webb v. Gorsuch, 699 F.2d 157 (4th Cir. 1983).

[111] 40 C.F.R. § 1508.18.

power over the outcome of a project or its involvement enables the project to proceed, then it is likely that the project will be characterized as a federal action.[112]

Regulations of the Council on Environmental Quality (CEQ) identify a number of categories of federal actions. These include—

1. Adoption of official policy, such as rules, regulations, and interpretations adopted pursuant to the Administrative Procedure Act, 5 U.S.C. 551 et seq.; treaties and international conventions or agreements; formal documents establishing an agency's policies which will result in or substantially alter agency programs.

2. Adoption of formal plans, such as official documents prepared or approved by federal agencies which guide or prescribe alternative uses of federal resources, upon which future agency actions will be based.

3. Adoption of programs, such as a group of concerted actions to implement a specific policy or plan; systematic and connected agency decisions allocating agency resources to implement a specific statutory program or executive directive.

4. Approval of specific projects such as construction or management activities located in a defined geographic area. Projects include actions approved by permit or other regulatory decision as well as federal and federally assisted activities.[113]

"Actions" include new and continuing activities, including projects and programs entirely or partly financed, assisted, conducted, regulated, or approved by federal agencies. New or revised agency rules, regulations, plans, policies or procedures, and legislative proposals can also be actions. Actions do not include funding assistance solely in the form of general revenue sharing funds, distributed under the State and Local Fiscal Assistance Act with no federal agency control over the subsequent use of those funds.[114]

Federal action may be present in a variety of transactions affecting historic buildings. For example, HUD approval of the sale of a National Register property has been determined to be a major federal action,[115] as has the federally funded construction of a medical center for prisoners in a historically significant area.[116] It has also been held that a demolition undertaken by a municipal redevelopment authority as part of a larger urban renewal project undertaken jointly with HUD is a

---

[112] Ely v. Velde, *supra*, 451 F.2d at 1137.

[113] 40 C.F.R. § 1508.18(b).

[114] *Id.* at (a).

[115] Hart v. Denver Urban Renewal Auth., 551 F.2d 1178 (9th Cir. 1977).

[116] Ely v. Velde, 451 F.2d 1130 (4th Cir. 1971).

major federal action requiring NEPA compliance.[117] In that situation, the reviewing court stated, "HUD cannot meet its statutory responsibility by ignoring demolition and other site clearance work performed by one clearly acting in partnership with the federal government nor can it shut its eyes to the demolition activity and look only to the environmental consequences of new construction which HUD is financing."[118]

Other cases, not all specifically in the context of historic preservation, suggest the range of actions by federal agencies that may trigger NEPA review. They include the lease of government-owned buildings,[119] construction by a federal agency,[120] construction by a nonfederal entity supported by federal funds,[121] construction supported by federal mortgage insurance and interest guarantees,[122] and permitting[123] and licensing[124] by a federal agency.

In considering whether an action is federal, a court may consider the relative extent of federal, as opposed to state and local, involvement. If the degree of federal involvement is relatively insubstantial, a federal environmental impact statement may not be required.[125] Moreover, the mere fact that an action is undertaken by a federally empowered entity such as a bank with a federal charter, may not be sufficient to make the action federal if no other federal involvement is present.[126]

Manipulation of the federal role to avoid EIS review of some or all of a project will generally not be countenanced by the courts. For example, segmenting federal and nonfederal funds for this purpose[127] or diverting federal funds from one project to another and replacing them with state funds[128] are both practices rejected by the courts. Proposed actions or parts of those actions which are in effect so closely related as to be a single course of action must be evaluated in one EIS.[129]

---

[117] Aertsen v. Harris, 467 F. Supp. 117 (D. Mass. 1979).

[118] *Id.* at 120.

[119] S. W. Neighborhood Assembly v. Eckard, 445 F. Supp. 1195 (D.D.C. 1978).

[120] City of Rochester v. U.S. Postal Service, 541 F.2d 967 (2d Cir. 1976).

[121] Ely v. Velde, 451 F.2d 1130 (4th Cir. 1971).

[122] Wilson v. Lynn, 372 F. Supp. 934 (D. Mass. 1974).

[123] Conservation Council of North Carolina v. Constanzo, 398 F. Supp. 653 (E.D.N.C.), *aff'd,* 528 F.2d 250 (4th Cir. 1975).

[124] Greene County Planning Board v. Federal Power Comm'n., 455 F.2d 412 (2d Cir.), *cert. denied,* 409 U.S. 849 (1972).

[125] Edwards v. First Bank of Dundee, 534 F.2d 1242 (7th Cir. 1976).

[126] Committee to Save the Fox Bldg. v. Birmingham Branch of Federal Reserve, 497 F. Supp. 504, 511 (Ala. 1980).

[127] Named Individual Members of San Antonio Conservation Society v. Texas Highway Dept., 446 F.2d 1013 (5th Cir. 1971), *cert. denied,* 406 U.S. 933 (1972).

[128] Ely v. Velde, 497 F.2d 252 (4th Cir. 1974).

[129] 40 C.F.R. § 1502.4(a).

If a proposed agency action is likely to be characterized as a major federal action, a full NEPA review is required if it is determined that the action will "significantly [affect] the quality of the human environment."[130]

The large majority of courts have recognized that alteration of the features of a historic building or demolition of a historic area or building affect the human environment.[131] But frequently the adverse effect is more indirect and therefore harder to categorize. Adverse change may pertain to the physical characteristics[132] or the aesthetics[133] of a neighborhood. It may also involve a change in the neighborhood's population[134] or traffic density.[135] To determine whether the action's effect on the environment is "significant" its context and intensity must be considered.[136] By "context" the regulations mean that an action's significance may vary with its setting. "Intensity" refers to the severity of the impact. Fators that should be considered in evaluating intensity include the "degree to which the action may adversely affect National Register eligible or listed properties or other significant scientific, cultural or historic resources."

6.3.2 The EIS Process

Each agency must adopt procedures to implement NEPA and ensure that decisions are made in accordance with NEPA's goals and comply with its procedural requirements.[137]

The procedures must be designed to determine whether preparation of an EIS is required and, if so, the procedures to be followed in its preparation. The EIS process is designed to provide a full and fair discussion of significant environmental impacts and to inform the public of the reasonable alternatives which would avoid or minimize adverse impacts or enhance the quality of the environment.[138] It is a planning tool, not merely a disclosure document, a means of assessing the

---

[130] 42 U.S.C. § 4332(2)(C).

[131] WATCH v. Harris, 603 F.2d 310 (2d Cir. 1979); Hanly v. Mitchell, 460 F.2d 640 (2d Cir.), *cert. denied*, 409 U.S. 990 (1972); Aluli v. Brown, 437 F. Supp. 602 (D. Hawaii 1977), *rev'd on other grounds*, 602 F.2d 876 (9th Cir. 1979); *contra*, St. Joseph Historical Soc. v. Land Clearance for Redevelopment Auth. of St. Joseph, Mo., 366 F. Supp. 605 (W.D. Mo. 1973).

[132] Goose Hollow Foothills League v. Romney, 334 F. Supp. 877 (D. Or. 1971).

[133] Maryland/National Capital Park & Planning Comm'n., v. U.S. Postal Service, 487 F.2d 1029 (D.C. Cir. 1973).

[134] Goose Hollow Foothills League v. Romney, 334 F. Supp. 877 (D. Or. 1971).

[135] City of Rochester v. U.S. Postal Service, 541 F.2d 967 (2d Cir. 1976).

[136] 40 C.F.R. § 1508.27.

[137] *Id.* § 1507.1.

[138] *Id.* § 1502.1.

environmental impact of proposed actions, rather than a document to justify decisions already made.

6.3.3   When Is an EIS Required?

The first step in the NEPA process is an agency's preparation of an environmental assessment. Based on the information acquired in preparing this public document, an agency determines whether an EIS is required. If, however, it is clear from the beginning that an EIS will be required, the agency may initiate that process immediately without first preparing an environmental assesment.

Based on the environmental assesment, the agency may decide that the action is not a major federal action or will not significantly affect the quality of the human environment. In either case, it would conclude that an EIS is not required and prepare a "Finding of No Significant Impact."[139] A determination that a full EIS is not required does not end an agency's duties under NEPA. The agency must demonstrate continuing sensitivity to environmental considerations in its actions, and must develop a system of decision-making which ensures the consideration of environmental values. It must adopt an interdisciplinary approach to ensure that nonagency interests are considered, particularly when there is a dispute over resource use. And the agency must continue to generate less environmentally intrusive alternatives to the proposed federal action.

If, however, the agency finds that the threshold requirements are satisfied, it must prepare an EIS.

6.3.4   Participants

Generally, the agency proposing the undertaking is responsible for preparing the EIS. In projects involving a number of agencies, a lead agency is designated to take primary responsibility for preparing the EIS.[140] But while one agency may bear that primary responsibility, a number of other federal agencies, state and local governments, and even private individuals may contribute to its preparation before the process is completed.

Other federal, state, and local agencies may become cooperating agencies in the preparation of the EIS. A cooperating agency is one that is requested to participate by the lead agency.[141] It is appropriate for a lead agency to delegate responsibility to another federal agency to draft portions of the EIS when a portion of the project falls within that

---

[139] *Id.* § 1508.13.
[140] *Id.* § 1501.5(c). This decision is hammered out by the agencies involved.
[141] *Id.* § 1501.6.

agency's jurisdiction or when that agency brings special expertise to the issue.

The lead agency must solicit comments from other federal agencies which because of their jurisdiction or special expertise are able to highlight any of the environmental consequences that can be anticipated, and those agencies have a duty to comment.[142] Also having a duty to comment are agencies created to develop and enforce environmental standards. For example, in historic preservation matters, the Department of Interior and the Advisory Council have special expertise and must comment on projects which may impact on historic resources.

Under some circumstances, a federal agency may also delegate responsibility for preparing an EIS to a state entity.[143] As a rule, federal agencies whose chief responsibility is funding of state projects (for example, the Federal Highway Administration) may rely upon an EIS prepared by a state agency. Federal entities with permitting responsibilities (for example, the Army Corps of Engineers) must undertake independent analyses, and reliance upon state efforts is not acceptable.[144] In any event, the federal agency retains ultimate responsibility for compliance with NEPA.

Under the Urban Development Action Grant program discussed in chapter 4, the Secretary of HUD is authorized to delegate environmental review procedures to local governments as grant applicants.[145] The applicant's request for a release of UDAG funds must be accompanied by a certification that the applicant has fully carried out its responsibilities under NEPA, and the Secretary's approval of that certification is deemed to satisfy his NEPA responsibilities.[146]

The regulations provide for public involvement at several steps in the EIS process. Citizen participation must be invited at the environmental assessment stage and with regard to both the draft and final versions of the EIS.[147] To facilitate this participation, hearings must be held and relevant documents made available for public review.

6.3.5 Timing

The EIS is intended to aid in an agency's decision-making process.[148] Preparation of the EIS must commence as close as possible to the time

---

[142] Id § 1503.2.

[143] 42 U.S.C. § 4332(2)(D).

[144] Sierra Club v. U.S. Army Corps of Engineers, 701 F.2d 1011, 1037–39 (2d Cir. 1983).

[145] 42 U.S.C. § 5304(f).

[146] Crosby v. Young, 512 F. Supp. 1363 (E.D. Mich. 1981); cf. Colony Federal Savings & Loan Ass'n. v. Harris, 482 F. Supp. 296 (W.D. Pa. 1980); Nat'l. Center for Preservation Law v. Landrieu, 496 F. Supp. 716 (D.S.C.), aff'd, 635 F.2d 324 (4th Cir. 1980).

[147] 40 C.F.R. § 1506.6.

[148] Id. § 1502.5.

the agency begins developing its proposal so that it is completed in time to be part of any recommendation or report on the proposal. For example, with respect to an action directly undertaken by a federal agency, such as demolition of a historic site for a highway project through a historic district, the EIS must be prepared at the feasibility (go -no go) stage. Where the federal agency must pass on an application of another entity that would affect a historic building (such as an application for federal funds for new construction in a historic district), the EIS must be prepared before granting the application.

6.3.6 Contents of the EIS

There are three main elements to the EIS.[149] It must describe the environment that will be affected, the proposed action and any alternatives, and the environmental consequences of the proposed action and its alternatives.

The purpose of describing the affected environment is to assist in understanding the effects of the proposed action and its alternatives. The agency, in satisfying this requirement, must keep its description and supporting data clear and concise to aid understanding, rather than cloud it in unnecessary verbiage.

A clear description of the proposed action and adequate discussion of the alternatives is critical. The discussion should be comparative and analytical. It should "rigorously explore and evaluate" all reasonable[150] alternatives, including the alternative of no action at all, in detail. The issues must be sharply defined and provide a clear basis for choosing among the options based on an evaluation of their comparative merits. The agency should indicate in the EIS which alternatives it prefers. The scope of the discussion of alternatives is determined by looking at the nature and scale of the proposed project and includes reasonable alternatives not within the jurisdiction of the lead agency.

The regulations delineate what the section on environmental consequences should discuss.[151] Most significant for historic preservation purposes is the requirement that this section discuss: "urban quality, historic and cultural resources, and the design of the built environment, including the reuse and conservation potential of various alternatives and mitigation measures."[152] It must discuss direct and indirect impacts of the action and its alternatives. Any unavoidable adverse effects of the proposal must be made clear.

---

[149] *Id.* § 1502.14–1502.16.

[150] *Id.* § 1502.14(a); Keith v. Volpe, 352 F. Supp. 1324, 1336 (C.D. Cal. 1972), *aff'd*, 506 F. 2d 696 (9th Cir. 1974), *cert. denied*, 420 U.S. 908 (1975); *cf.* Natural Resources Defense Council, Inc. v. Morton, 458 F.2d 827, 838 (D.C. Cir. 1972) (speculative alternatives are not required in the EIS).

[151] 40 C.F.R. § 1502.16.

[152] *Id.* at (g).

6.3.7   Judicial Review under NEPA

NEPA goals are substantive, but the rights it creates are only procedural.[153] Since the statute does not require a particular result or demand the preservation of a resource, a court is likely to defer to the discretion of the agency. Consequently, a claim disputing the substance of an agency's decision is not likely to succeed. A court is likely to limit its review to ensuring that all procedural requirements have been met and that the nature and extent of the environmental impact have been considered.[154]

However, CEQ's interpretation of NEPA in its regulations, which spell out agencies' procedural duties in great detail, is entitled to "substantial deference" from the courts.[155] This suggests that a challenge to the scope of the agency's deliberation may be more fruitful than a substantive challenge to the results of that deliberation. For example, the agency decision not to prepare an EIS based on its determination that no significant adverse environmental consequence will occur may be challenged as being inadequately considered.[156] Or the EIS itself may be attacked for neglecting to address necessary considerations such as direct or cumulative effects or reasonable alternatives.

The federal courts of appeals have utilized two different standards in reviewing challenges to the substance of agency decisions under NEPA. Several have held that NEPA determinations should be overturned only if they are shown to be "arbitrary and capricious."[157] Others apply a somewhat more rigorous test, and examine agency NEPA decisions to see if they meet a standard of reasonableness.[158] In both cases, the outcome will turn largely on the sufficiency of the information that the agency had available when it reached its decision. In practice, it is difficult to set aside an agency determination under either standard, and it is not enough to show that a better result might have been reached. Indeed, the fact that the court, upon *de novo* review, might not agree with the substance of an agency's determination on a NEPA question may well be immaterial to the outcome of a legal challenge.[159]

[153] Strycker's Bay Neighborhood Council, Inc. v. Karlen, 444 U.S. 223, 227 (1980).

[154] Vermont Yankee Nuclear Power Corp. v. Natural Resources Defense Council, Inc., 435 U.S. 519, 548 (1978).

[155] Andrus v. Sierra Club, 442 U.S. 347, 358 (1979).

[156] Preservation Coalition, Inc. v. Pierce, 667 F.2d 851, 855 (9th Cir. 1982), and cases cited therein.

[157] *E.g.* Nucleus of Chicago Homeowners Ass'n. v. Lynn, 524 F.2d 225, 229–30 (7th Cir. 1975), *cert. denied,* 424 U.S. 967 (1976); Hanly v. Kleindienst, 471 F.2d 823, 828-30 (2d Cir. 1972), *cert. denied,* 412 U.S. 908 (1973).

[158] Goodman Group, Inc. v. Dishroom, 679 F.2d 182, 186 (9th Cir. 1982); Minnesota Public Interest Research Group v. Butz, 498 F.2d 1314, 1320 (8th Cir. 1974) (en banc); Wyoming Outdoor Coordinating Council v. Butz, 484 F.2d 1244, 1249 (10th Cir. 1973); Save Our Ten Acres v. Kreger, 472 F.2d 463, 465–66 (5th Cir. 1973).

[159] Aertsen v. Landrieu, 637 F.2d 12, 19 (1st Cir. 1980).

The discussion of federal court jurisdiction to hear claims arising under the NHPA,[160] is fully applicable to NEPA as well.

6.4 COORDINATING NEPA AND THE NHPA FOR MAXIMUM FEDERAL STATUTORY PROTECTION

Both the NHPA and NEPA provide protection for historic, archaeological, and cultural resources. But the applicability of NEPA is not precisely coextensive with that of the NHPA. Like the NHPA, NEPA applies to federal agencies but not to their counterparts at other levels of government. Nor are private actions subject to regulation by either law.[161] Unlike the NHPA, though, the protection provided by NEPA may extend to historic property even if it has not been determined to be eligible for listing on the National Register.[162] In this respect, NEPA complements the NHPA, and expands federal statutory protection for historic resources.

Similarly, the NHPA may apply in some instances where NEPA does not. Because NEPA's threshold—requiring a major federal action significantly affecting the quality of the human environment—is higher than for the NHPA, there may be federal actions which are subject to review under the NHPA, but not NEPA.

Where the procedures of the two statutes overlap and steps taken to comply with one are the same steps required to comply with the other, the regulations attempt to avoid unnecessary duplication and expense by requiring that the procedures be integrated to the extent possible. Thus surveys, studies, and analyses prepared under the NHPA are incorporated into the EIS.[163] However, compliance with the NHPA does not assure compliance with the requirements of NEPA[164] and vice versa.[165] The requirements of each must be shown to have been met independently of the other.

6.5 STATE STATUTORY PROTECTION

Many states have sought to supplement the statutory protection available under both the NHPA and NEPA by adopting similar statutes which are applicable to the actions of state and, sometimes, local agencies. Typically, these state enactments seek to encourage protection of his-

---

[160] See Section 6.1.9.

[161] However, both laws do apply to the activities of these nonfederal or nongovernmental actors when their actions are subject to federal activity.

[162] Boston Waterfront Residents Ass'n. Inc. v. Romney, 343 F. Supp. 89 (D. Mass. 1972).

[163] 40 C.F.R. § 1502.25(a).

[164] Preservation Coalition, Inc. v. Pierce, 667 F.2d 851, 859 (9th Cir. 1982).

[165] Stop H-3 Ass'n. v. Coleman, 533 F.2d 434, 444–45 (9th Cir.), cert. denied sub nom. Wright v. Stop H-3 Ass'n., 429 U.S. 999 (1976).

toric and cultural resources by establishing a State Register, requiring governmental agencies to adhere to restrictions designed to protect these resources, and establishing an environmental review procedure.

6.5.1 State Historic Preservation Statutes

A number of states have enacted constitutional provisions, statutes, or administrative orders which offer direct protection to historical sites which exceeds that provided by federal law. These statutes often expand the scope of protection of the NHPA by imposing direct preservation duties upon state and often local government activities. A state may also create its own register of historic sites to complement the National Register, affording listed resources protection against state financed or licensed projects.

In New York State, the Historic Preservation Act of 1980[166] authorizes the creation of a Register of Sites and Structures "significant in the history, architecture, archaeology, or culture of the state, its communities or the nation."[167] The criteria for listing on the New York State Register mirror those for the National Register. This is in contrast to a number of states, which utilize eligibility criteria that are less stringent than those of the National Register.

The consequence of a listing in the New York State Register is that a state agency and, in some cases, local agencies, may not engage in any action that effects the listed property without considering the impact of its activity and taking steps to conserve the property. The statute makes it "the responsibility of every state agency . . . to avoid or mitigate adverse impacts to registered property or property determined eligible for listing" on the state register of historic places.[168]

The state of California has enacted legislation[169] which creates a State Historical Resources Commission to designate state historical landmarks.[170] This commission consists of members with broadly relevant expertise in the areas of history, architecture, and archaeology. The criteria used to designate property as state landmarks vary somewhat from those established in local ordinances and under the NHPA. Review and approval of a nomination are based on a finding of the presence of at least one of the following:

1. The property is the first, last, only, or most significant historic property of its type in the region.

---

[166] N.Y. Parks and Rec. Law § 14.01–14.09 (McKinney 1984).

[167] Id. § 14.07(a).

[168] Id. § 14.09(2).

[169] Cal. Pub. Res. Code §§ 5020–26 (West Supp. 1982).

[170] This commission also evaluates nominations to the National Register.

2. The property is associated with an individual or group having a profound influence in California.

3. The property is an outstanding example of a particular architectural movement or architect.

4. The property is listed on the National Register.

These statutes are illustrative of state efforts to protect historic properties on their own initiative. Like the federal statutes just discussed, they do not guarantee the preservation of historic structures, but they do signal an increasing commitment by states to consider the impacts of agency action on their cultural heritage and to require property owners and developers seeking state assistance to cooperate in this review.

6.5.2 State Environmental Protection Statutes

A majority of states have enacted environmental protection statutes which parallel the requirements of NEPA. As with their federal counterpart, it is the purpose of these state acts to mandate a full consideration of the environmental implications of planned state and, in some cases, local actions. Because NEPA's application is limited to federal actions, the role these statutes play in the preservation process is an important one particularly because a great number of actions which touch on land use are initiated at the state and local, rather than the federal, level.

The language of many state statutes is substantially similar to NEPA and their procedures are essentially the same. The New York State Environmental Quality Review Act (SEQRA)[171] is one example. It requires that state agencies prepare an EIS for any action they undertake which will have a significant effect on the environment. New York defines "environment" to include "objects of historic or aesthetic significance . . . and existing community or neighborhood character."[172] The agency must take the further step of choosing alternative actions which, to the extent practicable, minimize or avoid such adverse impacts. The "extent practicable" is influenced by a number of considerations including social and economic factors. The agency must choose a course of action based on findings reflecting consideration of these factors.

The SEQRA regulations identify two categories of actions for purposes of indicating the level of environmental scrutiny required. Type I actions are those which presumptively have a significant effect on the environment, and for which an EIS is likely to be required.[173] For all

---

[171] N.Y. Envtl. Conserv. Law §§ 8.0101 to 8.0117 (McKinney 1980).

[172] 6 N.Y.C.R.R. § 617.2(j).

[173] *Id.* § 617.12.

Type I actions, an EIS will be required unless the lead agency issues a determination that the specific action proposed will not have a significant effect on the environment (a so-called negative declaration), based upon a more limited Environmental Assessment Form which must be prepared in all cases. Type I actions include actions "occurring wholly or partially within, or contiguous to, any facility or site listed on the National Register of Historic Places, or any historic building, structure or site, or prehistoric site, that has been proposed by the Committee on the Registers for consideration by the New York State Board on Historic Preservation" for inclusion in the National Register.

Type II actions are those which have been generically determined not to have a significant effect on the environment.[174] They include such things as repaving of existing highways, construction of minor accessory facilities, and renewal of licenses or permits involving no material change in allowable activities. Neither an Environmental Assesment Form nor an EIS must be prepared for a Type II action.

The regulations also permit a local agency to designate as a "critical area of environmental concern" an area which has exceptional or unique "social, cultural, historic, archaeological, recreational, or educational" characteristics.[175] An action which takes place within or substantially contiguous to such a designated area automatically becomes a Type I action if it exceeds "locally established thresholds."

The California Environmental Quality Act (CEQA)[176] expresses the legislative intent that state agencies encourage activities of private citizens which enhance the environment, including historic resources. The statute expresses the state's policy that it is every citizen's responsibility to preserve and enhance the environment. The purpose of the CEQA is to prevent significant avoidable damage to the environment by requiring changes in projects through mitigation measures.

Any state agency must prepare an environmental impact report (EIR) when its activity will have a significant effect on the environment. In addition to ensuring that agencies assess all reasonable alternatives, the EIR is intended to inform the public and other agencies of the environmental impact of the project.

The concept of "project" is quite broad.[177] In addition to including direct governmental activities and discretionary projects such as zoning ordinances and variances, it also includes actions in which a public entity considers a permit, license, lease, certificate, or other entitlement for a private action that may potentially change the environment. The

[174] Id. § 617.13.

[175] Id. § 617.4(j)(3).

[176] Cal. Pub. Res. Code § 21000 et seq.; see Cal. Admin. Code tit. 14, R.15000 et seq. The CEQA applies to governmental agencies at all levels. State, regional, county, and local agencies must develop standards and procedures to protect environmental quality.

[177] See Cal. Pub. Res. Code § 21065.

CEQA is similar to many other statutes in distinguishing between discretionary and ministerial actions and applying only to discretionary ones. Whether an agency has discretionary or ministerial control over a project depends on the authority granted by the law which gives the agency control over the activity.[178] A discretionary project is one that requires the exercise of judgment, deliberation, or decision on the part of the public agency or body in the process of approving or disapproving a particular activity. In contrast, a ministerial project, as a general rule, includes those activities which are undertaken or approved by a governmental agency which entail a decision which a public officer or public agency makes upon a given set of facts in a prescribed manner in obedience to the mandate of legal authority. With these projects, the officer or agency must act upon the given facts without regard to his own judgemnt or opinion concerning the propriety or wisdom of the act. Similar projects may, then, be subject to discretionary controls—and environmental review requirements—in one city or county and only ministerial controls—and no such review—in another.

A project "significantly" affects the environment when it will potentially cause a substantial adverse change.[179] The "environment" is defined to mean the physical conditions which exist within the area which will be affected by the proposed project, including objects of historic and aesthetic significance.[180]

All reasonable alternatives to an agency's proposed project must be considered. This includes the alternative of not going forward with the project at all, and alternatives outside the expertise of the agency or which may require implementing legislation.[181] The California Supreme Court has stated, "Obviously if the adverse consequences to the environment can be mitigated, or if feasible alternatives are available, the proposed activity . . . should not be approved." However, the CEQA does not mandate the choice of the environmentally best feasible project.[182]

Significantly, some states provide substantive as well as procedural protection to historic resources by means of their environmental review statutes. These states create a binding protection that is much stronger than NEPA and other review laws which impose only procedural duties upon the agencies and which, even if complied with, may still result in the loss of the resource. The state of Washington is an example of one which imposes a substantive requirement.[183] Its State Environmental

---

[178] Cal. Admin. Code tit. 14, R.15357, 15369.

[179] Cal. Pub. Res. Code § 21068.

[180] Cal. Admin. Code tit. 14, R.15360.

[181] Resident Ad Hoc Stadium Comm. v. Board of Trustees of California State University, 89 Cal. App.3d 274 (Cal.Ct.App.1979).

[182] Friends of Mammoth v. Board of Supervisors, 8 Cal.3d 247, 502 P.2d 1049, 1059 n.8 (1972). This was legislatively ratified in 1976 amendments.

[183] *See* Polygon Corp. v. City of Seattle, 90 Wash.2d 59, 578 P.2d 1309 (1978) (en banc).

Protection Act (SEPA)[184] has been found to confer on localities the discretionary right to deny a building permit based on an EIS which concluded that an adverse environmental impact would result. The Minnesota Environmental Rights Act[185] has been interpreted to require the protection of resources unless the owner could show that there was no way that the resource could be preserved while meeting his needs.[186]

The purpose of the state environmental statute is to provide the fullest possible protection of the environment. To achieve this, the agencies and the courts must give its language the broadest interpretation.[187] Courts have, for instance, drawn assistance from NEPA cases in interpreting the CEQA. Thus, the more expansive case law developed under NEPA may be treated as persuasive authority in analyzing parallel state provisions.[188]

It is worthwhile for a state to make efforts to keep state requirements from duplicating those of federal law, in both the environmental and preservation contexts. New York is one state that has taken steps in this area. The state's Environmental Conservation Law provides that when a state agency is required to prepare or participate in preparing a federal EIS, any submissions required by state law "shall be coordinated with and made in conjunction with federal requirements in a single environmental reporting procedure."[189] Where both Section 106 review under the NHPA and state historic preservation review requirements coincide, the Section 106 process may be used to satisfy those elements of state law.[190]

[184] Wash. Rev. Code § 43.21C.

[185] Minn. Stat.c. 116B.

[186] State, by Powderly v. Erickson, 285 N.W.2d 84 (Sup.Ct. Minn. 1979).

[187] People ex rel. Younger v. Local Agency Formation Commission of San Diego County, 81 Cal. App.3d 464, 146 Cal.Rptr. 406 (Cal.Ct.App. 1978); Bozung v. Local Agency Formation Comm'n. of Ventura County, 13 Cal.3d 263, 529 P.2d 1017, 1024 (1973).

[188] Wildlife Alive v. Chickering, 18 Cal.3d 190, 201, 553 P.2d 537, 543 (1976).

[189] N.Y. Envtl. Conserv. Law. § 8-0111(1.).

[190] N.Y. Parks and Rec. Law § 14.09(2.).

Seven

Local Regulation of Historic Buildings

Introduction

Local regulation plays an increasingly important role in preservation law. Neither federal nor state laws prohibit actions that are adverse to the integrity of a historic structure. Rather, as explained in detail in chapter 6, they require government agencies to identify the potential impacts which a proposed action may have on historic resources, identify alternatives to an action that may have an adverse impact on such resources, and consider steps to mitigate the potential damage. Once these procedural steps are completed, the governmental agency may still proceed to take or allow the action which will adversely affect the historic resources. In contrast, local ordinances may provide direct protection for historic resources by regulating their maintenance, alteration, and demolition. It is not surprising, then, that approximately 900 communities throughout the United States have enacted local ordinances to provide additional protection for their built heritage.[1]

In general terms, the local preservation ordinance establishes the preservation framework for the community. Typically, the ordinance itself does not designate specific sites as landmarks. Instead, it articulates substantive and procedural standards to guide the community in identifying and protecting historic and architectural resources. The preservation ordinance may provide for the designation of individual structures as landmarks or of entire areas as historically, culturally, or aesthetically significant districts, or may do both. Since the purpose of a landmark ordinance is to provide a framework within which such decisions may be made, the ordinance must be flexible enough to be applied in a wide variety of circumstances and sufficiently specific to provide guidance in making individual designations.

This chapter is intended to assist communities which are drafting their own preservation ordinances, as well as building owners, developers, lawyers, architects, and citizens who appear before local preservation commissions.[2] The New York City ordinance, found in Appendix K, is an example of an ordinance which allows the designation of individual structures as well as districts. While specific reference is made to that ordinance at various places throughout this chapter, readers are encouraged to review it thoroughly and to consider its relative merits with regard to the needs of their own communities. Certification of a local ordinance by the Department of the Interior is often a necessary prerequisite to the rehabilitation tax credit discussed in chapter 1. Consequently, the final part of this chapter explains the process by which local ordinances are certified by state and federal agencies.

It should be noted that, unless expressly provided to the contrary, a preservation ordinance does not supersede or modify previously enacted zoning laws. Rather it is an overlay on existing land use regulation which focuses primarily on the preservation of existing buildings rather than the location, size, and use of new buildings.

---

[1] Christopher J. Duerksen, ed., *A Handbook on Historic Preservation Law* at 59 (The Conservation Foundation, et al. 1983) (hereinafter "Handbook"); *see* Stephen N. Dennis, *Directory of American Preservation Commissions* (National Trust for Historic Preservation, 1981). For a listing of local ordinances in New York State, *see* National Center for Preservation Law, *A Primer on Preservation Law in the State of New York* at Appendix A (Berle; Kass & Case, 1985).

[2] The Supreme Court has noted the concerns that give rise to these legislative efforts. *See* Penn Central Transportation Co. v. City of New York, 438 U.S. 104, 108 (1978).

7.1 STRUCTURAL PROVISIONS OF LOCAL PRESERVATION ORDINANCES

Local preservation ordinances share many common provisions. However, their precise language varies widely because each jurisdiction has tailored its ordinance to meet its own specific needs, preservation goals, and political realities. Each ordinance also reflects the scope of authority which the locality can exercise both under its state's enabling statute and the case law in its jurisdiction. Though ordinances will vary, much can be learned by surveying those features that are common and studying how various municipalities have approached the drafting of their ordinances. The discussion that follows outlines the provisions most commonly included in preservation ordinances throughout the country. Relevant examples are provided from a variety of local ordinances, in addition to the ordinance included in the appendix. Where appropriate, cases that interpret the language of these provisions are also surveyed.

7.1.1 Statement of Purpose

One of the most important provisions of the ordinance is its statement of purpose, a statement of public policy. Generally this provision will contain two components: a reference to the municipality's authority to use the police power in furtherance of the public welfare, and a statement of the interests the community seeks to protect. Together, these elements provide the legal basis upon which the ordinance rests.[3]

The public welfare interests advanced by the preservation ordinance should be stated as broadly as possible. Too often ordinances cite only the obvious cultural, historic, and aesthetic reasons for preservation and thereby miss the opportunity to place preservation in a broader context. It is now widely recognized that preservation may further education, commercial, and economic interests as well, and the statement of purpose should reflect the full range of public benefits which may be realized.

After the ordinance has been enacted, the statement of purpose will assist the local landmarks commission in interpreting other provisions of the ordinance. The purposes set forth in the statement will be considered by the commission and guide its deliberations. Moreover, the

---

[3] Reference to the public or general welfare should be included even though no court has yet struck down an ordinance solely for an inadequate statement of purpose. *See Handbook*, note 1 above, at 64. *See also* Berman v. Parker, 348 U.S. 26, 33 (1954); Figarsky v. Historic District Comm'n of Norwich, 171 Conn. 198, 368 A.2d 163 (1976); Bohannan v. City of San Diego, 106 Cal. Rptr. 333, 30 Cal. App.3d 416 (1973).

statement of purpose is one of the benchmarks against which courts of review can evaluate whether the commission has acted within the scope of its authority and has decided a particular case in accordance with the mandate of the ordinance.[4] Invoking a broad range of objectives to be served by the protection of qualifying properties provides a stronger basis for preservation activities, and may make it more likely that the ordinance will withstand judicial review, for the subjective nature of determinations which are solely aesthetic in nature has been a source of concern for some lower courts.[5]

[4]An example of a comprehensive Statement of Purpose is found in New York City's Ordinance:

Purpose and declaration of public policy.

a. The council finds that many improvements, as herein defined, and landscape features, as herein defined, having a special character or a special historical or aesthetic interest or value and many improvements representing the finest architectural products of distinct periods in the history of the city, have been uprooted, notwithstanding the feasibility of preserving and continuing the use of such improvements and landscape features, and without adequate consideration of the irreplaceable loss to the people of the city of the aesthetic, cultural and historic values represented by such improvements and landscape features. In addition distinct areas may be similarly uprooted or may have their distinctiveness destroyed, although the preservation thereof may be both feasible and desirable. It is the sense of the council that the standing of this city as a worldwide tourist center and world capital of business, culture and government cannot be maintained or enhanced by disregarding the historical and architectural heritage of the city and by countenancing the destruction of such cultural assets.

b. It is hereby declared as a matter of public policy that the protection, enhancement, perpetuation and use of improvements and landscape features of special character or a special historical or aesthetic interest or value is a public necessity and is required in the interest of the health, prosperity, safety and welfare of the people. The purpose of this chapter is to (a) effect and accomplish the protection, enhancement and perpetuation of such improvements and landscape features and of districts which represent or reflect elements of the city's cultural, social, economic, political and architectural history; (b) safeguard the city's historic, aesthetic and cultural heritage, as embodied and reflected in such improvements, landscape features and districts; (c) stabilize and improve property values in such districts; (d) foster civic pride in the beauty and noble accomplishments of the past; (e) protect and enhance the city's attractions to tourists and visitors and the support and stimulus to business and industry thereby provided; (f) strengthen the economy of the city; and (g) promote the use of historic districts, landmarks, interior landmarks and scenic landmarks for the education, pleasure and welfare of the people of the city. N.Y.C. Admin. Code § 205-1.0 (1976).

[5] See 21 A.L.R.3d 1222 for cases stating judicial concern about aesthetics being forwarded as the sole rationale for regulation. In some cases, state grounds (constitutional and statutory) have been invoked to strike down ordinances based on aesthetic objectives alone. Note, however, that cases which hold that zoning ordinances based on aesthetic criteria are invalid are not analogous to those involving ordinances designed to preserve historic architectural style. See A-S-P Associates v. City of Raleigh, 298 N.C. 207, 258 S.E.2d 444, 450 (1979), and cases cited therein.

The purposes clause should be drafted with reference not only to the U.S. Constitution but also the state constitution, case law relating to other preservation ordinances within the state and zoning and other forms of land use regulation.

7.1.2 Definitions

The definitions section which often follows the statement of purpose is a key section of the ordinance because the commission and the courts will refer to it to interpret the language, and thereby to determine the scope of the ordinance.[6]

The primary terms to be defined may include, *inter alia:*

☐ Historic district
  Landmark
  Alteration
  Maintenance
  Improvement
  Reasonable return

One example of a set of definitions for these terms can be found in the New York City Landmark and Historic District Ordinance.

Each definition should specify the word's meaning in the clearest possible terms. This will not only facilitate the administration of the ordinance, but is essential to its constitutionality, since due process requires that those subject to a law's regulation be able to understand it.[7] Persons with widely differing perspectives and backgrounds will be working with the ordinance, and it should be comprehensible to all. Unnecessary legal and technical terms should be avoided, for a definitions section which relies too heavily on legal and technical terms may be clear to a developer's attorney but incomprehensible to the small property owner who does not often read ordinances.[8]

While definitions should be drafted with precision, they should not be too narrow or rigid. The ordinance must be sufficiently flexible to be useful in the vast range of contexts in which it will be applied over time.

[6] *See* South of Second Assoc. v. Georgetown, 196 Col. 891, 580 P.2d 807 (1980).

[7] Rose v. Locke, 423 U.S. 48, 49–50 (1975).

[8] Dennis, "Do's and Don'ts in Drafting a Preservation Ordinance," in *Reusing Old Buildings: Preservation Law and the Development Process* at 315 (The Conservation Foundation, *et al.* 1982).

7.1.3 Establishing the Administrative Body

The local preservation commission designated by the ordinance bears overall responsibility for ongoing preservation planning and activities in the community. In addition to playing the key role in regulating the alteration or demolition of historic structures, its responsibilities may range from conducting surveys by which historic properties are identified and designated, to ensuring that other municipal agencies consider the goals of preservation in conducting their activities, to providing information to other levels of government in their preservation activities, such as nominations of properties to the National Register.

7.1.4 Creation and Makeup

The administrative bodies created by preservation ordinances are called a variety of names, including landmarks preservation commissions and boards of architectural review.[9] Whatever the administrative body is called (in this chapter, it will be called the commission), the ordinance usually provides for the appointment of its members by the community's mayor or by an elected body such as the city council.

Sometimes the composition of the commission is mandated in the state enabling law; in most cases it is left to the discretion of the locality. Although some communities opt to use existing municipal bodies, such as the planning board, most ordinances require the creation of a separate board composed of experts in relevant fields. Frequently the ordinance specifies the professions or interests to be represented, such as a local historian, architect, urban planner, lawyer, real estate professional, preservation group officer, or community resident.[10] Composition specifications should reflect the nature of the ordinance and of the tasks to be performed by the commission, some of which may require specialized skills. For example, the Palo Alto Historic Resources Committee consists of five members who must include an owner-occupant of a designated historic structure or historic district; at least two architects, landscape architects, building designers or other design professionals; and two representatives nominated by the Palo Alto Historical Association.[11]

Such diversity of perspective and expertise enhances the likelihood of a full airing of the issues and thorough consideration of the implications of approving a designation, alteration, or demolition. Courts have observed that a broad-based commission membership "curb[s] the

---

[9] See generally Dennis, ed., Directory of American Preservation Commissions, note 1 above.

[10] For an illustration of such a provision, see Model Preservation Ordinance prepared by the National Trust for Historic Preservation at § 54.10.2.

[11] Palo Alto Municipal Code Ch. 16.49(3)(a) (Ord. #3197).

possibility for abuse,"[12] and tends to protect a commission's decision from the allegation of arbitrary enforcement.[13] In addition, as noted, representation on the commission of a broad spectrum of relevant expertise is required if the local ordinance is to qualify for certification to receive federal funding.[14]

In small communities which do not have residents with the specialized skills needed for the commission, it may be possible to pool resources with neighboring communities and establish an intermunicipal commission.

In addition to specifying the categories of individuals to serve as the members of the commission and the method of their selection, the ordinance should establish procedures for their appointment and specify the length of their terms and amount of compensation, if any.

7.1.5   Scope of Authority

The authority which the ordinance grants to the administrative body will depend on two factors: first, the extent of authority granted to the locality in the state enabling statute and case law interpreting that statute; and second, the preservation goals of the community. Since 1976 every state has had enabling legislation for local preservation ordinances.[15] Under these empowering statutes, a municipality may choose whether or not to establish a commission to advance preservation purposes. Once it does choose to establish one, however, its discretion as to the methods and the standards by which the ordinance is to be administered may be limited by the enabling statute.[16]

If there is no state enabling statute specifically relating to preservation, or if the statute is not sufficiently encompassing for the locality's purposes, authority may be found elsewhere. In some states, the state constitution or a statute grants localities broad home rule powers. Alternatively, the general zoning power of localities may serve as a basis for preservation ordinances,[17] although assertions of authority based on the general zoning power may be viewed with less favor by the courts than those based on specific state enabling statutes.[18]

[12] Maher v. City of New Orleans, 516 F.2d 1051, 1062 (5th Cir.), *cert. denied,* 426 U.S. 905 (1975).

[13] *See* South of Second Assoc. v. Georgetown, 580 P.2d 807, 808–09 n.1 (1978).

[14] 16 U.S.C. § 470w(13).

[15] National Trust Guide to State Historic Preservation Programs (1976). *See, e.g.,* N.Y. Gen. Mun. Law §§ 96-a, 119-dd.

[16] *See* A-S-P Assoc. v. City of Raleigh, 298 N.C. 207, 258 S.E.2d 444, 452 (1979).

[17] *See* City of Santa Fe v. Gamble-Skogmo, Inc., 73 N.M. 410, 389 P.2d 13, 17 (1964).

[18] *See* Tierney v. Norwalk Planning & Zoning Comm'n, 3 Preservation L. Rept. 3016 (Conn. 1984).

Once the enabling law has been studied to determine the extent of the authority delegated to the municipality, it must be determined whether the municipality wishes to exercise the full extent of the power available to it. Since a landmark ordinance generally provides the framework within which decisions will be made, there is only one reason for a community not to reserve for itself the full range of power it can claim: It may be politically infeasible to gain passage of the stronger ordinance, because of fear of this additional—and to some, unfamiliar—form of regulation. Nonetheless, given the benefits to be gained from an effective preservation effort, local legislators often decide to enact an ordinance that grants to the commission the broadest possible authority to protect historic resources.

As outlined in more detail in this chapter, the commission's powers should include the designation of landmarks, districts, or sites (after nomination by a third party or on the commission's own initiative) and the review of applications for proposed physical changes—that is, alterations, demolitions, or new construction. The commission should also have the power to ensure maintenance of landmarks and, when necessary, to acquire property rights, particularly easements, to ensure that preservation goals are achieved. Typically, the ordinance also will grant the commission the power to adopt and amend the rules of procedure under which it will work.

7.1.6 Administrative Structure

The administrative structure required to carry out a landmark ordinance can be tailored to fit the needs of the particular community. For example, in some communities, the landmarks commission is responsible for designation and alteration determinations, but its decisions are subject to review by another city agency or legislative body. In New York City, for example, the Board of Estimate has the power to modify or disapprove a landmark designation made by the Landmarks Preservation Commission.[19]

In other cities, the commission may be advisory only, and will report and make recommendations to another municipal official, board, or legislature.[20] In Washington D.C., for instance, preservation proceedings are conducted by the mayor or by a designated agent with the advice of the Historic Preservation Review Board. If the permit application affects a landmark in the Old Georgetown District, it is subject to review

---

[19] N.Y.C. Admin. Code § 207-2.0.

[20] District of Columbia Historic Landmark and Historic District Protection Act of 1978, D.C. Code § 5-822(b) (Supp. 1980). *See e.g.,* San Francisco Planning Code Art. 10, which limits the authority of the landmarks board to nomination and review activities for the purpose of presenting recommendations to the planning commission.

by the Commission of Fine Arts and may be referred to the Historic Preservation Review Board for recommendation.

7.2   THE HISTORIC RESOURCES INVENTORY

The duties of the commission should include identifying the historic resources of the community and maintaining an inventory of them. Compiling a survey of the community's historical resources serves several purposes. This inventory is the groundwork upon which the community can base its preservation planning and justify its actions. For example, surveying properties in the community for their historic and architectural value, plotting them on a map, photographing them, and recording their significance may reveal clusters of properties that are worthy of inclusion in a historic district. Exhaustive recordkeeping will assist in formulating the boundaries of the historic district and in justifying the choices made should a dispute arise. The survey can also serve as a data base for nominations to the National Register, identification of properties that may be eligible for protection under environmental review statutes, and projects to promote public awareness through publications, tours, or historical identification plaques.

Finally, the survey should lead to the development of a comprehensive preservation plan which sets forth the community's goals and proposes actions to be undertaken to achieve them.[21]

---

[21] The Massachusetts Historical Commission has written an extremely helpful manual for local historical commissions which discusses the items that should be included in a community's plan. It is worth quoting here at length:

> This plan should summarize the information contained in the inventory, outline the local historical commission's policies and goals, and recommend actions to be taken to accomplish these objectives. The plan should discuss the methods the community will adopt to integrate the preservation of significant cultural resources with the development, growth and economic vitality of the city or town. The preservation plan should be directed to the entire community and should demonstrate to the residents of the city or town the significant beneficial impact which preservation activities can have on the community. The finished plan should be incorporated into the community planning process and should be used by all city or town planning agencies in their decision making processes. The plan should include the following components:
>
> a.   Summary of the history of the growth and development of the community.
>
> b.   Analysis of all of the architectural styles represented in the community.
>
> c.   Survey of the significant structures and areas in the community. This listing will usually include all of the properties listed in the inventory submitted to the Massachusetts Historical Commission.
>
> d.   Map showing all cultural resources (except archeological sites) surveyed.
>
> e.   Statement of local commission's preservation policy and overall objectives.
>
> f.   Recommendations for preservation measures for specific properties of areas. These might include but are not limited to the following: National Register nomination, local

A community may also wish to conduct a historical analysis to aid it in identifying how its physical resources reflect its heritage. The emphasis of the analysis should be to trace the forces that shaped the development of the area and gave it its essential character.[22] To a large extent, the inventory of historically prominent properties may be developed from such a historical analysis.

The identification of a community's historic resources is a critical first step in a local preservation program. The failure to gather information pertaining to potential landmarks can undermine later efforts

historic district designation, scenic road designation, acquisition of preservation restrictions, amendment of zoning ordinances.

g. Identification of funding sources for preservation projects in the community. These might include National Park Service grants-in-aid to National Register properties, Housing and Urban Development Community Development Block Grant Funds, National Endowment for the Arts architectural heritage grants, and the Architectural Conservation Trust for Massachusetts revolving fund.

h. Recommendations for preservation education projects, including publishing maps and pamphlets, walking tours, lecture series and adult education courses on preservation.

i. Recommendations for the integration of preservation goals and objectives into the community planning process.

[22] Some possible elements that may have contributed to a community's historical development include—

1. Geological formations, geographical features, and the natural environment.
2. Prehistoric life.
3. Historic Indian tribes and their culture.
4. The arrival of people from abroad, including explorers, missionaries, traders, early settlers, and later immigrants.
5. The reasons and incentives for settlement, such as land grants, mill sites, and mineral resources.
6. The patterns of transportation development, including rivers, canals, railroads, roads, and street patterns.
7. Military and political history, including the French and Indian War, American Revolution, War of 1812, local rebellions, impact of the Civil War, and later foreign wars.
8. Economic factors such as water power, technological inventions, turnpike, canal or railroad construction, war industries, mineral resources, recreation, and periods of boom and depression.
9. Social and cultural trends such as religious sects, reform movements, education, art, architecture, music, and literature.
10. Individuals or groups that played a part in shaping the character of the community, including farmers, industrialists, merchants, bankers, religious figures, judges, architects, and statesmen.

State of New York Division for Historic Preservation, *Historic Resources Survey Manual* at 27–28. (rev. ed., Albany, New York: New York State Office of Parks and Recreation 1974).

to save threatened buildings.[23] Moreover, the process can be time-consuming; starting it after a building or historic area is directly threatened may well be too late. For example, a neighborhood of Revolutionary war-era homes in Cambridge, Massachusetts recently completed a 2-year battle for approval as a historic district when the area was officially declared the Half Crown Neighborhood Conservation District. Residents had sought to preserve the area in the face of severe development pressures from the Harvard Square area, adjacent to the neighborhood. While this 2-year effort was in progress, nearby development included two office buildings, a hotel, a bank building, and more than 100 condominiums. The value of land in the area rose to as much as $1 million an acre.[24]

7.3 THE DESIGNATION PROCESS

7.3.1 Nomination

Most communities allow a nomination for designation as a landmark or historic district to be made by any of several methods. For example, in Boston, any ten registered voters, the mayor, or any Landmarks Commission member may submit a petition to the commission asking that a property be considered for designation.[25] A slightly different approach is taken by the Town of Pound Ridge, New York, which will initiate its process upon request from affected property owners or upon the commission's own motion.[26] Palo Alto, California, provides the broadest avenue for initiating the designation procedure. There, any individual or group may propose designation of a landmark building or district.[27]

By providing a variety of methods by which a property can be proposed for designation, a community takes a large step toward ensuring that no building that is potentially worthy of protection is overlooked.

7.3.2 Designation Standards

A local preservation ordinance should articulate clear standards for the designation of a structure or site as a landmark. These standards sometimes track the criteria provided in the enabling statute; if the

---

[23] See Committee to Save the Fox Bldg. v. Birmingham Branch of the Federal Reserve Bank of Atlanta, 497 F. Supp. 504 (N.D. Ala. 1980); Life of the Land, Inc. v. Honolulu, 61 Hawaii 390, 606 P.2d 866, 899–901 (1980).

[24] Boston Globe, Apr. 15, 1984 at 34.

[25] Mass. Gen. Laws Ch. 772, Acts of 1975 § 4.

[26] Pound Ridge, New York Local Law No. 4–1976, §§ II-1, and II-2.

[27] Palo Alto Code § 16.49.060.

statute is silent, they may be developed in the first instance for the ordinance. Other ordinances delegate the task of formulating designation standards to the commission, whose members can bring their expertise to bear on the question. Clear designation standards will provide fair notice to the owner or potential buyer of property as to the characteristics that may lead to designation. The owners can thereby better anticipate possible limitations upon alteration or development of the property. Carefully drafted standards will also serve as guidelines for the commission's deliberations. Finally, they provide a benchmark against which a reviewing court may judge whether the commission's actions were well grounded, arbitrary, or discriminatory.

7.3.3 General Criteria

Many communities, such as Rye, New York, adopt the list of characteristics which a structure, site, or area must possess to be included on the Department of Interior's National Register of Historic Places.[28] Other jurisdictions, such as Seattle, Washington, have drafted their own standards for identifying structures and districts of historical, cultural, geographic, or archaeological importance.[29]

---

[28] No preservation district or protected site or structure, as the case may be, shall be designated unless it is found to possess one or more of the following characteristics:

1. Association with persons or events of historic significance to the city, region, state or nation;
2. Illustrative of historic growth and development of the city, region, state, or nation;
3. In the case of structures, embodying distinctive characteristics of a type, period, or method of construction or representing the work of a master, or possessing unique architectural and artistic qualities, or representing a significant and distinguishable entity whose components may lack individual distinction;
4. In the case of districts, possessing a unique overall quality of architectural scale, texture, form, and visual homogeneity, even though certain structures within the district may lack individual distinction;
5. In the case of interiors, possessing one or more of the characteristics enumerated in (1), (2) or (3) above and, in addition, embodying distinctive characteristics of architectural scale, form, and visual homogeneity, which are an integral part of the character of the structure in which the space is contained

Rye, New York Landmarks Ordinance 7-5.5.

[29] Standards for Designation of Structures and Districts for Preservation. A structure, group of structures, site, or district may be designated for preservation as a landmark or landmark district if it has—

Historical, Cultural Importance

1. Has significant character, interest or value, as part of the development heritage; or
2. Is the site of an historic event with a significant effect upon society; or
3. Exemplifies the cultural, political, economic, social or historic heritage of the community; or

While the ordinance should be precise enough to provide adequate guidance to the commission and owners alike, language that is unreasonably and rigidly specific may preclude the designation of buildings that the community may wish to protect. The commission will be called on to apply the language of the ordinance in a wide range of situations, and the language must be sufficiently flexible to allow the commission to fulfill its task. It should also be noted that unless explicitly precluded, the property protected by a designation extends to structures and grounds which are necessary for the integrity of the landmark, but which may not be independently significant and worthy of protection.[30]

7.3.4 Historic Districts

The designation of boundaries of historic districts can present special difficulties. Frequently, the architectural or historical "theme" with which the buildings in an area may be associated pertains more strongly to some buildings than to others. Often, for example, the buildings in a proposed historic district will not be architecturally uniform. And while some buildings may not be worthy of individual designation, they may contribute significantly to the overall architectural or historical significance of a neighborhood. Accordingly, when defining boundaries, a commission must consider the unifying qualities of the structures to be included in the district. It should also evaluate the inclusion of structures and/or open space on the periphery of the district which are necessary to protect its historic properties. In addition, boundaries

Architectural, Engineering Importance

4.   Portrays the environment in an era of history characterized by a distinctive architectural style; or
5.   Embodies those distinguishing characteristics of an architectural-type or engineering specimen; or
6.   Is the work of a designer whose individual work has significantly influenced the development of Seattle; or
7.   Contains elements of design, detail, materials or craftsmanship which represent a significant innovation; or

Geographic Importance

8.   By being part of or related to a square, park or other distinctive area, should be developed or preserved according to a plan based on an historic, cultural or architectural motif; or
9.   Owing to its unique location or singular physical characteristic, represents an established and familiar visual feature of the neighborhood, community or city; or

Archeological Importance

10. Has yielded, or may be likely to yield, information important in pre-history or history.

Seattle, Washington Landmarks Ordinance No. 102229 § 6.

[30] *See* A-S-P Assoc. v. City of Raleigh, *supra*, 258 S.E.2d at 450–51 (1979).

should be clear and, where possible, should follow existing natural or mapped features.[31]

As a general matter, however, certain types of boundaries should be avoided. It is typically not a good idea, for example, to run the boundary of a historic district down the middle of a street. Doing so could result in incompatible building designs and facades on opposite sides, since construction and alteration on one side would be regulated while on the other it would not.

In order to reduce the potential for subjectivity in setting district boundaries, the ordinance should refer to the overall consistency of the proposed area and explicitly recognize that all buildings within the district may not be of the same significance. For example, the Rye ordinance refers to districts "possessing a unique overall quality of architectural scale, texture, form, and visual homogeneity, even though certain structures within the District may lack individual distinction." Other ordinances use such language as "congruity" or "consistency" in describing the relationship between the buildings in a district.

Generally, a court will uphold a commission's designation if it is based on clearly articulated standards.[32] In determining whether the discretion of the commission is limited, courts have gone fairly far to find that the commission acted under limiting factors.[33] Nevertheless, vague or nonexistent standards may result in a court invalidating the designation.[34] One court has said that discretion is sufficiently limited

---

[31] The Massachusetts Historical Commission has identified these factors to assist in evaluating historic district boundaries:

EVALUATION FOR SELECTION OF BOUNDARIES. In determining the boundaries of the historic district(s), the Study Committee should consider:

a. the significance of the architectural characteristics of each building and structure within the district and the degree to which they are visible from the public way;

b. the buildings and structures on the edges of the district which are an asset, as part of the setting or as protection to historic properties, or which are a detriment, because of incongruity of style, mass, use, condition;

c. the amount of open space within and on the edges of the district which can be justified as historic or necessary for protection to historic properties;

d. the distance up side street which district boundaries should extend to provide protection;

e. the amount of open space visible from the main public ways within the district;

f. the immediate surroundings of the district and the view into the district from the approaches to it.

[32] *See, e.g.,* Society for Ethical Culture v. Spatt, 416 N.Y.S.2d 246, 249–50 (1st Dep't, 1979), *aff'd,* 434 N.Y.S.2d 932 (1980).

[33] *See, e.g.,* A-S-P Assoc. v. City of Raleigh, *supra,* 258 S.E.2d at 453 (documents incorporated by reference in ordinance provide sufficient standards).

[34] *See* Historic Green Springs, Inc. v. Bergland, 497 F.Supp. 839 (E.D. Va. 1980); South of Second Assoc. v. Georgetown, 580 P.2d 807 (Colo. 1978); Texas Antiquities Comm'n. v. Dallas County Community College Dist., 554 S.W.2d 924 (Tex. 1977).

by a contextual standard found in the "readily identifiable, . . . predominant architectural style" of Victorian architecture in the district. This gave meaning to the general standard of "incongruity" in the ordinance.[35] Another court found that standards were provided to guide the commission in determining the effect of new construction on the character of a historic district when specific criteria to be considered, such as the architectural style of the new construction, texture, and material, were provided in the ordinance. The court found that such narrowing standards were "objective and easily discernible."[36]

For this reason, the standards section of a local ordinance assumes great importance. If the standards are found to be adequate, a court will be disinclined to substitute its judgment for that of the commission when it applies those standards to determine whether a particular building or district should be designated. Instead, the traditional deference extended by courts to the assumed expertise of administrative boards is reflected in the broad discretion allowed a commission in its designation determinations.

7.3.5 Procedural Standards

Proper procedural standards in the preservation context follow traditional constitutional and administrative law principles for agency decision-making. The purpose of these standards is to ensure fairness in the decision-making process and to provide a structural check on potential abuse of discretion by the commission.[37] Adequate procedural standards will include notice to the owner of the nominated property, an opportunity for the owner to present his views to the commission and the requirement that any decision include findings and embody a conclusion that is supported by the record. In addition to basic constitutional standards (derived from both the U.S. and applicable state constitutions), further procedural requirements may be found in a state's enabling act or administrative procedure act.

Designations have often been challenged on procedural grounds. As one court stated in requiring a city to rescind a landmark designation because of inadequate notice, rules concerning public notice, hearing, and procedure must be "spelled out and followed."[38] But while precise drafting is in the interests of due process, "[t]o satisfy due process, guidelines to aid a commission charged with implementing a public

---

[35] A-S-P Assoc. v. City of Raleigh, *supra,* 258 S.E.2d at 454.

[36] South of Second Assoc. v. Georgetown, 580 P.2d 807 (Colo. 1978).

[37] Maher v. City of New Orleans, *supra,* 516 F.2d at 1062–63.

[38] St. James United Methodist Church v. Kingston (N.Y. Ulster Cty. Sup. Ct. March 11, 1977).

zoning purpose need not be so rigidly drawn as to prejudge the outcome in each case, precluding reasonable administrative discretion."[39]

7.3.6   Notice

As noted earlier, many ordinances allow the designation process to be initiated by someone other than the owner of the property. To allow the process to continue without providing an opportunity for the owner's input would be fundamentally unfair. Since designation as a landmark may place restrictions on the use of the property and even impose affirmative duties on the owner, it seems beyond question that before any property is designated, due process requires notice which is "reasonably calculated, under all the circumstances to apprise the interested parties of the pendency of the action and afford them an opportunity to present their objections."[40]

The question remains open whether in the context of landmark designation personal notice or notice by publication is required. Although it is not a settled matter, a community would be well advised to provide personal notice by certified mail any time the number of property owners affected is not so large as to make that method impractical or economically prohibitive.

In circumstances where personal notice is appropriate, adequate notice should include a copy of the application, the time and location of the hearing, and either a copy of the rules and procedures that the commission will use or an explanation of where this can be obtained. Notice should be provided sufficiently in advance of the public hearing to allow proper preparation of testimony.

The designation of a site may affect a number of parties in addition to the applicant or property owner. These other parties may include adjoining landowners, tenants, preservation organizations, and relevant business organizations. Interested parties such as these should have the opportunity to have their views heard by the commission and the commission should encourage their participation by identifying them and by providing them with notice of hearing as well.[41]

7.3.7   Prehearing

After the designation of an individual building or district has been proposed, the commission must begin to develop the information which will be included in the record in support of its decision either to make

---

[39] Maher v. City of New Orleans, *supra,* 516 F.2d at 1062.

[40] Mullane v. Central Hanover Bank & Trust Co., 339 U.S. 306, 314 (1950).

[41] However, notice to these interested parties may not be required by the U.S. Constitution. *See* Zartman v. Reiser, 59 A.D.2d 237, 242, 399 N.Y.S.2d 506, 510 (4th Dep't 1977).

or reject the designation. Participants in the designation process should determine what practices the commission and its staff, if any, follow in evaluating the properties or districts prior to the hearing.

For example, in Boston the significance of a nominated property is evaluated in a formal study report which is prepared prior to the public hearing.[42] When a single landmark is proposed, the commission staff is responsible for preparing the study. In contrast, when a historic district is proposed, the study is performed by five of the nine commission members and six persons who have a "demonstrated interest in the district" and have been appointed by the mayor and confirmed by the City Council. The resulting report is made available three weeks prior to a public hearing on a proposed landmark designation and 60 days prior to a proposed district designation. The study, which makes recommendations as to the proposed designation, serves as an important reference document in all further deliberations of the Boston Commission.

7.3.8 Hearing

The due process clause of the Fifth Amendment of the U.S. Constitution generally requires the opportunity for a hearing prior to the deprivation of a constitutionally protected interest, such as a property right,[43] but it need not be analogous to a formal trial with cross-examination and a verbatim record. Instead, the hearing may be informal, more similar to a legislative hearing than a trial. At a minimum, however, some adjudicative procedure must be provided.

Most jurisdictions utilize the legislative hearing format rather than the triallike setting of the judicial approach. The municipality's ordinance will normally specify which is to be used. If the ordinance is silent on this point, counsel should refer to the jurisdiction's administrative procedures act which governs general agency procedures.[44] It is important that all parties involved with a designation proceeding be familiar with applicable procedural requirements.

Both proponents and opponents, if any, of the proposed designation should identify the particular concerns of the commission with regard to the property at issue and determine how best to satisfy these concerns. A number of methods may be helpful, including presenting testimony by experts, offering demonstrative evidence such as slides, and presenting a comparative analysis of prior approvals and denials and de-

[42] Boston, Massachusetts, Landmarks Ordinance § 4.

[43] *See* Board of Regents v. Roth, 408 U.S. 564, 570 n.7 (1972).

[44] *See* Society for Ethical Culture v. Spatt, 68 A.D.2d 112, 416 N.Y.S.2d 246, 250 (1st Dep't), *aff'd,* 434 N.Y.S.2d 932 (1980).

scribing how the proposal under consideration fits into that broader context.

7.3.9 Decision

The commission should develop a comprehensive written record upon which it can base its decision. The record should include written submissions, transcripts of oral testimony, and whatever documents the staff or the commission itself has produced. Based on this record, the commission should issue a written statement of its findings of fact and its determination and should include an explanation of how the designation relates to the standards set forth in the ordinance. A decision based on a repetition of the language of the ordinance and the mere conclusory statement that these requirements have been met may well not be found to be sufficient if the decision is challenged judicially.[45]

The ordinance should require that notice of a designation be given to the owner, published generally, and provided to municipal agencies with an interest in the outcome, such as those administering taxes and building permits. Notice to other municipal agencies is critical to the effective administration of the ordinance. The buildings department, for example, must know which buildings are subject to special procedures in connection with the grant of demolition or alteration permits by virtue of their status as landmarks. The designation should also be recorded in the appropriate title records so that any potential buyer of the property is on notice that it is subject to regulation under the ordinance.

7.3.10 Appeal

An ordinance should provide a mechanism for appealing a decision by the commission. Some ordinances allow review of the decision by another municipal body such as the city council. Others provide for immediate judicial review.

The question of standing—that is, of who may bring the appeal—should be dealt with in this section of the ordinance. In some communities only the owner of the designated property may appeal the decision. In other cases, standing is extended to any member of the community in recognition of the interest which each may have in preserving the community's historic resources.

---

[45] *See, e.g.,* Historic Green Springs, Inc. v. Bergland, *supra*, 497 F.Supp. at 850–51; Don't Tear it Down, Inc. v. D.C. Dep't of Housing & Community Dev., 428 A.2d 369 (D.C. App. 1981); Broadview Apartments Co. v. Comm'n for Historical & Architectural Preservation, 290 Md. 49, 433 A.2d 1214 (App. 1981).

7.4 APPLICATIONS RELATING TO ALTERATION, NEW CONSTRUCTION, AND DEMOLITION

7.4.1 Alteration and New Construction

The fundamental provision of the local ordinance is the requirement that the commission review and approve proposed alterations to any individual landmark or building in a historic district. This is the heart of the ordinance's substantive protection. Under the typical alterations section, no building permit for physical change to a designated landmark or property in an historic district can be issued without approval of the commission evidenced by its issuance of a certificate of appropriateness.

The authority of the commission to restrict alterations is found in the police power.[46] This authority extends to all changes to the building's exterior and, in some instances, interior.[47] In a historic district, the commission has the power to approve or deny proposed changes not only to historic buildings but to all of the property within the district.[48] This would include nonhistoric buildings and new construction on undeveloped lots. It is necessary that the commission's authority encompass all property within the district, since it would be impossible to protect the integrity of the district if unregulated development of any property were allowed.

7.4.2 Standards

A clear definition of the term "alteration" is crucial to the effective administration of the ordinance. The definition should assist the property owner in distinguishing between routine maintenance, which does not require review by the commission, and alterations, which do. An ordinance that does not make this distinction clear may create confusion regarding the scope of the commission's authority and ultimately undermine public support for the ordinance.

Prior to the commencement of any proposed change, the owner should be required to submit all plans and related materials to the

[46] See Section 7.1.1 above.

[47] For example, the inside of Radio City Music Hall in New York City is protected interior space.

[48] See, e.g., Sleeper v. Old King's Highway Regional Historic Dist. Comm'n, 417 N.E.2d 987, 989 (Mass. App. 1981); Faulkner v. Town of Chestertown, 290 Md. 214, 428 A.2d 879 (1981).

commission.[49] This information must be sufficient to allow the commissioners to determine the compatibility of the change with the existing structure or district. Determining compatibility can be problematic, since it necessarily involves some element of subjectivity. To circumscribe that subjectivity, the ordinance should establish criteria for review. Many ordinances instruct commissioners to look for compatibility, consistency, or appropriateness in the relationship between the existing landmark or district and the proposed change in such factors as mass, material, color, proportions, architectural style, and detailing.[50] Alternatively, when a change within an architecturally uniform district, rather than a single landmark, is proposed, the "observable character of the district"[51] can guide the commission.

Some ordinances contain specific formulas. For example, the Boston ordinance contains a provision devoted to the historic integrity of the doorways found on Beacon Hill.[52] Others, such as the Madison, Wisconsin, ordinance are more encompassing but contain less detail than the Boston ordinance. The Madison ordinance requires *inter alia:*

> (d) The proportions and relationships between doors and windows in the street facade(s) should be visually compatible with the buildings and environment with which it is visually related. . . .

[49] The preservation ordinance of Dallas, Texas, for example, requires the owner to submit "copies of all detailed plans, elevations, perspectives, specifications and other documents pertaining to the work." Ordinance No. 14012 of Dallas City Code, as amended, § 19A-9(b).

[50] New York City's Administrative Code §§ 207-6.0(b)(1) and (2) is an example:

> 1. In making [the] determination [as to whether the proposed work would be appropriate for and consistent with the effectuation of the purposes of this chapter] with respect to any such application for a permit to construct, reconstruct, alter or demolish an improvement in an historic district, the commission shall consider (a) the effect of the proposed work in creating, changing, destroying or affecting the exterior architectural features of the improvement upon which such work is to be done and (b) the relationship between the results of such work and the exterior architectural features of other, neighboring improvements in such district.

> 2. In appraising such effects and relationship, the commission shall consider, in addition to any other pertinent matters, the factors of aesthetic, historical and architectural values and significance, architectural style, design, arrangement, texture, material and color.

In Seattle the Board must consider

> among other things, the purposes of this ordinance, the historical and architectural value and significance of the landmark or landmark district, the texture, material and color of the building or structure in question or its appurtenant fixtures, including signs and the relationship of such features to similar features of other buildings within a landmark district, and the position of such building or structure in relation to the street or public way and to other buildings and structures.

Seattle, Washington Landmark Preservation Ordinance No. 102229, § 8(c).

[51] Town of Deering ex rel. Bittenbender v. Tibbetts, 105 N.H. 481, 202 A.2d 232 (1964).

[52] Mass. Gen. Laws, Ch. 616, Acts of 1955, § 1, as amended.

(1) All street facade(s) should blend with other buildings via directional expression. When adjacent buildings have a dominant horizontal or vertical expression, this expression should be carried over and reflected.[53]

Drafters of preservation ordinances should pay close attention to the case law in their state to ensure that the criteria adhere to minimum levels of explicitness.

New construction, like alterations, can be designed to be compatible with a historic building or district and therefore may be found to be appropriate. It is generally recognized that good "contemporary architecture per se is not incompatible with historic buildings. . . . "[54] For example, the U.S. Supreme Court in *Penn Central* noted that the New York City Landmarks Preservation Commission had approved construction of a new building which, "though completely modern in idiom, respects the qualities of its surroundings and will enhance the Brooklyn Heights Historic District. . . ."[55]

7.4.3 Certificate of Appropriateness

In contrast to the standards for designation which must be written into the ordinance itself, the standards to be applied when considering certificates of appropriateness may also be developed as separate detailed guidelines to be adopted by the commission.[56] While commission action taken without the benefit of narrowing guidelines is generally looked upon with disfavor by the courts, decisions based on incorporated materials have been upheld. These materials may include, *inter alia,* city and historic documents.[57]

The commission considering an application for a certificate of appropriateness should take care to establish an adequate record, make findings, and base its conclusion on those findings. Typically, a public hearing is held to receive oral and written testimony from the owner and any opponents to the alteration or new construction. When reviewing the decision of a commission that has failed to take these steps, a court may remand the case with an order that this be done.[58]

---

[53] Madison, Wisconsin General Ordinances 33,01.

[54] *Handbook,* note 1 above, at A-71; *see also* Hayes v. Smith, 167 A.2d 546, 549–50 (R.I. 1961).

[55] 438 U.S. at 118, n. 18.

[56] Dennis, "Do's and Don'ts in Drafting a Preservation Ordinance," in *Reusing Old Buildings,* note 8 above, at 315.

[57] Maher v. City of New Orleans, *supra,* 516 F.2d at 1063; A-S-P Assoc. v. City of Raleigh, *supra,* 258 S.E. 2d at 453 n.4, 454.

[58] *See* Fout v. Frederick Historic District Comm'n., Misc. No. 4005 (Md. Cir. Ct. Feb. 5, 1980); Equitable Funding Corp. v. Spatt, No. 12832/77 (N.Y. Sup. Ct. Feb. 8, 1978).

Although local commissions have the authority to deny certificates of appropriateness,[59] this is not the only action the commission may take. The ordinance may grant the commission the power to modify or amend an unacceptable proposal pending the applicant's agreement to such modification. If the modification is not acceptable to the property owner, then the effective result of the commission's decision would be a denial of the application. The ordinance may also permit the commission to suspend final action to allow the owner to make his proposal more consistent with the ordinance.[60]

7.4.4   Demolition

The presence of a landmark on a parcel of property can prevent the property from being used as the site of a new, larger, and presumably more economic building. Particularly in dense urban environments, this can create severe pressure to demolish the landmark to make way for a building capable of generating greater income to its owner.

Accordingly, to preserve the integrity of landmarks and districts, the ordinance should provide protection against demolition. The authority of landmark commissions to prohibit demolitions is settled.[61] Many communities make full use of this authority by granting their commission the power to refuse a permit for a landmark's demolition. However, some ordinances only allow the commission to comment prior to demolition. To the extent that the commission's authority to block demolition is decreased, protection of the city's landmarks is dramatically weakened.

An example of one city's approach is the Austin, Texas, preservation ordinance, which grants the commission the power to prohibit demolition:

> If an application is received for demolition or removal of a designated historic landmark, the building official shall immediately forward the application to the Landmark Commission. The Landmark Commission shall hold a public hearing on the application within thirty (30) days after the application is initially filed with the building official. The applicant shall be given ten (10) days written notice of the hearing. The Landmark Commission shall consider the state of repair of the building, the reasonableness of the cost of restoration or repair, the existing and/or potential

---

[59] See Penn Central Transportation Co. v. New York City, 438 U.S. 104 (1978).

[60] E.g., § 8(d) of the Seattle, Washington, ordinance states:

> In the event of a determination to deny a Certificate of Approval, the Board shall request consultation with the owner for a period not to exceed 90 days for the purpose of considering means of preservation in keeping with the criteria.

[61] See, e.g., Maher v. City of New Orleans, 516 F.2d 1051 (5th Cir.), cert. denied, 426 U.S. 905 (1975).

usefulness, including economic usefulness, of the building, the purpose behind preserving the structure as a historic landmark, the character of the neighborhood, and all other factors it finds appropriate. If the Landmark Commission determines that in the interest of preserving historical values, the structure should not be demolished or removed, it shall notify the building official that the application has been disapproved, and the building official shall so advise the applicant within five (5) days therefrom."[62]

This ordinance is a good example of one attempt to provide a commission with guidelines to aid in the difficult task of evaluating the economic impact of the restriction on demolition. This ordinance also indicates the relationship between the restrictions on demolition and other legal considerations. Although an owner does not have the right to receive the highest possible economic return from the use of his property, he must be able to make some economically viable use of his property. As discussed more fully in the following section on Economic Hardship,[63] a commission must consider whether denial of a demolition permit precludes any reasonable economic return. Thus, the demolition provision of the ordinance should be considered in relation to the ordinance's economic hardship clause.

Typically, a commission only has jurisdiction to consider applications to permit demolition of designated historic buildings or buildings within historic districts. Thus, unless the community has identified and designated significant buildings in the first instance, the effectiveness of the demolition provision will be curtailed. Since the designation process can be very time-consuming, a number of ordinances ban demolition (or alteration) until the designation process has been completed.[64]

Least effective from the standpoint of preservation goals are ordinances which only delay demolition for a period of time while alternatives are considered. If this approach is employed, the length of the delay period should be sufficient to allow meaningful negotiations to occur between the property owner, the commission, and other interested parties in an attempt to arrange an alternative to demolition.[65] Judicial interpretation of a provision granting a commission the power to delay has indicated that the power to grant a demolition permit implies the power to deny that permit.[66] Accordingly, an ordinance

[62] City of Austin, Texas Ordinance No. 740307-A § 45-51.

[63] See Section 7.4.5.

[64] See, e.g., Berkeley, California Ordinance § 15; District of Columbia Historic Landmark and Historic District Protection Act of 1978, D.C. Law 2–144, § 31(f).

[65] See Handbook, note 1 above, at A106; Dunbar High School Alumni Ass'n v. Dep't of Housing & Urban Dev., Civ. No. 8911-75 (D.C. Sup. Ct. April 1, 1977).

[66] San Diego Trust & Savings Bank v. Friends of Gill, 121 Cal. App.3d 203, 174 Cal. Rptr. 784 (1981).

which by its terms only authorized a demolition delay was interpreted
to give the commission authority to deny the permit if during that period
a feasible alternative to demolition was found. Despite this ruling, however, the property in question, San Diego's Melville/Klauber House,
was demolished while the case was being appealed. The lesson of this
case is that the power to deny demolition, if desired, should be explicitly
granted in the ordinance.

7.4.5 Economic Hardship Clause

By requiring that landmark buildings be retained and maintained,
preservation regulations have a direct economic impact on the owners
of such property. If the economic burden imposed is too severe, the
ordinance may be challenged on the grounds that the landmarked property has been constructively confiscated without just compensation. To
prevent such a challenge, the local ordinance should contain a hardship
clause.

A claim that property has been taken unlawfully may rest on several
alternative bases. Both the federal Constitution[67] and many state
constitutions[68] prohibit the taking of private property without just compensation to the owner. It is clear that a taking may occur without any
physical invasion of the property. A regulation which interferes with
the owner's use, possession or enjoyment of property, or his right to
dispose of it freely, may constitute a taking.[69] In practice, however, such
a de facto taking is extremely difficult to demonstrate. Courts do not
deem it adequate to show that the value of property has been reduced,
even drastically reduced, to make out a constitutional violation.[70]
Rather, the owner is generally required to show that he has been denied
any economically viable use of his land before a taking will be found.[71]

In the context of historic preservation, a taking claim may also be
based upon the state enabling legislation which empowers local government entities to preserve historic buildings, and to restrict their
alteration or demolition. New York's Town Law, for example, authorizes
town boards to enact measures which provide for the preservation and

---

[67] U.S. Const. amend. V.

[68] *See, e.g.,* N.Y. Const. art. 1 § 7.

[69] *See, e.g.,* Penn Central Transportation Co. v. City of New York, *supra,* 438 U.S. at 124;
City of Buffalo v. J. W. Clement Co., 28 N.Y.2d 241, 255, 321 N.Y.S.2d 345, 357 (1971).

[70] Federal courts have declined to find takings when government regulation has reduced
the value of property by as much as 90 percent. *See* Park Ave. Tower Assocs. v. City of
New York, 746 F.2d 135, 139–40 (2d Cir. 1984).

[71] Agins v. Tiburon, 447 U.S. 255, 260 (1980).

protection of historic buildings. The enabling provision, however, stipulates that "[a]ny such measures, if adopted in the exercise of the police power, shall be reasonable and appropriate to the purpose, or if constituting a taking of private property, shall provide for due compensation, which may include the limitation or remission of taxes."[72]

While a hardship clause may not be constitutionally required, it will help to provide flexibility where needed to prevent the application of the ordinance from unjustly burdening a particular property owner. If the owner can establish that the denial of a certificate of appropriateness (and the demolition or alteration it would have allowed) prevents realization of a given economic return, considering all possible alternatives including selling or leasing the property, then the clause may authorize the commission to devise a plan under which an appropriate return can be achieved. This may include tax relief or approval of a limited construction project.

The legal standard in the ordinance is usually expressed as one of reasonableness. In the District of Columbia, the mayor cannot issue a permit to demolish a designated landmark unless "failure to issue a permit will result in unreasonable economic hardship to the owner."[73] In some cases the ordinance itself will define "reasonable." For example, in New York City, a reasonable annual rate of return is defined as six percent of the valuation of the building and its site, including a depreciation allowance of two percent of the assessed value of improvements or the amount deducted for depreciation on the owner's latest federal tax return, whichever is less.[74] When the ordinance supplies a definition of reasonableness, the commission's inquiry in a given case can be limited largely to the economic situation of the particular building.[75]

When the party seeking alteration or demolition of its building is a charitable organization, use of the "reasonable rate of return" standard is not appropriate. Nor is the tax relief provided under many hardship provisions an ameliorative measure for tax-exempt institutions, which often enjoy tax-exempt status under local property tax statutes. Recognizing that the use of charitable property cannot be judged on the same basis as commercial property, the court may, instead of evaluating the availability of a reasonable return, ask whether the restrictions imposed by the ordinance prevent or seriously interfere with the organization's

---

[72] N.Y. Town Law § 64(17-a).

[73] D.C. Code § 5-1004(e) (1981); MB Assoc. v. D.C. Dep't of Licenses, Investigation & Inspection, 456 A.2d 344 (D.C. App. 1982).

[74] *See* N.Y.C. Admin. Code § 207-1.0(v.).

[75] For one set of factors to be considered in relation to economic hardship, *see* National Trust Model Ordinance, note 10 above, at § 54.10.29.

charitable purpose.[76] This approach is likely to be employed with respect to buildings used for religious purposes as well.[77]

The application of a hardship provision should be quite narrow, and the argument that the property could be put to a higher economic use should, by itself, be inadequate to secure relief.[78] Moreover, to further preservation purposes, the ordinance should contain the broadest possible array of alternatives for the commission to explore prior to permitting undesirable alteration or demolition of a historic building. The New York City ordinance is a good illustration.[79] If the commission determines that alteration of a given building would not be appropriate and that absent alteration a reasonable return cannot be obtained, it has several options. First, it must "endeavor to devise" a plan which would serve preservation goals and provide a reasonable return on the property. The plan may include tax exemption, tax remission, or authorization for appropriate alterations or construction. Once the plan is approved by the commission after public hearings, the property owner has the option to accept or reject it. If the owner accepts, and if the plan includes tax exemption or remission, the city tax commission must grant the approved tax relief upon the owner's application.

If the owner rejects the plan, the commission may recommend to the mayor that the city acquire a "specified appropriate protective interest" in the property, which would compensate the owner for his inability to undertake the work necessary to achieve a reasonable return. If the city does not take steps to condemn and acquire the recommended interest within ninety days, the commission must issue a notice to proceed with the proposed work.

A separate provision applies when an owner establishes that he can no longer economically use a property for the purposes to which it has been devoted, and seeks a certificate of appropriateness for alterations which are essential to enable him to enter into a contract of sale or long-term lease of the property. In that case, the commission must attempt to find a purchaser or tenant who will agree to take the property without the work for which the certificate of appropriateness is required. If a purchaser or tenant is found who will accept the property on "reasonably equivalent terms and conditions" as in the owner's original proposal, the certificate of appropriateness will be denied. If not,

[76] See Society for Ethical Culture v. Spatt, 51 N.Y.2d 449, 415 N.E.2d 922, 434 N.Y.S.2d 932 (1980); Trustees of Sailors' Snug Harbor v. Platt, 29 A.D.2d 376, 288 N.Y.S.2d 314, 316 (1st Dep't 1968).

[77] Lutheran Church in America v. City of New York, 35 N.Y.2d 121, 316 N.E.2d 305, 359 N.Y.S.2d 7 (1974).

[78] See Society for Ethical Culture v. Spatt, supra; Manhattan Club v. Landmarks Preservation Comm'n, 51 Misc.2d 556, 273 N.Y.S.2d 848 (Sup. Ct. 1966).

[79] N.Y.C. Admin. Code § 207-8.0.

the previously outlined procedure will be followed: The city must either condemn an appropriate protective interest, or the commission must issue the notice to proceed.

7.4.6 Maintenance

Issues relating to the maintenance of old buildings are growing in importance with the increase in the number of buildings which are landmarked and kept in active use for longer than had typically been the case in the past. Indeed, it has been suggested that in New York City issues relating to maintenance will be "[t]he greatest challenge facing the Landmarks Commission over the next twenty years."[80] A similar challenge will face preservation commissions of many other cities.

It is well settled that government has the constitutional authority to require owners to take affirmative steps to keep their landmarks in a state of good repair. *Maher v. City of New Orleans*[81] is the seminal case which upheld as constitutional a local ordinance's requirement that to achieve the legitimate end of preserving a landmark, "an owner may incidentally be required to make out-of-pocket expenditures" for the building's upkeep.

While *Maher* and other cases have upheld ordinances designed to prevent deterioration of historical structures, local governments do not have unfettered authority to require an owner to bear unlimited maintenance costs. In considering whether a commission has exceeded its authority, courts are likely to utilize a balancing test in which the public interest in preserving the landmark is weighed against the reasonableness of the economic burden to the owner.[82]

Some communities draft their maintenance provision in general terms while others specify the level of work which will be deemed to constitute adequate minimum upkeep. An example of a general minimum maintenance standard is the ordinance of the Town of Yorktown, New York, which states simply, "Every owner of property in a preservation district or a designated landmark shall keep it in good repair."[83] In contrast, the ordinance of Seattle, Washington, enumerates specific and quite detailed maintenance requirements, and a historic building is deemed "substandard" if it is permitted to deteriorate to the point at which occupant or public safety is endangered.[84]

---

[80] Gray, "Landmarks Preservation Comes of Age," *New York Affairs*, Nov. 3, 1980, at 52.

[81] 516 F.2d 1051, 1067 (5th Cir.), *cert. denied,* 426 U.S. 905 (1975).

[82] City of Chicago v. Kutil, 43 Ill. App.3d 826, 357 N.E.2d 200 (1976).

[83] Yorktown, New York Landmark Preservation Ordinance § 5-1 (1974).

[84] Section 6A of the Seattle Landmark Ordinance provides the following:

Minimum Maintenance Standards. The City Council finds that in order to accomplish the purposes of this ordinance, buildings and structures in the Historic District must be preserved against decay and deterioration occasioned by neglect. Any building or portion thereof or any structure appurtenant thereto in which there exists any of the following conditions to the degree that the preservation of the building or structure or the safety of its occupants, the occupants of adjacent buildings, or the public, is endangered is hereby declared for the purposes of this ordinance to be a "substandard historic district building":

A. Structural defects or hazards, including but not limited to the following:

1. Footings or foundations which are weakened, deteriorated, insecure or inadequate or of insufficient size to carry imposed loads with safety;

2. Floorings or floor supports which are defective, deteriorated, or of insufficient size or strength to carry imposed loads with safety;

3. Members of walls, partitions, or other vertical supports that split, lean, list, buckle, or are of insufficient size or strength to carry imposed loads with safety;

4. Members of ceilings, roofs, ceilings and roof supports, or other horizontal members which sag, split, buckle, or are of insufficient size or strength to carry imposed loads with safety;

5. Fireplaces or chimneys which list, bulge, settle, or are of insufficient size or strength to carry imposed loads with safety;

6. Exterior cantilever walls, or parapets, or appendages attached to or supported by an exterior wall of a building located adjacent to a public way or to a way set apart for exit from a building or passage of pedestrians, if such cantilever walls, parapets or appendages are not so constructed, anchored or braced as to remain wholly in their original position in event of an earthquake or wind capable of producing a lateral force equal to 0.2 of gravity;

7. Any exterior wall (or element thereof such as arches, keystones, lintels, etc.) located adjacent to a public way or to a way set apart for exit from a building or passage of pedestrians, if such wall is not so constructed, anchored or braced as to remain wholly in its original position in event of an earthquake or wind capable of producing a lateral force equal to 0.2 of gravity;

8. Any structure, chimney, flashing, antenna, air conditioner, or other appendage attached to or located on a roof of a building that is not so attached thereto so as to remain wholly in its original position in event of an earthquake or windstorm capable of producing a lateral force equal to 0.2 of gravity or an uplift force equal to the weight of the object plus gravity.

B. Defective or inadequate weather protection, including but not limited to the following:

1. Crumbling, broken, loose, or falling interior wall or ceiling covering;

2. Broken or missing doors and windows;

3. Deteriorated, ineffective, or lack of waterproofing of foundations or floors;

4. Deteriorated, ineffective, or lack of exterior wall covering, including lack of paint or other approved protective coating;

5. Deteriorated, ineffective, or lack of roof covering;

6. Broken, split, decayed or buckled exterior wall or roof covering;

7. Gutters and downspouts which have deteriorated.

C. Defects increasing the hazards of fire or accident, including, but not limited to the following:

1. Accumulations of rubbish and debris;

2. Any condition as to cause a fire of explosion or to provide a ready fuel to augment the spread or intensity of fire or explosion arising from any cause.

New York City's landmarks ordinance establishes fairly general maintenance standards, but contains separate provisions depending upon the nature of the landmark. Persons "in charge of" historic buildings in historic districts must keep in good repair all exterior portions of the buildings, and all interior portions which, if not maintained, could cause the exterior portions to "deteriorate, decay or become damaged or otherwise to fall into a state of disrepair." Persons controlling interior landmarks must keep those interior portions in good repair, as well as any other portions which, if not maintained, could cause damage of any kind to the interior landmark.[85]

Some communities provide financial assistance to property owners who face financial hardship caused by the requirement that certain repairs be made. For example, some ordinances establish a revolving fund for this purpose. Others allow the municipality to undertake directly maintenance which an owner fails to provide and to impose a lien against the property for the costs.

7.4.7 Exclusion for Routine Maintenance

It is important that ordinances distinguish between routine maintenance, which can be undertaken without commission approval, and repair work and alterations, which may require such review and approval. Ordinances which impose unduly burdensome approval procedures on owners who wish to undertake relatively minor maintenance will discourage the end they seek.

New York City's ordinance distinguishes among three types of work, each of which is subject to a difficult level of review. "Ordinary repairs and maintenance,"[86] for which no approval from the landmarks commission is required, is defined as:

> (1) work done on any improvement; or (2) replacement of any part of an improvement; for which a permit issued by the department of buildings is not required by law, where the purpose and effect of such work or replacement is to correct any deterioration or decay of or damage to such improvement or any part thereof and to restore same, as nearly as may be practicable, to its condition prior to the occurrence of such deterioration, decay or damage.

A second category, "minor work," consists of somewhat more extensive projects undertaken with respect to historic buildings, although still not so major as to require a building permit. Minor work includes "any change in, addition to or removal from the parts, elements or

---

[85] N.Y.C. Admin. Code § 207-10.0(a).
[86] *Id.* § 207-1.0(r.).

materials comprising an improvement including, but not limited to, the exterior architectural features or interior architectural features thereof . . ." The Landmarks Commission is authorized to incorporate by regulation within this category the "surfacing, resurfacing, painting, renovating, restoring, or rehabilitating of the exterior architectural features or interior architectural features or the treating of the same in any manner that materially alters their appearance. . . ."[87] Minor work may only be performed in accordance with a permit issued by the commission.[88]

The third and most intrusive category of work under the New York City ordinance, "alteration," includes "any of the acts defined as an alteration by the building code of the city."[89] Alterations may not be undertaken without the issuance of a certificate of no effect, certificate of appropriateness, or a notice to proceed.

The ordinance should also contain a provision for undertaking work necessary to remedy dangerous conditions relating to historic buildings on an emergency basis, without the issuance of a certificate of appropriateness or permit. The New York City ordinance permits essential steps to be taken, without commission approval, to remedy "conditions determined to be dangerous to life, health or property," if the work has been ordered by the Department of Buildings, the Fire Department, the Health Service Administration, or by a court upon application of any of those agencies.[90]

7.4.8 Compliance Inspections

The authority of local governments to require that affirmative steps be taken by the owner of a landmark to maintain the property suggests a corresponding right to evaluate the owner's diligence in doing so. However, Fourth-Amendment issues relating to the government's right to search private property are implicated when a landmark commission attempts to detect compliance through a schedule of periodic inspections. Therefore, an ordinance's program of inspections must meet certain minimum constitutional standards or risk successful challenge in court. In *Camara v. Municipal Court*,[91] for example, the Supreme Court held that a warrant based on probable cause was required prior to an administrative inspection.[92]

---

[87] *Id.* § 207-1.0(q.).

[88] *Id.* § 207-9.0.

[89] *Id.* § 207-1.0(a). *See also* National Trust Model Ordinance, note 10 above, at § 54.10.21.

[90] N.Y.C. Admin. Code § 207-11.0.

[91] 387 U.S. 523 (1967).

[92] *Camara,* 387 U.S. at 538.

Establishing probable cause for purposes of an administrative inspection, however, does not require concrete evidence that a violation of a city ordinance exists. Rather, it is sufficient to show that "reasonable legislative or administrative standards for conducting an . . . inspection are satisfied" in order to obtain a warrant to inspect.[93] If the preservation ordinance or regulations adopted by the commission outline a "general administrative plan" for enforcement of the ordinance's maintenance requirements, based upon "neutral" and nondiscriminatory principles, that should provide an adequate basis for the issuance of warrants.[94]

In addition to, or in lieu of, inspections to verify that routine maintenance is being performed, a municipality may wish to monitor work being done pursuant to a permit or certificate of appropriateness issued by the commission.[95] Permission for such inspections could be made a condition of issuance of the permit or certificate, thus avoiding any constitutional complications. In addition, it may be advisable, bureaucratic considerations permitting, to vest total or primary responsibility for inspection of historic buildings in the city buildings department, which will already have a trained inspection staff and an established program.[96] If this is done, of course, the commission must make sure the building inspectors are trained to identify those particular maintenance problems which are peculiar to historic preservation of buildings, as opposed to more general health and safety considerations.

7.4.9 Publicly Owned Property

Historical buildings are often publicly owned. Nevertheless, it should be noted that the extent to which a higher level of government, particularly a county, must comply with a local ordinance remains an unsettled question.[97] The use of state-owned property by state agencies is not generally within the jurisdiction of local preservation ordinances. It has been held that a city's police power may not extend to the designation of state-owned buildings as landmarks.[98]

---

[93] Marshall v. Barlow's, Inc., 436 U.S. 307, 320 (1978).

[94] *Id.* at 321.

[95] *See* National Trust Model Ordinance, note 10 above, at § 54.10.31-.32.

[96] *Id.* at 54.10.31; Seattle Landmark Ordinance § 6B.

[97] *Compare* City of Ithaca v. County of Tompkins, 77 Misc.2d 882, 355 N.Y.S.2d 275 (Sup. Ct. 1974) and Mayor of Annapolis v. Anne Arundel County, 271 Md. 265, 316 A.2d 807 (1974) (county property subject to city landmark ordinance) *with* Texas Antiquities Comm'n. v. Dallas County Community College Dist., 554 S.W.2d 924 (Tex. 1977) (county property not subject to local regulations).

[98] *See, e.g.,* City of Santa Fe v. Armijo, 96 N.M. 663, 634 P.2d 685 (N.M. 1981); State v. City of Seattle, 94 Wash.2d 162, 615 P.2d 461 (1980).

Even if the commission cannot control the use of historic buildings that are publicly owned, it should have the power to participate in decisions which affect such buildings. New York City's ordinance requires all agencies to refer to the landmarks commission plans for construction, reconstruction, alteration, or demolition of any buildings, owned by the city or situated on land owned by the city, which are landmarks themselves or are located within historic districts. No agency or official can approve such plans until the commission has had the opportunity to prepare a report detailing its views on the proposed work.[99] This procedure does not permit the commission to veto a project, but does insure that preservation concerns will be considered before irreversible action is taken.

7.4.10   Penalties

A city should provide the commission with the means to enforce its actions. Most ordinances provide for penalties to be imposed against those who violate their provisions. For example, in Syracuse, New York, those who fail to comply with the provisions of the preservation ordinance are subject to a penalty of $100 a day. Willful violators are subject to fines of $150 per day or up to 15 days' imprisonment.[100] New York City's ordinance provides varying penalties for those who violate its alteration and demolition provisions or its repair and minor work provisions, and those who withhold information or make false statements to the commission. It also provides for injunctive relief.[101]

The District of Columbia requires that any person who demolishes, alters, or constructs a structure in violation of the act must restore the property and its site to its appearance prior to the violation.[102] This remedy is, in practice, designed to be an economic deterrent to the destruction of historic property, since damage to a historic property is often irremediable.

7.5   CERTIFICATION OF LOCAL STATUTES AND HISTORIC DISTRICTS

In order to survive court challenges to the designation of historic districts and buildings, it is important that communities incorporate the appropriate substantive provisions in their preservation statutes, and

[99] N.Y.C. Admin. Code § 207-17.0(b)(1).

[100] Syracuse, N.Y., General Ordinance No. 11 § IX(A) (1975).

[101] N.Y.C. Admin. Code § 207-16.0.

[102] District of Columbia Historic Landmark and Historic District Protection Act of 1978, D.C. Law 2-144, § 11(b).

abide by the processes just outlined in making designations. There is, however, an important additional reason for complying with these requirements: Such compliance may determine whether local property owners are eligible for the federal tax incentives for historic preservation discussed in chapter 1.

As explained in chapter 2, to qualify for certain preservation tax benefits, a structure must either be listed individually on the National Register or located within a registered historic district. A district can achieve that status in only two ways. First, it may itself be listed on the National Register as a historic district. Alternatively, the district must be designated under a state or local statute which has been certified by the Secretary of the Interior as containing criteria which will substantially achieve the purpose of preserving and rehabilitating buildings of significance to the district,[103] and be certified as meeting the same requirements as apply to listing of districts on the National Register. Only when the locality has complied with this two-step certification process can individual building owners benefit from the federal tax incentive program.

7.5.1   Certification of Statutes

The National Park Service (NPS) will review requests for certification of state or local statutes only at the request of the chief elected official of the enacting jurisdiction or that official's duly authorized representative. A private citizen, therefore, cannot go directly to the federal government seeking certification of a local ordinance; he must prevail upon the local government to initiate the process. In evaluating the adequacy of a local statute, the NPS will decide whether the statute "contain[s] criteria which will substantially achieve the purpose of preserving and rehabilitating buildings of historic significance to the district." The regulations specify only two additional requirements: The statute must designate a landmarks review board or commission, and that entity must be empowered to review proposed alterations to historic structures within any districts designated under the statute.[104]

Past certification decisions indicate additional criteria which the NPS applies in reviewing local statutes.[105] Statutes have been deemed not to achieve preservation purposes if they permit owners to veto landmark designation of their own property. This power, according to the NPS, could threaten the integrity of an historic district, by withholding protection from important properties within it. For similar reasons, the

---

[103] 36 C.F.R. § 67.8(a)

[104] *Id.*

[105] *See* "Certification of State and Local Statutes," in *Reusing Old Buildings*, note 8 above, at 346–49 (1982).

NPS refused to certify a statute providing for review of alterations only to listed structures within a district, rather than to all structures. These "patchwork" concerns also led to denial of certification for a statute which excluded from landmark review any district structures less than 100 years old.

Conversely, however, the NPS has certified statutes which provide for district-wide review of designation. These have contained provisions that a majority of property owners within a proposed historic district must agree to designation, or even that 2/3 of such owners must consent to designation in writing. The NPS has indicated that even 100 percent owner concurrence would be acceptable.

Moreover, before the NPS will review local ordinances enacted under the authority of state enabling legislation, the empowering state law must itself be certified. The NPS generally will not review state enabling laws in isolation, unless those laws actually designate historic districts without the need for local action. Rather, state laws, when submitted to the NPS, must be accompanied by local statutes that implement them and carry out their purposes. If the state enabling legislation does not satisfy the intent of federal law, then no local statutes enacted under its authority will be certified.

7.5.2 Certification of Historic Districts

Once the enabling statute is certified, the district itself must also be certified before individual buildings can be certified for tax incentive purposes. Although designation under a state or local ordinance is an alternative to listing in the National Register, district certification depends upon a determination by the NPS that the locally created district "substantially" meets the requirements for district certification.[106] Thus, it is imperative that the local ordinance provide for full and reasoned consideration of district designations by the local commission, so that its ultimate decisions will survive NPS scrutiny.

Again, the certification request must come from an official of the relevant governmental entity, not from an individual property owner. The certification request should include the following documentation:

1. A description of the physical or historical qualities which define the district, an explanation of the choice of boundaries, and a description of the typical architectural styles and types of buildings;

2. A statement explaining why the district has significance and substantially meets the National Register criteria;

[106] 36 C.F.R. § 67.9(a). The National Register criteria are found in 36 C.F.R. Part 60, *see* Appendix I.

3. A definition of the types of buildings that do not contribute to the significance of the district, and an estimate of the percentage of buildings that do not;

4. A map showing all district buildings, preferably identifying contributing and noncontributing structures; and

5. Photographs of typical areas and buildings in the district.

Once a district itself is certified, individual property owners within it are eligible to begin the certification process for rehabilitation to their own buildings.

Eight

A Rehabilitation Sampler

Introduction

By its nature, each rehabilitation or preservation project is unique; there is no single formula for success. Nevertheless, there are common issues which arise in the course of rehabilitation efforts, and an understanding of these issues is important to virtually all preservation projects. Objectives must be defined, properties identified and inspected, proposed uses evaluated, financing obtained, applicable laws and regulations satisfied, and the project carried out and operated as planned. The preceding chapters have identified the major considerations which the sponsor or coordinator of a rehabilitation project must address. When the property to be rehabilitated is historic, the project takes on additional levels of complexity but, as previously mentioned, offers additional rewards as well.

The first portion of this chapter outlines an overall program for the rehabilitation of a historic property. The framework it provides is skeletal; it does not purport to describe all of the decisions that will confront those who undertake historic rehabilitations, nor will the issues necessarily arise in the order presented. It is intended to illustrate the breadth and complexity of the tasks which may be involved in a preservation project, and the range of expertise which may be required. No single individual should expect to possess all of the skills required to carry out a project of any significant size, and outside assistance—whether legal, financial, architectural, or otherwise—should be brought in as necessary.

The second portion of the chapter describes four preservation projects that illustrate both the common elements and the extraordinary diversity of the rehabilitation process. They are also examples of rehabilitation successfully undertaken with preservation objectives in mind and demonstrate greatly varying approaches to the challenge of adapting historic structures to contemporary use.

8.1   THE REHABILITATION PROGRAM

This section is intended to assist an individual or organization wishing to embark upon a preservation project which has not yet been defined in detail. The following discussion identifies the principal steps which must be taken on the path from initial project plan to completion. It must be emphasized, however, that this description is not exhaustive, and that the preservationist should obtain outside expertise as appropriate to insure that all components of the project are properly in place.

8.1.1   Identifying the Property

The first step in any project is to locate the historic structure or structures to be preserved. In some cases, this will be obvious: An organization may have been formed for the precise purpose of restoring a particular local landmark, designated or otherwise. In others, a project sponsor may have a certain objective in mind—creating new residential units, for example—but may not yet have identified a property which will serve that purpose.

In either event, a number of characteristics of the property should be analyzed. The condition of the property will determine the extent and cost of the required renovation. These in turn may affect eligibility for the rehabilitation tax credit, which is available only for a "substantial" rehabilitation—one whose cost must generally exceed the owner's basis in the building. At the same time, if tax credits are desired, the building must not require so much expansion or alteration that 75 percent of the existing external walls cannot be retained.

Eligibility for the tax credit will depend on other building characteristics as well. Age is an important criterion; a structure must be at least 30 years old to qualify for the 15 percent credit, 40 years old for the 20 percent credit, and, as a general rule, 50 years old for the maximum 25 percent credit. If the building is located in a National Register historic district, federal restrictions may govern the type of rehabilitation work which can be undertaken. Unless certified as not contributing to the significance of the district, a building within a historic district must undergo a certified rehabilitation for the tax credit to be available. If the structure is a locally designated landmark or situated within a locally created district, provisions of the local preservation ordinance may dictate the extent and nature of alterations which can be carried out under any circumstances.

In choosing a property for rehabilitation, then, the prospective developer should establish that the property can be adapted to its intended use in conformity with applicable federal and local regulations. In making this determination, he may wish to consult with an architect or structural engineer familiar with historic preservation work, or with the staffs of the local preservation commission, the state historic preservation office, or (particularly in states that do not participate in the certification process) the office of the National Park Service which administers the certification program.

8.1.2 Planning the Project

Having selected an appropriate property or properties, the next step is to formulate a plan for carrying out the rehabilitation project. In most cases, this will involve hiring an architect to design the project and, at a later date, to prepare the required plans for submission to governmental authorities and financing sources. It may also be useful at this point to retain an attorney experienced in preservation law (and other relevant fields) to identify, as early as possible in project planning, the likely regulatory reviews and legal standards to which the project will be subject. As indicated in preceding chapters, a complex rehabilitation project may require one or more levels of environmental and historic review and involve legal issues such as code compliance; zoning regulations; local, state, and federal preservation and environmental regulations; and the tax, securities, and partnership matters discussed in earlier portions of this book. A clear understanding of the impact of these issues on a particular project is an essential component of successful advance planning.

Once an overall project plan has been completed, the architect can proceed with the detailed drawings and plans. After the architect's plans have been completed, they will have to be submitted to local building authorities, who must certify that they comply with applicable building code standards. Depending upon the nature of the project, it may be necessary to take advantage of special building code provisions pertaining to historic buildings in order to achieve compliance without significantly violating the building's historic character; in some cases, where these do not exist or are not adequate, a project sponsor may wish to seek a variance or exemption from code requirements.

The architect should also prepare a zoning analysis to determine the permissible building size and configuration under local zoning ordinances. If the structure to be renovated does not utilize all of the available building bulk for the site on which it is located, the owner should explore its eligibility for the transfer of development rights to another site or sites, a step which could help raise funds to finance the project.

The architect should also prepare an estimated construction budget for the project. This will serve as the basis for the overall financing package discussed in the next section. If the property to be rehabilitated is a federally or locally designated landmark or located within a federally or locally designated historic district, additional reviews may be required beyond that of local building authorities. If the building is a local landmark, construction plans will typically have to be submitted to the landmarks preservation commission in order to obtain a certificate of appropriateness permitting alteration.

If the building is federally designated or within a federally certified district, and the owner intends to take a rehabilitation tax credit, the appropriate certification procedures must be followed. If the building is located within a federally designated district, and no rehabilitation credit is desired, the owner must first obtain a certification that the building does not contribute to the significance of that district. This will enable him to claim either the 15 percent or 20 percent tax credit, depending upon the age of the structure. Absent such certification, or in the case of a National Register landmark, a certified rehabilitation will be necessary if *any* tax credit is desired. In such cases it may also be useful, although not essential, to request preliminary certification of the rehabilitation plans, to provide some assurance to the owner and other potential investors that the benefits of the tax credit will be forthcoming when the project is completed. If a certified rehabilitation is to be undertaken, the project sponsor would be well advised to retain the services of an architect or other consultant who is well versed in the certification process.

8.1.3 Structuring the Financing Package

Once the overall cost of the project has been estimated, the sponsor must determine how it is to be financed. The simplest methods, which may be adequate for small-scale projects, involve self-financing of some or all project costs, with the remainder to be provided as a construction or mortgage loan from a bank or other lending institution. However, most preservation projects can benefit from financial aid targeted or available for historic preservation activities. Such aid may come from public or private sources, may be provided in the form of loans or grants, and may constitute either direct assistance or indirect benefits which have the effect of lowering project costs.

The latter category, consisting of a variety of tax incentives for rehabilitation, is potentially the most beneficial and should generally be explored first. The most powerful incentive is the rehabilitation tax credit, which, as just discussed, can offset 15, 20, or 25 percent of qualifying project costs. If a project can be structured to take advantage of one of these credits, it is generally advisable to do so. In addition, a number of approaches are available under the tax laws for depreciating project costs. Different depreciation formulas may be most beneficial in different instances, and the optimal combination of tax credits and depreciation deductions will depend upon the specifics of the project. The sponsor should seek tax advice from either an attorney or a financial adviser to determine the best possible utilization of such tax incentives.

An additional offset to project costs may be achieved, in certain instances, through the donation of a facade easement (technically, a qualified conservation contribution) on the building to be renovated. By agreeing not to alter the facade, the owner may be entitled to a tax deduction equal to the amount by which that restriction reduces the value of the building. The possibility of benefiting from such a donation should be explored. Finally, a variety of state and local tax incentives may also be available. These will generally take the form of a partial or total exemption from, or abatement of, property taxes for projects which meet specified criteria.

The advantages of using these cost offsets, however, depend upon the existence of a tax liability to offset. If the entity undertaking the rehabilitation cannot use tax credits or deductions at all (for example, if it is a nonprofit, tax-exempt organization), or if the entity or individual cannot make full use of the amount of tax benefits which will be generated by the project, it will probably be desirable to include other investors in financing the project, and to offer a portion of the tax savings as an incentive for them to invest. This is an additional advantage of syndication, and is discussed in this chapter.

In addition to those tax provisions, there may be other programs of direct governmental assistance for preservation. The federal government funds programs which can be used to facilitate preservation involving adaptive reuse, as, for example, through Urban Development Action Grants, Community Development Block Grants, or Federal Housing Administration loans. Other programs, such as those of the National Trust for Historic Preservation, may not typically provide direct project support but can provide technical assistance and other services. States and localities, either directly or through quasi-public corporations, may be able to issue tax-exempt bonds, the revenue from which can be devoted to certain kinds of rehabilitation or conversion projects. This may be especially helpful for large-scale commercial or residential projects with significant public benefits.

Private sources of funding may also exist. Many communities have revolving funds which will lend money to finance a portion of the cost of rehabilitating older buildings. Foundations, particularly community foundations, with a commitment to economic development and civic improvement in a particular locale, may also offer financial assistance. All of these potential funding sources should be explored in putting together a construction funding package.

A final method of obtaining financing, though by no means a last resort, involves putting together a syndicate to finance the rehabilitation. This may be desirable if the sources of public and private funding just discussed will not cover project costs; if sufficient bank financing cannot be obtained or is prohibitively expensive; if the sponsor wishes to share the risk of a project (as well as the potential benefits) with others; or if the project will produce tax savings of which the sponsor cannot take full advantage.

A project sponsor may wish to obtain the services of a professional syndicator to assist in structuring financing and securing investors.[1] On the basis of information provided by the developer of the project, the syndicator will typically prepare a proposal letter indicating the amount of equity he believes can be raised for the renovation and the terms on which solicitations for investment should be made. If the proposal is acceptable, the syndicator will put together a more comprehensive syndication agreement. This will specify the guarantees which the sponsor will be required to provide, such as a construction completion guarantee, a promise to repurchase investor interests in the event completion is not achieved, or even a price adjustment provision if the project ultimately does not qualify for an investment tax credit. The agreement will also outline the form of syndication, usually a limited partnership, the requirements and timetable for capital contributions from the limited partners, the apportioning of partnership items (income, loss, credits, and so on) among the general and limited partners, and the obligation of the developer to assume other costs associated with the project.

If the agreement is acceptable to the developer, the next step is to prepare a partnership agreement, which will contain a more detailed set of rules for the conduct of partnership business. In most instances, a private placement memorandum or prospectus will also be prepared and, depending upon relevant state requirements, submitted to state officials for a review of compliance with securities regulations. This document will describe the project, particularly its financial aspects, and will identify any risks associated with the project of which potential investors should be aware.

In particular, the memorandum or prospectus will usually contain

---

[1] *See* Oldham, "The Equity Syndication Process," in *Historic Preservation Law: Tax Planning for Old and Historic Buildings* at 371 (American Law Institute 1984).

a *pro forma* projection of financial data relating to the project for the duration of a prospective limited partner's investment. It will itemize the anticipated expenses of the project over time, as well as the sources of funds to meet rehabilitation and other project costs. The memorandum will also break down projected cash flow from the project by category, including direct income from the rental, sale, or use of the rehabilitated property, and indirect returns in the form of tax benefits. Finally, it will project the annual capital contributions which will be required of potential investors, and indicate the expected return of capital and tax benefits which will be allocated to each limited partnership share on an annual basis. While these projections cannot be made with certainty, and will be accompanied by appropriate disclaimers, they do provide a basis to help investors evaluate the merits of a proposed investment. Appendix G contains an illustration of such a *pro forma* projection for a typical rehabilitation project. Once the appropriate disclosure document has been approved, it will be used as the basis for soliciting the participation of limited partners.

When sufficient funds have been raised for a project through the methods just described, and after all required permits and approvals have been secured, the project can begin. However, as with any construction project, the hiring of a competent and knowledgeable contractor is essential to the successful completion of rehabilitation work. If historic features of the building are to be preserved, it may be necessary to find a contractor or subcontractor particularly skilled in restoration. Depending upon the scale of the project, it may also be desirable to employ a supervisor to oversee the entire project, and to make sure the work is completed as planned, on time and as close to budget as possible. The project's architect may be willing to provide this construction supervision.

Once the renovated building is placed in service in its new (or rejuvenated) function, the partnership agreement should govern the allocation of revenues among investors, as well as the ultimate disposition of the building. In some cases it will be retained as an income-producing asset; in others, it will be sold when rehabilitation is finished to generate an immediate cash flow. If the project is eligible for the 25 percent investment tax credit, it will be necessary to seek certification of the completed work, regardless of whether preliminary certification was sought or obtained.

8.2 REHABILITATION PROJECTS

The foregoing discussion outlines the basic elements which must come together for a successful rehabilitation project. The remainder of this chapter describes four widely varying rehabilitation projects which have combined these elements in different, but ultimately successful, ways. These illustrations are not intended as models for undertaking historic preservation work, nor will the financing packages used to underwrite these projects necessarily be appropriate in other circumstances. Nevertheless, each of these projects demonstrates the range of benefits possible, as well as the leadership and perseverance required, in the rehabilitation field and the need for project sponsors both to assemble a completed project team and to be familiar with the essential components of real estate development and finance. The detailed financial feasibility analysis which must be undertaken before initiating a project is beyond the scope of this book.[2]

8.2.1 Small Residential Building (78 Hudson Street, Providence, Rhode Island)

Small-scale rehabilitation projects are often undertaken for reasons which are not solely financial, and in those cases a range of nonprofit and governmental entities may have to work together to make the projects feasible. An example of this type of project is the rehabilitation of a 2½ story building within the Broadway-Armory National Register Historic District in Providence, Rhode Island.[3] Although the project resulted in the creation of only four apartments, it was thought important as a catalyst to further rehabilitation activity in the district, and money from several sources, including a revolving fund, was made available to supplement the funds obtained from private investors.

The costs associated with the creation of a limited partnership and the sale of the limited partnership interests may cast doubt on the advisability of a partnership syndication in situations, such as this one, where the amount raised from the limited partners is relatively small. However, there are often situations in which other forms of financial assistance are not available unless private funds are committed, or where the nonprofit and public funds are not sufficient. If preservation rather than profit alone is the objective, and the limited funds available from the syndication are critical, the costs of the syndication may be necessary transaction costs. In any event, the structure of this trans-

---

[2] A number of useful publications are available to the prospective sponsor of a rehabilitation project, which will help in analyzing the investment prospects of a particular project and in choosing among various financing options. Among those worth consulting are J. Canestaro, *Real Estate Financial Feasibility Analysis Handbook: A Guide to Project Cost-Benefit Evaluation* (Virginia Polytechnic Institute and State University, 1982); A. Ring and J. Dasso, *Real Estate: Principles and Practices* (1981); R. Lifton, *Practical Real Estate: Legal, Tax and Business Strategies* (1979).

[3] The information in this section is taken primarily from an offering memorandum for limited partnership interests in Hudson Street Associates, dated June 30, 1982, generously provided by the Providence Preservation Society Revolving Fund.

action can serve as a useful model for the rehabilitation of buildings that are more expensive to acquire or rehabilitate, and which therefore require more substantial capital contributions from the limited partners.

The Armory District, part of the Broadway-Armory National Register Historic District, is five minutes from downtown Providence. The district is a neighborhood of late nineteenth and early twentieth-century homes constructed in a variety of Victorian styles, which have sheltered successive influxes of new Americans, including most recently Spanish, Portuguese, and Southeast Asian immigrants.

The character of the neighborhood has reflected the changing economic climate of Providence; the last few decades have not been kind to the district. Large-scale migration from the city has led to disinvestment, and a 1980 study revealed a 16 percent vacancy rate in the district. At the time this project was initiated, the district contained a number of abandoned properties and many others in varying states of disrepair.

Nonetheless, the neighborhood's housing stock escaped the large-scale clearance and demolition activities of the urban renewal programs of the 1960s, and its rows of predominantly two- and three-family wood-frame houses have remained basically unchanged. With the renewed popularity of urban residential areas, these buildings are becoming more attractive, and there have recently been signs of reinvestment by both new and older residents of the area. To this end, neighborhood residents have established a variety of organizations to assist in improving housing conditions in the district. The city of Providence has also lent its support. The city's Community Development Office has funded a program to provide financial assistance to district property owners for home improvements. The city has also planted 300 trees around the perimeter of the Dexter Training Field, a park which is a focal point of the neighborhood.

By early 1985, in fact, the character of the district had been substantially upgraded as a result of the rehabilitation of over 30 properties in the area, including the residential structure located at 78 Hudson Street. This building is a Second Empire house built in 1875, apparently as a two-family dwelling with servants' quarters in the basement and on the top floor. This large, 2½ story wood-frame, mansard-roofed structure has retained most of its unusually fine exterior detailing. The acquisition and rehabilitation of 78 Hudson Street was coordinated by the Providence Preservation Society Revolving Fund, Inc. (Fund), a nonprofit organization incorporated in 1980 as an affiliate of the Providence Preservation Society. The society set up the Fund to work in areas where urban blight threatens buildings of historic significance, and to undertake projects of high visibility in an effort to spur other neighborhood revitalization efforts.

Hudson Street Associates, a limited partnership, was created to help finance and carry out the project. The general partner, who is responsible for the administration of the partnership and management of the property, is a former carpenter who has refitted large sailing yachts and worked on a variety of construction projects, including two rehabilitation projects within the Armory District. He also has experience in managing an apartment building in Providence.   •

The rehabilitation of 78 Hudson Street consisted of restoring the exterior of the building, part of which had been damaged by fire, to its original condition and converting the interior to four residential units. New heating, electrical, and plumbing systems were installed. Through the use of paint scrapings, the original colors of the building's exterior were approximated and the entire building was repainted. Clapboards which had been custom-shaped to duplicate the originals were installed, and deteriorated wooden detail was replaced.

The total purchase and rehabilitation effort cost approximately $90,000, of which $56,673 qualified for the 25-percent investment tax credit. Funds were obtained from several sources. The general partner borrowed $25,000 from a commercial lending institution to purchase the property and to finance a portion of the rehabilitation. The loan is secured by a first mortgage on the property. The general partner then conveyed the property to the partnership, subject to the bank loan and first mortgage. He remained personally liable on the loan.

The Fund loaned the partnership $15,000, secured by a second mortgage on the property. This loan, made possible by funds loaned to the Fund by the National Trust for Historic Preservation, was in turn partially secured by the Fund's assignment of its second mortgage to the National Trust. The Fund also loaned the partnership an additional $15,000 from its own resources for renovation work undertaken prior to the receipt of revenues from the partnership's planned syndication.

The Providence Office of Community Development and the West Broadway Incentive Corporation provided a grant of approximately $8,200 to be applied to the exterior renovation of the building. The remainder of the financing came from an offering of 10 limited partnership shares of $2,700 each. Under the agreement, the limited partners are to share 98 percent of the profits, losses, and tax credits from normal operation of the partnership, with the remaining 2 percent to go the general partner. The partners shared the 25-percent investment tax credit of $14,168 and will also share losses, expected to total $11,629 during 1983–87. In addition, they shared a charitable deduction based on the appraised value of a historic easement granted to the Fund. Under the terms of this easement, without the authorization of the Fund no construction or modification may be undertaken that would affect the appearance or structural integrity of the building's exterior, alter the composition of the exterior surfaces, or increase the building's height. The easement agreement specifies that the building's exterior is to be maintained in its restored condition in perpetuity and that the building shall contain no more than four dwelling units. The amount of the deduction available as a result of the donation of this easement to a charitable organization was expected to be about 12–15 percent of the unrestricted fair market value of the renovated property.

For its services and expenses in organizing the partnership and preparing the offering memo, the Fund received a fee of $5,000 upon admission of the limited partners. The general partner, for his work in managing the construction and rehabilitation work, as well as actually doing much of the construction, received $225 per week during the year-long construction period. In addition, the general partner receives a management fee equal to the rent for the apartment he occupies within the building, plus 2 percent of the net cash flow from the apartment rentals.

Upon the sale of the property, and after the limited partners recover their capital contributions, the general partner will be reimbursed for his capital contribution and for the loans he made to the partnership to help finance the rehabilitation. The Fund will then receive 10 percent of the balance for its services as the representative of the limited partners. Of the remainder, 55 percent will go to the general partner and 45 percent will be distributed among the limited partners.

In addition, the general partner received a 2-year option, to begin 5 years after the completion of rehabilitation, to purchase the building for its fair market value, so long as that price is sufficient to satisfy the debts of the partnership and to repay the limited partners' capital contributions. The general partner also received a right of first refusal should the partnership choose to sell the property at any time between 5 and 15 years after the completion of rehabilitation.

8.2.2 Low-income Housing (Kensington Square, New Haven, Connecticut)

Large-scale low-income housing projects generally involve governmental assistance at the federal, state, and local levels, as well as support from nonprofit organizations. These projects, which have been notoriously complex, have become even more difficult over the past few years since the forms of government assistance available for low-income housing projects have been in a state of flux. Some of the types of financial assistance used for the Kensington Square project described later in this chapter (most notably Section 8 rent guarantees) are no longer available. Nevertheless, since this project was undertaken new programs, such as the Housing Development Action Grant (HODAG) program described in chapter 4, have been created. Developers should be aware of all of the funding sources available at any given time and combine them as creatively as possible.

Dwight-Edgewood is a neighborhood of single-family and small multifamily buildings, many of which are 50 to 100 years old.[4] It is situated
two blocks from the western edge of the Yale University campus in New
Haven, Connecticut. These residences range in condition from well
maintained to newly rehabilitated to badly deteriorated; several commercial properties are intermingled among them. A major retail district
is located a few blocks away.

A local nonprofit group, Neighborhood Housing, Inc. (NHI), was
designated by the city of New Haven to redevelop the neighborhood.
NHI was organized in 1971 by a group of area residents who gathered
together local architects, bankers, lawyers, and builders to discuss the
problems of substandard housing in the neighborhood and to create
home ownership opportunities for low and moderate-income families.
The city chose NHI for this project (which was to include the rehabilitation and rental to low-income tenants of 225 housing units in more
than 40 scattered sites under the Neighborhood Strategies Area program
administered by HUD) because the group had previously developed
some 86 dwelling units under a "buy, rehabilitate and sell" program, it
was an established grass roots organization with broad community support, and it had a highly respected board of directors.

In order to avoid jeopardizing the tax-exempt status of NHI, a new
organization called Neighborhood Housing Developers, Inc. (NHDI)
was formed to serve as the cogeneral partner of NHP Kensington Square
Associates, the limited partnership formed for this project. The other
general partner was the Washington, D.C.-based National Housing Partnership (NHP), which in turn has as its general partner the National
Corporation for Housing Partnerships (NCHP). Both the NCHP and the
NHP are private, profit-making organizations created by Congress in
1968. The purpose of each is to encourage the widest possible participation by private enterprise in the provision of housing for low and
moderate-income families. The NHP and the NCHP were initially capitalized in 1970 by means of a public securities offering that raised over
$42 million. The NHP thus brought to the project the financial strength
to enable it to offer, among other guarantees, to repurchase each investor's interest if rehabilitation were not completed or if the project did
not qualify for the anticipated government assistance by the date of
final closing.

Because its aim was the creation of low-income housing, the project
received government assistance at the federal, state, and local levels.
This assistance included a Section 8 rental payment guarantee for 20
years, as well as tax-exempt financing through the state public housing

---

[4]The information in both this section and section 3 is taken, with the permission of
Preservation Action, from *Another Revolution in New England: A Case Study of the Historic
Rehabilitation Tax Incentives*, Deborah Dunning and Nellie Longsworth, (Boston: Pres-
ervation Action and Boston University Preservation Studies Program).

agency and mortgage loan insurance from the Federal Housing Administration (FHA). In addition, the city of New Haven granted a 15-year freeze on property tax assessments and helped to secure the federal commitment to provide assistance under Section 8. Despite this backing, and the availability at that time of 60-month depreciation for low-income housing under Section 167(k) of the Internal Revenue Code, plans for the project were complicated by higher-than-anticipated cost projections until the 25-percent investment tax credit provided the possibility of substantial additional tax benefits, the precise amount of which depends upon the number of buildings certified as contributing to the historic district in which they are located.

The 21 buildings (120 units overall) included in the project consist of 2½ and 3-story structures, some of which are wood frame buildings with wood siding, others of which have masonry bearing walls with brick or masonry exterior finishes. Some were constructed during New Haven's carriage building boom of the early and mid-nineteenth century; others date from the city's era of industrial prosperity during the late nineteenth and early twentieth century. They are all located within what, in September 1983, became the Dwight Street National Register Historic District.

The bulk of the project financing came from tax-exempt construction loan notes and from mortgage revenue bonds issued by the Kensington Square Housing Development Corporation, created by the Connecticut State Housing Authority. All proceeds from the sale of the tax-exempt securities were held by the Connecticut Bank and Trust Company (CBT) as trustee. A mortgage loan from the development corporation to the partnership is evidenced by a note in the amount of $5,738,900 from the partnership to CBT. The note is secured by a mortgage on the properties that constitute the project, a security interest in all personal property to be used in connection with the properties and an assignment to CBT, effective in the event of a default, of all tenant leases and all subsidy payments to be received by the partnership under the Section 8 program. The NHP was also required by HUD to post both a working capital letter of credit (to secure performance of the partnership's obligations during the construction period and unanticipated costs of equipping the properties, rent-up expenses, taxes, mortgage insurance premiums, and hazard insurance premiums) and a debt service letter of credit as additional security for the bonds issued by the corporation.

The partnership's agreement with the general contractor provided for liquidated damages for each day of delay beyond a specified date for the completion of construction. Under the partnership's agreement with HUD, as well as the Section 8 program, the partnership was required to deposit monthly payments in a reserve fund for replacements, beginning upon initial occupancy of the properties. Whenever rents are increased in future years by a HUD-established annual automatic adjustment factor, the amount of the monthly deposit is to be increased by the same factor.

Since the project's estimated costs amounted to more than $7 million, it was necessary to supplement the $5.7 million tax-exempt bond issue through the sale of limited partnership interests. Thirty such partnership interests, valued at $74,400 each, were sold to raise a total of $2,232,000. Together with the tax-exempt bond proceeds, this covered the project's anticipated $7.9 million cost.

To assist the partnership offering the NHP offered to repurchase each investor's interest in the partnership if the rehabilitation of the properties was not completed by the date scheduled for final closing, or if the properties failed to qualify for rent subsidy or other contemplated government assistance. The NHP also agreed that, if the partnership were unable to cover costs of construction, the NHP or the NCHP would advance or lend the amounts necessary to complete rehabilitation within the required time (or the NHP would repurchase the investors' interests at cost). The purchase price for each investment unit included a $3,720 fee for providing this construction completion/repurchase guarantee. The NHP provided a similar guarantee, for the first several years of operation, with respect to operating expenditures and debt service payments, at a cost of $5,207 per unit.

Since the purchase price for each investment unit was based upon the assumption that 17 of the buildings involved in this project would be certified by the Department of the Interior and hence eligible for the rehabilitation tax credit, the NHP offered to reduce by $84 the cost of each investment unit for every dwelling unit that did not qualify for the tax credit. The NHP could also, in the event some buildings were not certified, elect to repurchase all units at the investors' cost. (In fact, 19 buildings were ultimately submitted for certification.)

Each partner was to be allocated partnership losses incurred after his or her acquisition of an interest in the partnership. The 25-percent investment tax credit was also allocated among the partners. The tax losses of the partnership are expected to decline over time and, unless the properties are sold or otherwise disposed of, after 16 years the partnership is expected to generate income in excess of capital returned to the investors.

It is encouraging to note that plans are under way for a second Kensington Square project.

8.2.3 Commercial Building (Hildreth Building, Lowell, Massachusetts)

The key role that the investment tax credits can play in conjunction with industrial development bonds is indicated in the following example, which focuses on the rehabilitation of an office building in Lowell, Massachusetts, for commercial use. New Internal Revenue Code limits on the amount of industrial development bonds that states are authorized to issue on a tax-free basis may make such transactions more difficult in certain states in the future. However, the Lowell project also emphasizes the importance of a building's facade, not simply as a conspicuous architectural feature, but as a potential financing source through either direct restoration grants or tax deductions for the donation of facade easement to a qualifying organization.

Lowell, Massachusetts, is located thirty-three miles northwest of Boston, on the banks of the Merrimack River. It is one of the many New England riverside mill towns that enjoyed their commercial heyday before this century began. The Hildreth Building, a product of that era of growth and vitality, is one of the largest buildings in Lowell's old, and in parts deteriorated, commercial and retail center. It lies within the Merrimack-Middle Street National Register Historic District. Because the building, which has suffered from some deferred maintenance, was found to contribute to the historic significance of the district, federal tax incentives were available for a rehabilitation that conformed to the Secretary of the Interior's standards.

In September 1980, Ted Trivers, a Lowell native and area businessman, was looking for space to relocate a family retail clothing store when he became interested in the Hildreth Building, in which both F. W. Woolworth and S. S. Kresge stores were located. Trivers was impressed with the building's architecture, low vacancy rate, central location, and the potential income from its four floors of office space and the retail space on the ground floor. The city was also eager to see the building rehabilitated, if possible by a local resident, and both the city manager, B. Joseph Tully, and then-Massachusetts senator Paul Tsongas threw their weight behind the project. The Lowell Historic Preservation Commission also let its support, and in March 1981, the building's owner agreed to sell the property.

Financing the rehabilitation proved problematic since interest rates were high. A financial consultant, New England Communities, Inc. (NEC), was brought in to oversee the project and structure its financing. Trivers, NEC and NEC president Mark Slotnick became general partners of Hildreth Associates, a limited partnership created to undertake the project.

The initial rehabilitation work involved only the facade, lobby, and common areas, allowing many existing tenants to remain and providing a needed income stream throughout construction. Although a Boston syndicator was approached, initial efforts to syndicate the project to raise the necessary equity proved unsuccessful. And, notwithstanding the support of the city and the Lowell Historic Preservation Commission, local financial institutions showed limited interest in providing industrial revenue bond financing. In the midst of these difficulties, however, Congress passed the Economic Recovery Tax Act (ERTA) of

1981. The 25 percent investment tax credit for certified historic rehabilitations provided by ERTA suddenly made the project salable to limited partner investors and, as a result, financially feasible.

In an effort to expedite the project, a request for the Secretary of the Interior to certify that the Hildreth Building contributed to the character of the historic district was submitted immediately upon closing. In preparing this application, the partners and their architect worked closely with the Lowell Historic Preservation Commission, the technical staff of the National Park Service, and the review director at the State Historical Commission, all to ensure that the plans submitted could be approved as consistent with the historic character of the structure and the district.

The bulk of the project's financing was provided by a consortium of nine local banks, which purchased $1,250,000 worth of industrial revenue bonds issued by the Lowell Industrial Development Finance Agency. The proceeds of that sale were loaned to the partnership at an interest rate of 11.5 percent for the first four years. A variable rate of 75 percent of a local bank's prime rate, with a floor of 11.5 percent and a ceiling of 17 percent, applies to the remaining 21 years of the loan. The partners, with the help of the Lowell Historic Preservation Commission, also secured a $75,000, 25-year loan from the Lowell Development Financing Corporation at an interest rate of 6.5 percent for the first four years, adjusted semiannually to 40 percent of a Boston bank's prime rate, with a floor of 5 percent. Grants totaling $28,000 for the rehabilitation of the building's facade were obtained from the preservation commission and from the Lowell Planning and Development Department (under the terms of the commission's grant, the partnership must maintain the historic character of the facade for at least 15 years). The last major piece of the financing came from the project's 15 limited partners, each of whom paid $50,000 per unit (for a total of $750,000). As a result of these efforts, a total financing package amounting to some $2.1 million was successfully assembled.

With financing in place, actual rehabilitation could begin. A photograph of the Hildreth Building taken in 1884 revealed that the upper stories of the structure had changed little over the past century. Routine repointing and stabilizing of some decorative elements were necessary. The retail ground level and main entry, however, had been completely changed during 1930 and 1963 renovations. Most of the nineteenth-century columns, piers, and ornaments on the ground floor had been removed, and very little of the original lobby survived. Many of the offices in the building were partially or totally renovated in the 1950s and 1960s.

The missing ground floor elements were replaced and painted wooden storefronts replaced, including signs and awnings. Several original features were uncovered when paneling, floor tiles, and the hung ceiling were removed from the lobby. The uncovered plaster ceiling was repaired, the wall replastered, and the floor and walls retiled. Columns were cast in place, with original capitals recreated in metal.

In the upper story offices, many original features were also found hidden behind newer wall panels. The original wooden doors were reused and wood trim preserved. For the facade of the upper stories, a matching mortar formula was developed for spot pointing. A bay window on the second floor was removed, and the original arched windows it had replaced were restored. Rotted window sashes were replaced with wood rather than with more modern materials.

The cost of this work amounted to some $883,440 and yielded an investment tax credit of $220,860. It is expected that income from the building will provide a rate of return on equity of over 23 percent, including tax benefits.[5]

Complying with the Secretary of the Interior's Standards for Rehabilitation added somewhat to the construction costs of the project. Whether the incremental costs of such compliance exceeded the marginal 5-percent difference between the 25 percent certified rehabilitation credit and the 20 percent tax credit for a noncertified rehabilitation is difficult to determine. The owners did, of course, consider the latter option, but decided that there was little likelihood that the Interior Department would decertify the building, a prerequisite where, as here, the structure is situated in a historic district. In addition, they believed their own rehabilitation standards and the requirements of the preservation commission grant would meet or exceed the mandated standards, and that a first-class renovation would give them a significant marketing advantage.

The techniques used to rehabilitate the Hildreth Building have proven to be broadly adaptable to other projects in Lowell, and dozens of older buildings have been rehabilitated using similar techniques. These include both downtown commercial buildings and a number of the massive old mills that once fueled Lowell's economy along the city's canals. These rehabilitated mills have been revived as homes for high tech industry—Lowell's recently acquired source of economic strength and a major new source of employment for its residents.

Lowell's historic rehabilitation has been accompanied by a boom in new construction. In 1983 new building permits were issued for projects valued at almost $70 million, an increase of 169 percent over 1982. Lowell's young people, who once left the city in search of jobs, now tend to remain in an area that is presently considering plans to bus in

---

[5]This rate of return would be slightly different were this project begun now, after the passage of the Tax Equity and Financial Responsibililty Act of 1982 (TEFRA). This difference seems sufficiently small, however, that it would not mandate any different structure for the transaction. The rate of return and the tax benefits, other than the 25 percent investment tax credit, would be slightly smaller under TEFRA. The gain on the sale of the building, however, would be slightly greater.

workers to remedy a labor shortage. In summary, Lowell has experienced a dramatic and successful turnaround, and one in which the rehabilitation of its remarkable architectural and industrial heritage played a major role.

8.2.4 Comprehensive Public Development Project (Forty-second Street Development Project, New York, New York)

At one time comprehensive public development projects were generally urban renewal programs that entailed the wholesale clearance of inner-city areas. In the past few years, however, there has been increased sensitivity to the importance of retaining structures of historic or architectural merit and integrating them into projects designed to revitalize their surrounding areas. Given the scope and complexity of such projects, they often require active participation by several governmental bodies, as well as major commitments of private capital.

The presence of government actors in the rehabilitation process brings with it the possibility of utilizing powers, such as eminent domain, property tax abatements, and zoning or building code exemptions, which are not available to private entities. At the same time, this requires that the environmental and historic review processes that are triggered by government activities be satisfied. The Forty-second Street Development Project, jointly undertaken by the State and City of New York, illustrates the complexities of area-wide redevelopment which includes the preservation and restoration of historic structures in a major downtown area.

The Forty-second Street Development Project is a collaborative undertaking of the City of New York and the New York State Urban Development Corporation (UDC) that is intended to eliminate the persistent blight which has characterized West Forty-second Street near Times Square, and to return the area to productive use.[6] The project proposes to rehabilitate and redevelop West Forty-second Street between Broadway and Eighth Avenue with a mix of major office buildings, a hotel, a wholesale mart, restaurants and other retail uses, major subway improvements and, as a centerpiece for the development, ten renovated theaters. Most of these theaters are currently used for "action" (violence), pornographic, and occasional first-run movies, as well as for cut-rate retail and fast food outlets. The theaters occupy nine separate buildings, constructed between 1899 and 1918, that once formed the heart of New York's theater district. Despite their present use and the "exotic seediness" of their surroundings, the theaters have been remarkably

---

[6] For a comprehensive description and analysis of the Forty-second Street Development Project, *see* the Final Environmental Impact Statement issued by the UDC in August 1984, from which much of the information contained in this section is drawn.

well preserved because, apart from movies, no alternative use has appeared economically viable in the context of Forty-second Street's current condition.

The theaters included within the project are as follows:

New Amsterdam Theater.   Built in 1902 and listed on the National Register of Historic Places (as well as a New York City landmark), this is one of the few public structures in the United States designed entirely in the Art Nouveau style. The New Amsterdam is most famous as the home of the Ziegfeld Follies; the careers of Fannie Brice, Ed Wynn, Will Rogers, Eddie Cantor and Marilyn Miller were also launched there. In addition to the main theater, the New Amsterdam includes a smaller rooftop theater once used for private shows by Florenz Ziegfield.

Harris Theater.   Built along with the Candler Building, an office structure, this theater is well-known as the site of John Barrymore's *Hamlet*, which played there in 1922.

Liberty Theater.   Designed at the same time as the New Amsterdam and by the same firm, this theater's facade makes a strong contribution to the Forty-second Street streetscape; its interior still contains a grand proscenium, carved arches, boxes, and an unusual printed asbestos curtain. George M. Cohan and Fred Astaire both played there.

Empire Theater.   Built in 1912, the 80-foot facade of this theater is one of the most distinctive elements on Forty-second Street. It housed the Eltinge Follies, a burlesque, from its inception until 1930.

Victory.   Designed in 1899 for Oscar Hammerstein, the Victory is the oldest of Forty-second Street's theaters, though its splendid interior is extremely well-preserved. Its most famous play was *Abie's Irish Rose*, whose record of 2,327 performances was only recently broken.

Lyric Theater.   Built shortly after 1900, this theater specialized in musical comedy until 1925 (when the Marx Brothers appeared in *Coconuts*). Its 119-foot Beaux Arts Facade on Forty-third Street, of red and white brick and terra cotta, is largely intact.

Times Square and Apollo Theaters.   The 100-foot common facade of these two theaters, with its 3-story colonnade, is a major architectural feature on Forty-second Street. The Times Square was first constructed in 1910 for movies and vaudeville and was later reconstructed and joined to the new Apollo Theater.

The Selwyn Theater.   In its 15 years as a legitimate theater, the Selwyn housed plays by Rudolph Friml, Cole Porter, Noel Coward, and George S. Kaufman. Its Beaux Arts facade is largely intact.

In addition to the Forty-second Street theaters, the project area includes the 24-story Candler Building, an office structure constructed in 1912 that is rich in decorative detail and clad in white terra cotta with a stepped-back roof tower. Like the New Amsterdam Theater, the Candler Building is listed on the National Register. The area also contains the former Times Tower (now known as One Times Square), which is widely known for its New Year's Eve celebrations and for its electric news sign (currently unused) which brought New Yorkers news of both V-E Day and V-J Day in 1945. The original facade of the Times Tower was completely stripped in 1965, and as a consequence the building lost much of its architectural character. However, the site is still regarded as important by many, and continues to serve as a visual focal point for the southern end of Times Square.

Committed to restoring the Forty-second Street theaters to legitimate theater, theater-related, and upgraded retail use, the city joined forces with the UDC in 1980 to develop a comprehensive redevelopment plan which relied almost entirely on private investment. Not only were the individual theaters to be preserved and renovated, but the low scale of Forty-second Street's mid-blocks was to be maintained and the area revitalized as a entertainment area attractive to a broad cross-section of New Yorkers and tourists. In addition, the city was eager to overcome the blight and crime that have plagued West Forty-second Street for many years and deterred many visitors (and New Yorkers) from patronizing the area. Finally, the city sought to carry out this development in a way which would contribute to the city's long-term tax revenues, help finance needed subway improvements, and be a positive influence on the neighboring garment center, the Clinton residential area, and the Broadway theater district.

With these objectives in mind, the city and the UDC undertook an economic feasibility study of the overall development plan (of which the project area theaters were to be a part) and commenced a detailed analysis of the architectural and historic significance of each of the theaters. In addition, the project team carried out a careful assessment of land use, social and street conditions, traffic, mass transit, pedestrian, air quality, waste disposal, and other characteristics of the project area. As part of these studies, the project team, with assistance from outside consultants, also reviewed the eligibility of each of the theaters for inclusion on both the National and New York State Registers of Historic Places, as well as their eligibility for local landmark designation. From these studies grew a series of specific guidelines for both the use and exterior and interior treatment of each theater.

After completing this review, the city and the UDC concluded that it would be possible to carry out the project (including acquisition and renovation of the theaters) by "transferring the theaters" unused development rights to the intersections of Forty-second Street with Broadway (on the east) and Eighth Avenue (on the west). These corner sites were judged to be appropriate for large-scale commercial development, which could in turn generate sufficient revenues to pay for both the theaters' acquisition and renovation and significant subway improvements. After extensive discussion of this concept with local officials and civic and business organizations, the city and the UDC requested developer proposals for four separate office towers to be located at the Forty-second Street/Broadway intersection and a hotel and wholesale mart to be located at the Forty-second Street/Eighth Avenue intersection. At the same time, the city and the UDC sought developers to renovate and operate the theaters (with initial financial assistance from the commercial developers, who would also be required to pay for the renovation of the Times Square subway station). Under the plan, the Candler Building was to be left in private ownership as part of the renovated mid-block and the former Times Tower substantially modified to enhance its visual contribution to the redeveloped southern end of Times Square.

Since the project's economic feasibility studies had shown that it would be necessary, in order to preserve both the theaters and the low scale of the Forty-second Street midblocks, to "transfer" unused development potential from the midblocks to the corner development sites, an appropriate technique had to be identified to effect that transfer. As noted previously in chapter 5, New York City's zoning and landmark regulations authorize only limited transfers of unused development rights within the city's Special Theater Subdistrict. Moreover, to make the project economically feasible, it would be necessary for the sponsoring agency to proceed by eminent domain in completing assembly of the complex project site and, once site assembly had been completed, to offer phased property tax abatements to the developers designated to carry out individual components of the project. For these reasons, among others, the city requested the UDC to join in carrying out the project. Under the New York State Urban Development Corporation Act (UDC Act), the UDC has the power to acquire all necessary sites, to hold the acquired sites free of local property taxes and to construct the project in accordance with its project plan without being bound by local zoning requirements.

Use of the UDC to implement the project meant that the UDC would have "lead agency" status under New York's State Environmental Quality Review Act (SEQRA) and State Historic Preservation Act (SHPA), both of which are discussed in chapter 6. It also meant that the project would be subject to review and approval under the UDC Act and under the State's Eminent Domain Procedure Law (EDPL). On the other hand, the project would not be subject to the city's Uniform Land Use Review Procedure, which governs approval of privately sponsored development projects within the city.

Just as NEPA and the NHPA (both discussed in chapter 6) require comprehensive environmental and historic reviews for proposed federal actions, SEQRA and the SHPA require a comparable review of the environmental and historic impacts of actions by state agencies, including the UDC, likely to have significant impacts on their surroundings. In addition, SEQRA, the UDC Act and the EDPL each required public hearings, followed by specific statutory findings by the UDC with respect to the project's purpose and impacts. In August 1981 the UDC held its first public hearings on the project, after which the agency commenced (with the cooperation of the city) an exhaustive environmental review process that drew upon not only the UDC's and the city's own staffs, but independent experts in the relevant disciplines, including historic preservation, as well as special environmental and preservation counsel. This process included extensive consultation with the state commissioner of parks, recreation, and historic preservation under the SHPA and the city's Landmarks Preservation Commission under SEQRA.

All this culminated in the issuance, in February 1984, of a draft environmental impact statement for the entire project, after which extensive public hearings were again held on the project's likely impacts and public benefits. A final environmental impact statement, containing more than 1000 pages, was then issued by the UDC in August 1984. Following yet another round of public comments, the UDC formally approved the project in October 1984, and the city's Board of Estimate took similar action in November 1984.

The City and the UDC have now completed conditional designation of developers for each of the project's major components, including the mid-block theaters, and lease negotiations are currently under way with such developers. Although a number of court actions challenging the project have been commenced by project area property owners, the UDC and the city expect to commence site acquisition in 1985, after which rehabilitation of the theaters along the lines originally proposed by the city is expected to begin in 1986. Discussions are continuing concerning the ultimate treatment of the former Times Tower, which may either be modified, as originally proposed, or replaced by an entirely new structure designed to provide more light and vitality at the Times Square crossroads.

As finally approved, the project plan contemplates that the Lyric, Apollo, Selwyn, Harris, and New Amsterdam theaters will be used for "legitimate" musical or dramatic productions, the Victory and, possibly, the Liberty theaters will be used as nonprofit "institutional theaters" and the Times Square and Empire (and, possibly, the Liberty) will be used for upgraded restaurant, retail, or mart-related uses. In accordance with the project's original objective, the cost of acquiring and renovating these theaters will be borne largely by developers of the project's commercial sites. Experienced theater operators have been designated with the aim of producing quality shows intended to help restore Forty-second Street to the place it once held as the heart of New York's theater district. If this goal is realized, it will be in significant part because the city and the UDC succeeded in structuring their efforts in a way which combined public condemnation powers, property tax abatements, and exemption from local zoning controls with private funding, comprehensive historic, environmental and economic impact analyses, and a concerted effort to integrate the project's historic resources into an overall redevelopment plan for Forty-second Street and the Times Square area.

Appendix A

National Historic Preservation Act

16 U.S.C. § § 470–470w-6

§ **470.** [NHPA § 1] Short title; Congressional finding and declaration of policy
§ **470-1.** [NHPA § 2] Declaration of policy of the Federal Government
§ **470a.** [NHPA § 101] Historic preservation program

- (a) National Register of Historic Places; designation of properties as historic landmarks; properties deemed included; criteria; nomination of properties by states, local governments or individuals; regulations
- (b) Regulations for State Historic Preservation Programs; periodic evaluations and fiscal audits of state programs; administration of state programs; contracts and cooperative agreements with nonprofit or educational institutions; treatment of state programs as approved programs
- (c) Certification of local governments by State Historic Preservation Officer; transfer of portion of grants; certification by Interior Secretary; nomination of properties by local governments for inclusion in National Register
- (d) Matching grants-in-aid to states for programs and projects; matching grants-in-aid to National Trust for Historic Preservation in the United States; program of direct grants for preservation of properties included in National Register; grants or loans to Indian tribes and ethnic or minority groups for preservation of cultural heritage
- (e) Prohibition of use of funds for compensation of intervenors in preservation program
- (f) Guidelines for federal agency responsibility for agency-owned or controlled historic properties
- (g) Professional standards for preservation of federally owned or controlled historic properties
- (h) Dissemination of information concerning professional methods and techniques for preservation of historic properties

§ **470a-1.** World Heritage Convention

- (a) United States participation

215

§ 470w-3. [NHPA § 304]   Information relating to location or character of
historic resources, disclosure to public

§ 470w-4. [NHPA § 305]   Attorneys' fees and costs to prevailing parties in civil
actions

§ 470w-5. [NHPA § 306]   National Museum for the Building Arts

    (a) Cooperative agreement between Interior Secretary, administrator of
General Services Administration and Committee for National Museum
of the Building Arts; purposes

    (b) Provisions of cooperative agreement

    (c) Matching grants-in-aid to committee; limitation on amounts

    (d) Renovation of site

    (e) Annual committee report to secretary and administrator

    (f) Definition of "building arts"

§ 470w-6. [NHPA § 307]   Effective date of regulations

    (a) Copy to Congress prior to publication in Federal Register; effective date
of final regulations

    (b) Effective date of final regulation in case of emergency

    (c) Disapproval of regulation by resolution of Congress

    (d) Failure of Congress to adopt resolution of disapproval of regulation

    (e) Sessions of Congress

    (f) Congressional inaction or rejection of resolution of disapproval not
deemed approval of regulation

§ 470. Short title; Congressional finding and declaration of policy

(a)   Sections 470 to 470a, 470b, and 470c to 470w-6 of this title may be cited
as the "National Historic Preservation Act".

(b)   The Congress finds and declares that—

(1)   the spirit and direction of the Nation are founded upon and reflected
in its historic heritage;

(2)   the historical and cultural foundations of the Nation should be
preserved as a living part of our community life and development in order to give
a sense of orientation to the American people;

(3)   historic properties significant to the Nation's heritage are being lost
or substantially altered, often inadvertently, with increasing frequency;

(4)   the preservation of this irreplaceable heritage is in the public interest
so that its vital legacy of cultural, educational, aesthetic, inspirational, economic,
and energy benefits will be maintained and enriched for future generations of
Americans;

(5)   in the face of ever-increasing extensions of urban centers, highways,
and residential, commercial, and industrial developments, the present governmental
and nongovernmental historic preservation programs and activities are inadequate
to insure future generations a genuine opportunity to appreciate and enjoy the
rich heritage of our Nation;

(6)   the increased knowledge of our historic resources, the establishment
of better means of identifying and administering them, and the encouragement of
their preservation will improve the planning and execution of Federal and federally
assisted projects and will assist economic growth and development; and

(7)   although the major burdens of historic preservation have been borne and major efforts initiated by private agencies and individuals, and both should continue to play a vital role, it is nevertheless necessary and appropriate for the Federal Government to accelerate its historic preservation programs and activities, to give maximum encouragement to agencies and individuals undertaking preservation by private means, and to assist State and local governments and the National Trust for Historic Preservation in the United States to expand and accelerate their historic preservation programs and activities.

§ 470-1. Declaration of policy of the Federal Government

It shall be the policy of the Federal Government, in cooperation with other nations and in partnership with the States, local governments, Indian tribes, and private organizations and individuals to—

(1)   use measures, including financial and technical assistance, to foster conditions under which our modern society and our prehistoric and historic resources can exist in productive harmony and fulfill the social, economic, and other requirements of present and future generations;

(2)   provide leadership in the preservation of the prehistoric and historic resources of the United States and of the international community of nations;

(3)   administer federally owned, administered, or controlled prehistoric and historic resources in a spirit of stewardship for the inspiration and benefit of present and future generations;

(4)   contribute to the preservation of nonfederally owned prehistoric and historic resources and give maximum encouragement to organizations and individuals undertaking preservation by private means;

(5)   encourage the public and private preservation and utilization of all usable elements of the Nation's historic built environment; and

(6)   assist State and local governments and the National Trust for Historic Preservation in the United States to expand and accelerate their historic preservation programs and activities.

§ 470a. Historic preservation program—National Register of Historic Places; designation of properties as historic landmarks; properties deemed included; criteria; nomination of properties by States, local governments or individuals; regulations

(a)   (1)   (A) The Secretary of the Interior is authorized to expand and maintain a National Register of Historic Places composed of districts, sites, buildings, structures, and objects significant in American history, architecture, archeology, engineering, and culture.

(B)   Properties meeting the criteria for National Historic Landmarks established pursuant to paragraph (2) shall be designated as "National Historic Landmarks" and included on the National Register, subject to the requirements of paragraph (6). All historic properties included on the National Register on December 12, 1980 shall be deemed to be included on the National Register as of their initial listing for purposes of sections 470 to 470a, 470b to 470w-6 of this title. All historic properties listed in the Federal Register of February 6, 1979, as "National Historic Landmarks" or thereafter prior to the effective date of this Act are declared by Congress to be National Historic Landmarks of national historic significance as of their initial listing as such in the Federal Register for purposes of sections 470 to 470a, 470b, and 470c to 470w-6 of this title and the Act of August 21, 1935; except that in cases of National Historic Landmark districts for which no boundaries have been established, boundaries must first be published in the Federal Register and submitted to the Committee on Energy and National Resources of the United States Senate and to the Committee on Interior and Insular Affairs of the United States House of Representatives.

(2)   The Secretary in consultation with national historical and archaeological associations, shall establish or revise criteria for properties to be included on the National Register and criteria for National Historic Landmarks, and shall also promulgate or revise regulations as may be necessary for—

(A)   nominating properties for inclusion in, and removal from, the National Register and the recommendation of properties by certified local governments;

(B)   designating properties as National Historic Landmarks and removing such designation;

(C)   considering appeals from such recommendations, nominations, removals, and designations (or any failure or refusal by a nominating authority to nominate or designate);

(D)   nominating historic properties for inclusion in the World Heritage List in accordance with the terms of the Convention concerning the Protection of the World Cultural and Natural Heritage;

(E)   making determinations of eligibility of properties for inclusion on the National Register; and

(F)   notifying the owner of a property, any appropriate local governments, and the general public, when the property is being considered for inclusion on the National Register, for designation as a National Historic Landmark or for nomination to the World Heritage List.

(3)   Subject to the requirements of paragraph (6), any State which is carrying out a program approved under subsection (b) of this section, shall nominate to the Secretary properties which meet the criteria promulgated under subsection (a) of this section for inclusion on the National Register. Subject to paragraph (6), any property nominated under this paragraph or under section 470h-2 (a) (2) of this title shall be included on the National Register on the date forty-five days after receipt by the Secretary of the nomination and the necessary documentation, unless the Secretary disapproves such nomination within such forty-five day period or unless an appeal is filed under paragraph (5).

(4)   Subject to the requirements of paragraph (6) the Secretary may accept a nomination directly from any person or local government for inclusion of a property on the National Register only if such property is located in a State where there is no program approved under subsection (b) of this section. The Secretary may include on the National Register any property for which such a

nomination is made if he determines that such property is eligible in accordance with the regulations promulgated under paragraph (2). Such determination shall be made within ninety days from the date of the nomination unless the nomination is appealed under paragraph (5).

(5)   Any person or local government may appeal to the Secretary a nomination of any historic property for inclusion on the National Register and may appeal to the Secretary the failure or refusal of a nominating authority to nominate a property in accordance with this subsection.

(6)   The Secretary shall promulgate regulations requiring that before any property or district may be included on the National Register or designated as a National Historic Landmark, the owner or owners of such property, or a majority of the owners of the properties within the district in the case of an historic district, shall be given the opportunity (including a reasonable period of time) to concur in, or object to, the nomination of the property or district for such inclusion or designation. If the owner or owners of any privately owned property, or a majority of the owners of such properties within the district in the case of an historic district, object to such inclusion or designation, such property shall not be included on the National Register or designated as a National Historic Landmark until such objection is withdrawn. The Secretary shall review the nomination of the property or district where any such objection has been made and shall determine whether or not the property or district is eligible for such inclusion or designation, and if the Secretary determines that such property or district is eligible for such inclusion or designation, he shall inform the Advisory Council on Historic Preservation, the appropriate State Historic Preservation Officer, the appropriate chief elected local official and the owner or owners of such property, of his determination. The regulations under this paragraph shall include provisions to carry out the purposes of this paragraph in the case of multiple ownership of a single property.

(7)   The Secretary shall promulgate, or revise, regulations—

(A)   ensuring that significant prehistoric and historic artifacts, and associated records, subject to section 470h-2 of this title, the Act of June 27, 1960, and the Archaeological Resources Protection Act of 1979 are deposited in an institution with adequate long-term curatorial capabilities;

(B)   establishing a uniform process and standards for documenting historic properties by public agencies and private parties for purposes of incorporation into, or complementing, the national historical architectural and engineering records within the Library of Congress; and

(C)   certifying local governments, in accordance with subsection (c) (1) of this section and for the allocation of funds pursuant to section 470c (c) of this title.

Regulations for State Historic Preservation Programs; periodic evaluations and fiscal audits of State programs; administration of State programs; contracts and cooperative agreements with nonprofit or educational institutions; treatment of State programs as approved programs

(b)   (1)   The Secretary, in consultation with the National Conference of State Historic Preservation Officers and the National Trust for Historic Preservation, shall promulgate or revise regulations for State Historic Preservation Programs. Such regulations shall provide that a State program submitted to the Secretary under this section shall be approved by the Secretary if he determines that the program—

(A)   provides for the designation and appointment by the Governor of a "State Historic Preservation Officer" to administer such program in accordance with paragraph (3) and for the employment or appointment by such officer of such professional qualified staff as may be necessary for such purposes;

(B)   provides for an adequate and qualified State historic preservation review board designated by the State Historic Preservation Officer unless otherwise provided for by State law; and

(C)   provides for adequate public participation in the State Historic Preservation Program, including the process of recommending properties for nomination to the National Register.

(2)   Periodically, but not less than every four years after the approval of any State program under this subsection, the Secretary shall evaluate such program to make a determination as to whether or not it is in compliance with the requirements of sections 470 to 470a, 470b, and 470c to 470w-6 of this title. If at any time, the Secretary determines that a State program does not comply with such requirements, he shall disapprove such program, and suspend in whole or in part assistance to such State under subsection (d)(1) of this section, unless there are adequate assurances that the program will comply with such requirements within a reasonable period of time. The Secretary may also conduct periodic fiscal audits of State programs approved under this section.

(3)   It shall be the responsibility of the State Historic Preservation Officer to administer the State Historic Preservation Program and to—

(A)   in cooperation with Federal and State agencies, local governments, and private organizations and individuals, direct and conduct a comprehensive statewide survey of historic properties and maintain inventories of such properties;

(B)   identify and nominate eligible properties to the National Register and otherwise administer applications for listing historic properties on the National Register;

(C)   prepare and implement a comprehensive statewide historic preservation plan;

(D)   administer the State program of Federal assistance for historic preservation within the State;

(E)   advise and assist, as appropriate, Federal and State agencies and local governments in carrying out their historic preservation responsibilities;

(F)   cooperate with the Secretary, the Advisory Council on Historic Preservation, and other Federal and State agencies, local governments, and organizations and individuals to ensure that historic properties are taken into consideration at all levels of planning and development;

(G)   provide public information, education, and training and technical assistance relating to the Federal and State Historic Preservation Programs; and

(H)   cooperate with local governments in the development of local historic preservation programs and assist local governments in becoming certified pursuant to subsection (c) of this section.

(4)   Any State may carry out all or any part of its responsibilities under this subsection by contract or cooperative agreement with any qualified nonprofit organization or educational institution.

(5)   Any State historic preservation program in effect under prior authority of law may be treated as an approved program for purposes of this subsection until the earlier of—

(A)   the date on which the Secretary approves a program submitted by the State under this subsection, or

(B)   three years after December 12, 1980.

Certification of local governments by State Historic Preservation Officer; transfer of portion of grants; certification by Secretary; nomination of properties by local governments for inclusion on National Register

(c)   (1)   Any State program approved under this section shall provide a mechanism for the certification by the State Historic Preservation Officer of local governments to carry out the purposes of section 470 to 470a, 470b, and 470c to 470w-6 of this title and provide for the transfer, in accordance with section 470c (c) of this title, of a portion of the grants received by the States under sections 470 to 470a, 470b, and 470c to 470w-6 of this title, to such local governments. Any local government shall be certified to participate under the provisions of this section if the applicable State Historic Preservation Officer, and the Secretary, certifies that the local government—

(A)   enforces appropriate State or local legislation for the designation and protection of historic properties;

(B)   has established an adequate and qualified historic preservation review commission by State or local legislation;

(C)   maintains a system for the survey and inventory of historic properties that furthers the purposes of subsection (b) of this section;

(D)   provides for adequate public participation in the local historic preservation program, including the process of recommending properties for nomination to the National Register; and

(E)   satisfactorily performs the responsibilities delegated to it under sections 470 to 470a, 470b, and 470c to 470w-6 of this title.

Where there is no approved State program, a local government may be certified by the Secretary if he determines that such local government meets the requirements of subparagraphs (A) through (E); and in any such case the Secretary may make grants-in-aid to the local government for purposes of this section.

(2)   (A)   Before a property within the jurisdiction of the certified local government may be considered by the State to be nominated to the Secretary for inclusion on the National Register, the State Historic Preservation Officer shall notify the owner, the applicable chief local elected official, and the local historic preservation commission. The commission, after reasonable opportunity for public comment, shall prepare a report as to whether or not such property, in its opinion, meets the criteria of the National Register. Within sixty days of notice from the

State Historic Preservation Officer, the chief local elected official shall transmit the report of the commission and his recommendation to the State Historic Preservation Officer. Except as provided in subparagraph (B), after receipt of such report and recommendation, or if no such report and recommendation are received within sixty days, the State shall make the nomination pursuant to subsection (a) of this section. The State may expedite such process with the concurrence of the certified local government.

(B)   If both the commission and the chief local elected official recommend that a property not be nominated to the National Register, the State Historic Preservation Officer shall take no further action, unless within thirty days of the receipt of such recommendation by the State Historic Preservation Officer an appeal is filed with the State. If such an appeal is filed, the State shall follow the procedures for making a nomination pursuant to subsection (a) of this section. Any report and recommendations made under this section shall be included with any nomination submitted by the State to the Secretary.

(3)   Any local government certified under this section or which is making efforts to become so certified shall be eligible for funds under the provisions of section 470c (c) of this title, and shall carry out any responsibilities delegated to it in accordance with such terms and conditions as the Secretary deems necessary or advisable.

Matching grants-in-aid to States for programs and projects; matching grant-in-aid to National Trust for Historic Preservation in the United States; program of direct grants for preservation of properties included on National Register; grants or loans to Indian tribes and ethnic or minority groups for preservation of cultural heritage

(d)   (1)   The Secretary shall administer a program of matching grants-in-aid to the States for historic preservation projects, and State historic preservation programs, approved by the Secretary and having as their purpose the identification of historic properties and the preservation of properties included on the National Register.

(2)   The Secretary shall administer a program of matching grant-in-aid to the National Trust for Historic Preservation in the United States, chartered by Act of Congress approved October 26, 1949 for the purposes of carrying out the responsibilities of the National Trust.

(3)   (A)   In addition to the programs under paragraphs (1) and (2), the Secretary shall administer a program of direct grants for the preservation of properties included on the National Register. Funds to support such program annually shall not exceed 10 per centum of the amount appropriated annually for the fund established under section 470h of this title. These grants may be made by the Secretary, in consultation with the appropriate State Historic Preservation Officer—

(i) for the preservation of National Historic Landmarks which are threatened with demolition or impairment and for the preservation of historic properties of World Heritage significance,

(ii) for demonstration projects which will provide information concerning professional methods and techniques having application to historic properties,

(iii) for the training and development of skilled labor in trades and crafts, and in analysis and curation, relating to historic preservation; and

(iv) to assist persons or small businesses within any historic district included in the National Register to remain within the district.

(B)   The Secretary may also, in consultation with the appropriate State Historic Preservation Officer, make grants or loans or both under this section to Indian tribes and to nonprofit organizations representing ethnic or minority groups for the preservation of their cultural heritage.

(C)   Grants may be made under subparagraph (A) (i) and (iv) only to the extent that the project cannot be carried out in as effective a manner through the use of an insured loan under section 470d of this title.

Prohibition of use of funds for compensation of intervenors in preservation program

(e)   No part of any grant made under this section may be used to compensate any person intervening in any proceeding under sections 470 to 470a, 470b, and 470c to 470w-6 of this title.

Guidelines for Federal agency responsibility for agency-owned historic properties

(f)   In consultation with the Advisory Council on Historic Preservation, the Secretary shall promulgate guidelines for Federal agency responsibilities under section 470h-2 of this title.

Professional standards for preservation of Federally owned or controlled historic properties

(g)   Within one year after December 12, 1980, the Secretary shall establish, in consultation with the Secretaries of Agriculture and Defense, the Smithsonian Institution, and the Administrator of the General Services Administration, professional standards for the preservation of historic properties in Federal ownership or control.

Dissemination of information concerning professional methods and techniques for preservation of historic properties

(h)   The Secretary shall develop and make available to Federal agencies, State and local governments, private organizations and individuals, and other nations and international organizations pursuant to the World Heritage Convention, training in, and information concerning, professional methods and techniques for the preservation of historic properties and for the administration of the historic preservation program at the Federal, State, and local level. The Secretary shall also develop mechanisms to provide information concerning historic preservation to the general public including students.

§ 470a-1. World Heritage Convention—United States participation

(a)   The Secretary of the Interior shall direct and coordinate United States participation in the Convention Concerning the Protection of the World Cultural and Natural Heritage, approved by the Senate on October 26, 1973, in cooperation with the Secretary of State, the Smithsonian Institution, and the Advisory Council on Historic Preservation. Whenever possible, expenditures incurred in carrying out activities in cooperation with other nations and international organizations shall be paid for in such excess currency of the country or area where the expense is incurred as may be available to the United States.

Nomination of property to World Heritage Committee

(b)   The Secretary of the Interior shall periodically nominate properties he determines are of international significance to the World Heritage Committee on behalf of the United States. No property may be so nominated unless it has previously been determined to be of national significance. Each such nomination shall include evidence of such legal protections as may be necessary to ensure preservation of the property and its environment (including restrictive covenants, easements, or other forms of protection). Before making any such nomination, the Secretary shall notify the Committee on Interior and Insular Affairs of the United States House of Representatives and the Committee on Energy and Natural Resources of the United States Senate.

Nomination of non-Federal property to World Heritage Committee

(c)   No non-Federal property may be nominated by the Secretary of the Interior to the World Heritage Committee for inclusion on the World Heritage List unless the owner of the property concurs in writing to such nomination.

§ 470a-2. Federal undertakings outside United States; mitigation of adverse effects

Prior to the approval of any Federal undertaking outside the United States which may directly and adversely affect a property which is on the World Heritage List or on the applicable country's equivalent of the National Register, the head of a Federal agency having direct or indirect jurisdiction over such undertaking shall take into account the effect of the undertaking on such property for purposes of avoiding or mitigating any adverse effects.

§ 470b. Requirements for awarding of grant funds

(a)   No grant may be made under sections 470 to 470a, 470b, and 470c to 470w-6 of this title—
(1)   unless application therefor is submitted to the Secretary in accordance with regulations and procedures prescribed by him.

(2)   unless the application is in accordance with the comprehensive state-wide historic preservation plan which has been approved by the Secretary after considering its relationship to the comprehensive statewide outdoor recreation plan prepared pursuant to the Land and Water Conservation Fund Act of 1965 (78 Stat. 897);

(3)   for more than 50 per centum of the aggregate cost of carrying out projects and programs specified in section 470a (d) (1) and (2) of this title in any one fiscal year, except that for the costs of State or local historic surveys or inventories the Secretary shall provide 70 per centum of the aggregate cost involved in any one fiscal year;

(4)   unless the grantee has agreed to make such reports, in such form and containing such information as the Secretary may from time to time require;

(5)   unless the grantee has agreed to assume, after completion of the project, the total cost of the continued maintenance, repair, and administration of the property in a manner satisfactory to the Secretary; and

(6)   until the grantee has complied with such further terms and conditions as the Secretary may deem necessary or advisable.

Except as permitted by other law, the State share of the costs referred to in paragraph (3) shall be contributed by non-Federal sources. Notwithstanding any other provision of law, no grant made pursuant to sections 470 to 470a, 470b, and 470c to 470w-6 of this title shall be treated as taxable income for purposes of the Internal Revenue Code of 1954.

(b)   The Secretary may in his discretion waive the requirements of subsection (a), paragraphs (2) and (5) of this section for any grant under sections 470 to 470b and 470c to 470t of this title to the National Trust for Historic Preservation in the United States, in which case a grant to the National Trust may include funds for the maintenance, repair, and administration of the property in a manner satisfactory to the Secretary.

(c)   Repealed. Pub.L. 96-515, Title II, § 202 (c), Dec. 12, 1980, 94 Stat. 2993.

(d)   No State shall be permitted to utilize the value of real property obtained before October 15, 1966, in meeting the remaining cost of a project for which a grant is made under sections 470 to 470b and 470c to 470t of this title.

§ 470b-1. Grants to National Trust for Historic Preservation— Authority of Secretary of Housing and Urban Development; renovation or restoration costs; terms and conditions; amounts

(a)   The Secretary of Housing and Urban Development is authorized to make grants to the National Trust for Historic Preservation, on such terms and conditions and in such amounts (not exceeding $90,000 with respect to any one structure) as he deems appropriate, to cover the costs incurred by such Trust in renovating or restoring structures which it considers to be of historic or architectural value and which it has accepted and will maintain (after such renovation or restoration) for historic purposes.

Authorization of appropriations

(b)   There are authorized to be appropriated such sums as may be necessary for the grants to be made under subsection (a) of this section.

§ 470c. Apportionment of grant funds; reapportionment

(a)   The amounts appropriated and made available for grants to the States for comprehensive statewide historic surveys and plans under sections 470 to 470b and 470c to 470t of this title shall be apportioned among the States by the Secretary on the basis of needs as determined by him.

(b)   The amounts appropriated and made available for grants to the States for projects and programs under sections 470 to 470a, 470b, and 470c to 470w-6 of this title for each fiscal year shall be apportioned among the States by the Secretary in accordance with needs as disclosed in approved statewide historic preservation plans.

The Secretary shall notify each State of its apportionment under this subsection within thirty days following the date of enactment of legislation appropriating funds under sections 470 to 470a, 470b, and 470c to 470w-6 of this title. Any amount of any apportionment that has not been paid or obligated by the Secretary during the fiscal year in which such notification is given, and for two fiscal years thereafter, shall be reapportioned by the Secretary in accordance with this subsection.

(c)   A minimum of 10 per centum of the annual apportionment distributed by the Secretary to each State for the purposes of carrying out sections 470 to 470a, 470b, and 470c to 470w-6 of this title shall be transferred by the State, pursuant to the requirements of sections 470 to 470a, 470b, and 470c to 470w-6 of this title, to local governments which are certified under section 470a (c) of this title for historic preservation projects or programs of such local governments. In any year in which the total annual apportionment to the States exceeds $65,000,000, one half of the excess shall also be transferred by the States to local governments certified pursuant to section 470a (c) of this title.

(d)   The Secretary shall establish guidelines for the use and distribution of funds under subsection (c) of this section to insure that no local government receives a disproportionate share of the funds available, and may include a maximum or minimum limitation on the amount of funds distributed to any single local government. The guidelines shall not limit the ability of any State to distribute more than 10 per centum of its annual apportionment under subsection (c) of this section, nor shall the Secretary require any State to exceed the 10 per centum minimum distribution to local governments.

§ 470d. Loan insurance program for preservation of property included on National Register—Establishment

(a)   The Secretary shall establish and maintain a program by which he may, upon application of a private lender, insure loans (including loans made in accordance with a mortgage) made by such lender to finance any project for the preservation of a property included on the National Register.

Loan qualifications

(b)   A loan may be insured under this section only if—

(1)   the loan is made by a private lender approved by the Secretary as financially sound and able to service the loan property;

(2)   the amount of the loan, and interest rate charged with respect to the loan, do not exceed such amount, and such a rate, as is established by the Secretary, by rule;

(3)   the Secretary has consulted the appropriate State Historic Preservation Officer concerning the preservation of the historic property;

(4)   the Secretary has determined that the loan is adequately secured and there is reasonable assurance of repayment;

(5)   the repayment period of the loan does not exceed the lesser of forty years or the expected life of the asset financed;

(6)   the amount insured with respect to such loan does not exceed 90 per centum of the loss sustained by the lender with respect to the loan; and

(7)   the loan, the borrower, and the historic property to be preserved meet other terms and conditions as may be prescribed by the Secretary, by rule, especially terms and conditions relating to the nature and quality of the preservation work.

The Secretary shall consult with the Secretary of the Treasury regarding the interest rate of loans insured under this section.

Limitation on amount of unpaid principal balance of loans

(c)   The aggregate unpaid principal balance of loans insured under this section and outstanding at any one time may not exceed the amount which has been covered into the Historic Preservation Fund pursuant to section 470h of this title and subsections (g) and (i) of this section, as in effect on December 12, 1980, but which has not been appropriated for any purpose.

Assignability of insurance contracts; contract as obligation of United States; contestability

(d)   Any contract of insurance executed by the Secretary under this section may be assignable, shall be an obligation supported by the full faith and credit of the United States, and shall be incontestable except for fraud or misrepresentation of which the holder had actual knowledge at the time it became a holder.

Conditions and methods of payment as result of loss

(e)   The Secretary shall specify, by rule and in each contract entered into under this section, the conditions and method of payment to a private lender as a result of losses incurred by the lender on any loan insured under this section.

Protection of financial interests of Federal Government

(f)   In entering into any contract to insure a loan under this section, the Secretary shall take steps to assure adequate protection of the financial interests of the Federal Government. The Secretary may—

(1)   in connection with any foreclosure proceeding, obtain, on behalf of the Federal Government, the property securing a loan insured under sections 470a, 470b, and 470c to 470h-3 of this title, and

(2)   operate or lease such property for such period as may be necessary to protect the interest of the Federal Government and to carry out subsection (g) of this section.

Conveyance to governmental or nongovernmental entity of property acquired by foreclosure

(g)   (1)   In any case in which a historic property is obtained pursuant to subsection (f) of this section, the Secretary shall attempt to convey such property to any governmental or nongovernmental entity under such conditions as will ensure the property's continued preservation and use; except that if, after a reasonable time, the Secretary, in consultation with the Advisory Council on Historic Preservation, determines that there is no feasible and prudent means to convey such property and to ensure its continued preservation and use, then the Secretary may convey the property at the fair market value of its interest in such property to any entity without restriction.

(2)   Any funds obtained by the Secretary in connection with the conveyance of any property pursuant to paragraph (1) shall be covered into the historic preservation fund, in addition to the amounts covered into such fund pursuant to section 470h of this title and subsection (i) of this section, and shall remain available in such fund until appropriated by the Congress to carry out the purposes of sections 470 to 470a, 470b, and 470c to 470w-6 of this title.

Assessment of fees in connection with loans

(h)   The Secretary may assess appropriate and reasonable fees in connection with insuring loans under this section. Any such fees shall be covered into the Historic Preservation Fund, in addition to the amounts covered into such fund pursuant to section 470h of this title and subsection (g) of this section, and shall remain available in such fund until appropriated by the Congress to carry out purposes of sections 470 to 470a, 470b, and 470c to 470w-6 of this title.

Treatment of loans as non-Federal funds

(i)   Notwithstanding any other provision of law, any loan insured under this section shall be treated as non-Federal funds for the purposes of satisfying any requirement of any other provision of law under which Federal funds to be used for any project or activity are conditioned upon the use of non-Federal funds by the recipient for payment of any portion of the costs of such project or activity.

Authorization of appropriations for payment of losses

(j)   Effective after the fiscal year 1981 there are authorized to be appropriated, such sums as may be necessary to cover payments incurred pursuant to subsection (e) of this section.

Eligibility of debt obligation for purchase, etc., by Federal Financing Bank

(k)   No debt obligation which is made or committed to be made, or which is insured or committed to be insured, by the Secretary under this section shall be eligible for purchase by, or commitment to purchase by, or sale or issuance to, the Federal Financing Bank.

§ 470e. Required record keeping by recipients of assistance

The beneficiary of assistance under sections 470 to 470b and 470c to 470n of this title shall keep such records as the Secretary shall prescribe, including records which fully disclose the disposition by the beneficiary of the proceeds of such assistance, the total cost of the project or undertaking in connection with which such assistance is given or used, and the amount and nature of that portion of the cost of the project or undertaking supplied by other sources, and such other records as will facilitate an effective audit.

§ 470f. Effect of Federal undertakings upon property listed in National Register; comment by Advisory Council on Historic Preservation

The head of any Federal agency having direct or indirect jurisdiction over a proposed Federal or federally assisted undertaking in any State and the head of any Federal department or independent agency having authority to license any undertaking shall, prior to the approval of the expenditure of any Federal funds on the undertaking or prior to the issuance of any license, as the case may be, take into account the effect of the undertaking on any district, site, building, structure, or object that is included in or eligible for inclusion in the National Register. The head of any such Federal agency shall afford the Advisory Council on Historic Preservation established under sections 470i to 470t of this title a reasonable opportunity to comment with regard to such undertaking.

§ 470g. White House, United States Supreme Court building, and United States Capitol not included in program for preservation of historical properties

Nothing in sections 470 to 470b and 470c to 470n of this title shall be construed to be applicable to the White House and its grounds, the Supreme Court Building and its grounds, or the United States Capitol and its related buildings and grounds.

§ 470h. Historic Preservation Fund; establishment; appropriations; source of revenue

To carry out the provisions of sections 470 to 470a, 470b, and 470c to 470w-6 of this title, there is hereby established the Historic Preservation Fund (hereafter referred to as the "fund") in the Treasury of the United States. "There shall be covered into such fund $24,400,000 for fiscal year 1977, $100,000,000 for fiscal year 1978, $100,000,000 for fiscal year 1979, $150,000,000 for fiscal year 1980, and $150,000,000 for fiscal year 1981 and $150,000,000 for each of fiscal years 1982 through 1987, from revenues due and payable to the United States under the Outer Continental Shelf Lands Act (67 Stat. 462, 469), as amended, and/or under the Act of June 4, 1920 (41 Stat. 813), as amended, notwithstanding any provision of law that such proceeds shall be credited to miscellaneous receipts of the Treasury. Such moneys shall be used only to carry out the purposes of section 470 to 470a, 470b, and 470c to 470w-6 of this title and shall be available for expenditure only when appropriated by the Congress. Any moneys not appropriated shall remain available in the fund until appropriated for said purposes: *Provided,* That appropriations made pursuant to this paragraph may be made without fiscal year limitation.

§ 470h-1. Acceptance of privately donated funds by Secretary— Authorization; use of funds

(a)   In furtherance of the purposes of sections 470 to 470b and 470c to 470t of this title, the Secretary may accept the donation of funds which may be expended by him for projects to acquire, restore, preserve, or recover data from any district, building, structure, site, or object which is listed on the National Register of Historic Places established pursuant to section 470a of this title, so long as the project is owned by a State, any unit of local government, or any nonprofit entity.

Consideration of factors respecting expenditure of funds

(b)   In expending said funds, the Secretary shall give due consideration to the following factors: the national significance of the project; its historical value to the community; the imminence of its destruction or loss; and the expressed intentions of the donor. Funds expended under this subsection shall be made available without regard to the matching requirements established by section 470b of this title, but the recipient of such funds shall be permitted to utilize them to match any grants from the Historic Preservation Fund established by section 470h of this title.

Transfer of unobligated funds

(c)   The Secretary is hereby authorized to transfer unobligated funds previously donated to the Secretary for the purposes of the National Park Service, with the consent of the donor, and any funds so transferred shall be used or expended in accordance with the provisions of sections 470 to 470b and 470c to 470t of this title.

§ 470h-2. Historic properties owned or controlled by Federal agencies—
Responsibilities of Federal agencies; program for location, inventory and
nomination

(a)   (1)   The heads of all Federal agencies shall assume responsibility for the
preservation of historic properties which are owned or controlled by such agency.
Prior to acquiring, constructing, or leasing buildings for purposes of carrying out
agency responsibilities, each Federal agency shall use, to the maximum extent feasible, historic properties available to the agency. Each agency shall undertake, consistent with the preservation of such properties and the mission of the agency and
the professional standards established pursuant to section 470a (f) of this title, any
preservation, as may be necessary to carry out this section.
        (2)   With the advice of the Secretary and in cooperation with the State
historic preservation officer for the State involved, each Federal agency shall establish a program to locate, inventory, and nominate to the Secretary all properties
under the agency's ownership or control by the agency, that appear to qualify for
inclusion on the National Register in accordance with the regulations promulgated
under section 470a (a) (2) (A) of this title. Each Federal agency shall exercise
caution to assure that any such property that might qualify for inclusion is not
inadvertently transferred, sold, demolished, substantially altered, or allowed to
deteriorate significantly.

Records on historic properties to be altered or demolished; deposit
in Library of Congress or other appropriate agency

(b)   Each Federal agency shall initiate measures to assure that where, as a result of Federal action or assistance carried out by such agency, an historic property
is to be substantially altered or demolished, timely steps are taken to make or have
made appropriate records, and that such records then be deposited, in accordance
with section 470a (a) of this title, in the Library of Congress or with such other
appropriate agency as may be designated by the Secretary, for future use and
reference.

Agency Preservation Officer; responsibilities; qualifications

(c)   The head of each Federal agency shall, unless exempted under section
470v, designate a qualified official to be known as the agency's "preservation
officer" who shall be responsible for coordinating that agency's activities under
sections 470 to 470a, 470b, and 470c to 470w-6 of this title. Each Preservation
Officer may, in order to be considered qualified, satisfactorily complete an appropriate training program established by the Secretary under section 470a (g) of
this title.

Agency programs and projects

(d)   Consistent with the agency's missions and mandates, all Federal agencies
shall carry out agency programs and projects (including those under which any

Federal assistance is provided or any Federal license, permit, or other approval is required) in accordance with the purposes of sections 470 to 470a, 470b, and 470c to 470w-6 of this title and, give consideration to programs and projects which will further the purposes of sections 470 to 470a, 470b, and 470c to 470w-6 of this title.

Review of plans of transferees of surplus federally owned historic properties

(e)   The Secretary shall review and approve the plans of transferees of surplus federally owned historic properties not later than ninety days after his receipt of such plans to ensure that the prehistorical, historical, architectural, or culturally significant values will be preserved or enhanced.

Planning and actions to minimize harm to National Historic Landmarks

(f)   Prior to the approval of any Federal undertaking which may directly and adversely affect any National Historic Landmark, the head of the responsible Federal agency shall, to the maximum extent possible, undertake such planning and actions as may be necessary to minimize harm to such landmark, and shall afford the Advisory Council on Historic Preservation a reasonable opportunity to comment on the undertaking.

Costs of preservation as eligible project costs

(g)   Each Federal agency may include the costs of preservation activities of such agency under sections 470 to 470a, 470b, and 470c to 470w-6 of this title as eligible project costs in all undertakings of such agency or assisted by such agency. The eligible project costs may also include amounts paid by a Federal agency to any State to be used in carrying out such preservation responsibilities of the Federal agency under sections 470 to 470a, 470b, and 470c to 470w-6 of this title, and reasonable costs may be charged to Federal licensees and permittees as a condition to the issuance of such license or permit.

Annual preservation awards program

(h)   The Secretary shall establish an annual preservation awards program under which he may make monetary awards in amounts of not to exceed $1,000 and provide citations for special achievement to officers and employees of Federal, State, and certified local governments in recognition of their outstanding contributions to the preservation of historic resources. Such program may include the issuance of annual awards by the President of the United States to any citizen of the United States recommended for such award by the Secretary.

Environmental impact statement

(i)   Nothing in sections 470 to 470a, 470b, and 470c to 470w-6 of this title shall be construed to require the preparation of an environmental impact statement where such a statement would not otherwise be required under the National Environment Policy Act of 1969, and nothing in sections 470 to 470a, 470b, and 470c to 470w-6 of this title shall be construed to provide any exemption from any requirement respecting the preparation of such a statement under such Act.

Waiver of provisions in event of natural disaster or imminent threat to national security

(j)   The Secretary shall promulgate regulations under which the requirements of this section may be waived in whole or in part in the event of a major natural disaster or an imminent threat to the national security.

§ 470h-8. Lease or exchange of historic property—Authorization; consultation with Advisory Council on Historic Preservation

(a)   Notwithstanding any other provision of law, any Federal agency may, after consultation with the Advisory Council on Historic Preservation, lease an historic property owned by the agency to any person or organization, or exchange any property owned by the agency with comparable historic property, if the agency head determines that the lease or exchange will adequately insure the preservation of the historic property.

Proceeds of lease for administration, etc., of property; deposit of surplus proceeds into Treasury

(b)   The proceeds of any lease under subsection (a) of this section may, notwithstanding any other provision of law, be retained by the agency entering into such lease and used to defray the costs of administration, maintenance, repair, and related expenses incurred by the agency with respect to such property or other properties which are on the National Register which are owned by, or are under the jurisdiction or control of, such agency. Any surplus proceeds from such leases shall be deposited into the Treasury of the United States at the end of the second fiscal year following the fiscal year in which such proceeds were received.

Contracts for management of historic property

(c)   The head of any Federal agency having responsibility for the management of any historic property may, after consultation with the Advisory Council on Historic Preservation, enter into contracts for the management of such property. Any such contract shall contain such terms and conditions as the head of such agency deems necessary or appropriate to protect the interests of the United States and insure adequate preservation of the historic property.

§ 470i.   Advisory Council on Historic Preservation—Establishment; membership; chairman

(a)   There is established as an independent agency of the United States Government an Advisory Council on Historic Preservation (hereinafter referred to as the "Council") which shall be composed of the following members:

    (1)   a Chairman appointed by the President selected from the general public;

    (2)   the Secretary of the Interior;

    (3)   the Architect of the Capitol;

    (4)   the Secretary of Agriculture and the heads of four other agencies of the United States (other than the Department of the Interior) the activities of which affect historic preservation, appointed by the President;

    (5)   one Governor appointed by the President;

    (6)   one mayor appointed by the President;

    (7)   the President of the National Conference of State Historic Preservation Officers;

    (8)   the Chairman of the National Trust for Historic Preservation;

    (9)   four experts in the field of historic preservation appointed by the President from the disciplines of architecture, history, archeology, and other appropriate disciplines; and

    (10)   three at-large members from the general public, appointed by the President.

Designation of substitutes

(b)   Each member of the Council specified in paragraphs (2) through (8) (other than (5) and (6)) of subsection (a) of this section may designate another officer of his department, agency, or organization to serve on the Council in his stead, except that, in the case of paragraphs (2) and (4), no such officer other than an Assistant Secretary or an officer having major department-wide or agency-wide responsibilities may be so designated.

Term of office

(c)   Each member of the Council appointed under paragraph (1), and under paragraphs (9) and (10) of subsection (a) of this section shall serve for a term of four years from the expiration of his predecessor's term; except that the members first appointed under that paragraph shall serve for terms of one to four years; as designated by the President at the time of appointment, in such manner as to insure that the terms of not more than two of them will expire in any one year. The members appointed under paragraphs (5) and (6) shall serve for the term of their elected office but not in excess of four years. An appointed member may not serve more than two terms. An appointed member whose term has expired shall serve until that member's successor has been appointed.

Vacancies; term of office of members already appointed

(d)   A vacancy in the Council shall not affect its powers, but shall be filled, not later than sixty days after such vacancy commences, in the same manner as the original appointment (and for the balance of any unexpired terms). The members of the Advisory Council on Historic Preservation appointed by the President under sections 470 to 470a, 470b, and 470c to 470w-6 of this title as in effect on the day before December 12, 1980 shall remain in office until all members of the Council, as specified in this section, have been appointed. The members first appointed under this section shall be appointed not later than one hundred and eighty days after December 12, 1980.

Designation of Vice Chairman

(e)   The President shall designate a Vice Chairman, from the members appointed under paragraph (5), (6), (9), or (10). The Vice Chairman may act in place of the Chairman during the absence or disability of the Chairman or when the office is vacant.

Quorum

(f)   Nine members of the Council shall constitute a quorum.

§ 470j. Functions of Council; annual report to President and Congress; recommendations

(a)   The Council shall—
(1)   advise the President and the Congress on matters relating to historic preservation; recommend measures to coordinate activities of Federal, State, and local agencies and private institutions and individuals relating to historic preservation; and advise on the dissemination of information pertaining to such activities;
(2)   encourage, in cooperation with the National Trust for Historic Preservation and appropriate private agencies, public interest and participation in historic preservation;
(3)   recommend the conduct of studies in such areas as the adequacy of legislative and administrative statutes and regulations pertaining to historic preservation activities of State and local governments and the effects of tax policies at all levels of government on historic preservation;
(4)   advise as to guidelines for the assistance of State and local governments in drafting legislation relating to historic preservation;
(5)   encourage, in cooperation with appropriate public and private agencies and institutions, training and education in the field of historic preservation;
(6)   review the policies and programs of Federal agencies and recommend to such agencies methods to improve the effectiveness, coordination, and consistency of those policies and programs with the policies and programs carried out under sections 470 to 470a, 470b, and 470c to 470w-6 of this title; and

(7) inform and educate Federal agencies, State and local governments, Indian tribes, other nations and international organizations and private groups and individuals as to the Council's authorized activities.

(b) The Council shall submit annually a comprehensive report of its activities and the results of its studies to the President and the Congress and shall from time to time submit such additional and special reports as it deems advisable. Each report shall propose such legislative enactments and other actions as, in the judgment of the Council, are necessary and appropriate to carry out its recommendations and shall provide the Council's assessment of current and emerging problems in the field of historic preservation and an evaluation of the effectiveness of the programs of Federal agencies, State and local governments, and the private sector in carrying out the purposes of sections 470 to 470a, 470b, and 470c to 470w-6 of this title.

§ 470k. Cooperation between Council and instrumentalities of Executive Branch of Federal government

The Council is authorized to secure directly from any department, bureau, agency, board, commission, office, independent establishment or instrumentality of the executive branch of the Federal Government information, suggestions, estimates, and statistics for the purpose of sections 470i to 470n of this title; and each such department, bureau, agency, board, commission, office, independent establishment or instrumentality is authorized to furnish such information, suggestions, estimates, and statistics to the extent permitted by law and within available funds.

§ 470l. Compensation of members of Council

The members of the Council specified in paragraphs (2), (3), and (4) of section 470i (a) of this title shall serve without additional compensation. The other members of the Council shall receive $100 per diem when engaged in the performance of the duties of the Council. All members of the Council shall receive reimbursement for necessary traveling and subsistence expenses incurred by them in the performance of the duties of the Council.

§ 470m. Administration—Executive Director of Council; appointment; functions and duties

(a) There shall be an Executive Director of the Council who shall be appointed in the competitive service by the Chairman with the concurrence of the Council. The Executive Director shall report directly to the Council and perform such functions and duties as the Council may prescribe.

General Counsel; appointment; functions and duties

(b) The Council shall have a General Counsel, who shall be appointed by the Executive Director. The General Counsel shall report directly to the Executive

Director and serve as the Council's legal advisor. The Executive Director shall appoint such other attorneys as may be necessary to assist the General Counsel, represent the Council in courts of law whenever appropriate, including enforcement of agreements with Federal agencies to which the Council is a party, assist the Department of Justice in handling litigation concerning the Council in courts of law, and perform such other legal duties and functions as the Executive Director and the Council may direct.

Appointment and compensation of officers and employees

(c)   The Executive Director of the Council may appoint and fix the compensation of such officers and employees in the competitive service as are necessary to perform the functions of the Council at rates not to exceed that now or hereafter prescribed for the highest rate for grade 15 of the General Schedule under section 5332 of Title 5: *Provided,* however, That the Executive Director, with the concurrence of the chairman, may appoint and fix the compensation of not to exceed five employees in the competitive service at rates not to exceed that now or hereafter prescribed for the highest rate of grade 17 of the General Schedule under section 5332 of Title 5.

Appointment and compensation of additional personnel

(d)   The Executive Director shall have power to appoint and fix the compensation of such additional personnel as may be necessary to carry out its duties, without regard to the provisions of the civil service laws and chapter 51 and subchapter III of chapter 53 of Title 5.

Expert and consultant services; procurement

(e)   The Executive Director of the Council is authorized to procure expert and consultant services in accordance with the provisions of section 3109 of Title 5.

Financial and administrative services; Department of Interior

(f)   Financial and administrative services (including those related to budgeting, accounting, financial reporting, personnel and procurement) shall be provided the Council by the Department of the Interior, for which payments shall be made in advance, or by reimbursement, from funds of the Council in such amounts as may be agreed upon by the Chairman of the Council and the Secretary of the Interior: *Provided,* That the regulations of the Department of the Interior for the collection of indebtedness of personnel resulting from erroneous payments (section 5514 (b) of Title 5) shall apply to the collection of erroneous payments made to or on behalf of a Council employee, and regulations of said Secretary for the administrative control of funds (section 665 (g) of Title 31) shall apply to appropriations of the Council:

And provided further, That the Council shall not be required to prescribe such regulations.

Use of funds, personnel, facilities, and services of Council members

(g)   The members of the Council specified in paragraphs (2) through (4) of section 470i (a) of this title shall provide the Council, with or without reimbursement as may be agreed upon by the Chairman and the members, with such funds, personnel, facilities, and services under their jurisdiction and control as may be needed by the Council to carry out its duties, to the extent that such funds, personnel, facilities, and services are requested by the Council and are otherwise available for that purpose. To the extent of available appropriations, the Council may obtain, by purchase, rental, donation, or otherwise, such additional property, facilities, and services as may be needed to carry out its duties and may also receive donations of moneys for such purpose, and the Executive Director is authorized, in his discretion, to accept, hold, use, expend, and administer the same for the purposes of sections 470 to 470a, 470b, and 470c to 470w-6 of this title.

§ 470n. International Centre for Study of Preservation and Restoration of Cultural Property; United States participation; Council recommendations; authorization of appropriations

(a)   The participation of the United States as a member in the International Centre for the Study of the Preservation and Restoration of Cultural Property is hereby authorized.

(b)   The Council shall recommend to the Secretary of State, after consultation with the Smithsonian Institution and other public and private organizations concerned with the technical problems of preservation, the members of the official delegation which will participate in the activities of the Centre on behalf of the United States. The Secretary of State shall appoint the members of the official delegation from the persons recommended to him by the Council.

(c)   For the purposes of this section there is authorized to be appropriated an amount equal to the assessment for United States membership in the Centre for fiscal years 1979, 1980, 1981, and 1982: *Provided,* That no appropriation is authorized and no payment shall be made to the Centre in excess of 25 per centum of the total annual assessment of such organization. Authorization for payment of such assessments shall begin in fiscal year 1981, but shall include earlier costs.

§ 470o. Transfer of personnel, property, etc., by Department of Interior to Council; time limit

So much of the personnel, property, records, and unexpended balances of appropriations, allocations, and other funds employed, held, used, programed, or available or to be made available by the Department of the Interior in connection with the functions of the Council, as the Director of the Office of Management and Budget shall determine, shall be transferred from the Department to the Council within 60 days of the effective date of this Act.

§ 470p. Rights, benefits, and privileges of transferred employees

Any employee in the competitive service of the United States transferred to the Council under the provisions of this section shall retain all the rights, benefits, and privileges pertaining thereto held prior to such transfer.

§ 470q. Operations of Council; exemption

The Council is exempt from the provisions of the Federal Advisory Committee Act (86 Stat. 770), and the provisions of subchapter II of chapter 5 and chapter 7 of Title 5 shall govern the operations of the Council.

§ 470r. Transmittal of legislative recommendations, or testimony, or comments to President, Office of Management and Budget, and congressional committees; prohibition

No officer or agency of the United States shall have any authority to require the Council to submit its legislative recommendations, or testimony, or comments on legislation to any officer or agency of the United States for approval, comments, or review, prior to the submission of such recommendations, testimony, or comments to the Congress. In instances in which the Council voluntarily seeks to obtain the comments or review of any officer or agency of the United States, the Council shall include a description of such actions in its legislative recommendations, testimony, or comments on legislation which it transmits to the Congress.

§ 470s. Rules and regulations; participation by local governments

The Council is authorized to promulgate such rules and regulations as it deems necessary to govern the implementation of section 470f of this title. The Council shall, by regulation, establish such procedures as may be necessary to provide for participation by local governments in proceedings and other actions taken by the Council with respect to undertakings referred to in section 470f of this title which affect such local governments.

§ 470t. Budget; authorization of appropriations

(a)   The Council shall submit its budget annually as a related agency of the Department of the Interior. To carry out the provisions of sections 470i to 470t of this title, there are authorized to be appropriated not more than $1,500,000 in fiscal year 1977, $1,750,000 in fiscal year 1978, and $2,000,000 in fiscal year 1979. There are authorized to be appropriated not to exceed $2,250,000 in fiscal year 1980, $2,500,000 in fiscal year 1981, $2,500,000 in fiscal year 1982, and $2,500,000 in fiscal year 1983.

(b)   Whenever the Council submits any budget estimate or request to the President or the Office of Management and Budget, it shall concurrently transmit copies of that estimate or request to the House and Senate Appropriations Com-

mittees and the House Committee on Interior and Insular Affairs and the Senate
Committee on Energy and Natural Resources.

§ 470u. Report by Secretary to Council.

To assist the Council in discharging its responsibilities under sections 470 to 470a,
470b, and 470c to 470w-6 of this title, the Secretary at the request of the Chairman,
shall provide a report to the Council detailing the significance of any historic property, describing the effects of any proposed undertaking on the affected property,
and recommending measures to avoid, minimize, or mitigate adverse effects.
Pub. L. 89-665, Title II, § 213, as added Pub. L. 96-515, Title III, § 302 (a),
Dec. 12, 1980, 94 Stat. 3000.

§ 470v. Exemption for Federal programs or undertakings; regulations

The Council, with the concurrence of the Secretary, shall promulgate regulations
or guidelines, as appropriate, under which Federal programs or undertakings may
be exempted from any or all of the requirements of sections 470 to 470a, 470b,
and 470c to 470w-6 of this title when such exemption is determined to be consistent with the purposes of sections 470 to 470a, 470b, and 470c to 470w-6 of this
title, taking into consideration the magnitude of the exempted undertaking or program and the likelihood of impairment of historic properties.
Pub.L. 89-665, Title II, § 214, as added Pub.L. 96-515, Title III, § 302 (a),
Dec. 12, 1980, 94 Stat. 3000.

§ 470w. Definitions

As used in sections 470 to 470a, 470b, and 470c to 470w-6 of this title, the
term—
   (1)   "Agency" means agency as such term is defined in section 551 of
Title 5, except that in the case of any Federal program exempted under section
470v of this title, the agency administering such program shall not be treated as
an agency with respect to such program.
   (2)   "State" means any State of the United States, the District of
Columbia, the Commonwealth of Puerto Rico, Guam, the Virgin Islands, American
Samoa, the Commonwealth of the Northern Mariana Islands, and the Trust Territories of the Pacific Islands.
   (3)   "Local government" means a city, county, parish, township, municipality, or borough, or any other general purpose political subdivision of any State.
   (4)   "Indian tribe" means the governing body of any Indian tribe, band,
nation, or other group which is recognized as an Indian tribe by the Secretary of
the Interior and for which the United States holds land in trust or restricted status
for that entity or its members. Such term also includes any Native village corporation, regional corporation, and Native Group established pursuant to the Alaska
Native Claims Settlement Act.

(5)   "Historic property" or "historic resource" means any prehistoric or historic district, site, building, structure, or object included in, or eligible for inclusion on the National Register; such term includes artifacts, records, and remains which are related to such a district, site, building, structure, or object.

(6)   "National Register" or "Register" means the National Register of Historic Places established under section 470a of this title.

(7)   "Undertaking" means any action as described in section 470f of this title.

(8)   "Preservation" or "historic preservation" includes identification, evaluation, recordation, documentation, curation, acquisition, protection, management, rehabilitation, restoration, stabilization, maintenance and reconstruction, or any combination of the foregoing activities.

(9)   "Cultural park" means a definable urban area which is distinguished by historic resources and land related to such resources and which constitutes an interpretive, educational, and recreational resource for the public at large.

(10)   "Historic conservation district" means an urban area of one or more neighborhoods and which contains (A) historic properties, (B) buildings having similar or related architectural characteristics, (C) cultural cohesiveness, or (D) any combination of the foregoing.

(11)   "Secretary" means the Secretary of the Interior except where otherwise specified.

(12)   "State historic preservation review board" means a board, council, commission, or other similar collegial body established as provided in section 470a (b) (1) (B) of this title—

    (A)   the members of which are appointed by the State Historic Preservation Officer (unless otherwise provided for by State law),

    (B)   a majority of the members of which are professionals qualified in the following and related disciplines: history, prehistoric and historic archaeology, architectural history, and architecture, and

    (C)   which has the authority to—

        (i) review National Register nominations and appeals from nominations;

        (ii) review appropriate documentation submitted in conjunction with the Historic Preservation Fund;

        (iii) provide general advice and guidance to the State Historic Preservation Officer, and

        (iv) perform such other duties as may be appropriate.

(13)   "Historic preservation review commission" means a board, council, commission, or other similar collegial body which is established by State or local legislation as provided in section 470a (c) (1) (B) of this title, and the members of which are appointed, unless otherwise provided by State or local legislation, by the chief elected official of the jurisdiction concerned from among—

    (A)   professionals in the disciplines of architecture, history, architectural history, planning, archaeology, or related disciplines, to the extent such professionals are available in the community concerned, and

    (B)   such other persons as have demonstrated special interest, experience, or knowledge in history, architecture, or related disciplines, and as will provide for an adequate and qualified commission.

§ 470w-1. Authorization for expenditure of appropriated funds

Where appropriate, each Federal agency is authorized to expend funds appropriated for its authorized programs for the purposes of activities carried out pursuant to sections 470 to 470a, 470b, and 470c to 470w-6 of this title, except to the extent appropriations legislation expressly provides otherwise.

§ 470w-2. Donations and bequests of money, personal property and less than fee interests in historic property

(a)   The Secretary is authorized to accept donations and bequests of money and personal property for the purposes of sections 470 to 470a, 470b, and 470c to 470w-6 of this title and shall hold, use, expend, and administer the same for such purposes.

(b)   The Secretary is authorized to accept gifts or donations of less than fee interests in any historic property where the acceptance of such interests will facilitate the conservation or preservation of such properties. Nothing in this section or in any provision of sections 470 to 470a, 470b, and 470c to 470w-6 of this title shall be construed to affect or impair any other authority of the Secretary under other provision of law to accept or acquire any property for conservation or preservation or for any other purposes.

§ 470w-3. Information relating to location or character of historic resources, disclosure to public

The head of any Federal agency, after consultation with the Secretary, shall withhold from disclosure to the public, information relating to the location or character of historic resources whenever the head of the agency or the Secretary determines that the disclosure of such information may create a substantial risk of harm, theft, or destruction to such resources or to the area or place where such resources are located.

§ 470w-4. Attorneys' fees and costs to prevailing parties in civil actions

In any civil action brought in any United States district court by any interested person to enforce the provisions of sections 470 to 470a, 470b, and 470c to 470w-6 of this title, if such person substantially prevails in such action, the court may award attorneys' fees, expert witness fees, and other costs of participating in such action, as the court deems reasonable.

§ 470w-5. National Museum for the Building Arts—Cooperative agreement between Secretary, Administrator of General Services Administration and Committee for National Museum of the Building Arts; purposes

(a)   In order to provide a national center to commemorate and encourage the building arts and to preserve and maintain a nationally significant building which exemplifies the great achievements of the building arts in the United States, the Secretary and the Administrator of the General Services Administration are authorized and directed to enter into a cooperative agreement with the Committee for a National Museum of the Building Arts, Incorporated, a nonprofit corporation organized and existing under the laws of the District of Columbia, or its successor, for the operation of a National Museum for the Building Arts in the Federal Building located in the block bounded by Fourth Street, Fifth Street, F Street, and G Street, Northwest in Washington, District of Columbia. Such museum shall—

(1)   collect and disseminate information concerning the building arts, including the establishment of a national reference center for current and historic documents, publications, and research relating to the building arts;

(2)   foster educational programs relating to the history, practice and contribution to society of the building arts, including promotion of imaginative educational approaches to enhance understanding and appreciation of all facets of the building arts;

(3)   publicly display temporary and permanent exhibits illustrating, interpreting and demonstrating the building arts;

(4)   sponsor or conduct research and study into the history of the building arts and their role in shaping our civilization; and

(5)   encourage contributions to the building arts.

Provisions of cooperative agreement

(b)   The cooperative agreement referred to in subsection (a) of this section shall include provisions which—

(1)   make the site available to the Committee referred to in subsection (a) of this section without charge;

(2)   provide, subject to available appropriations, such maintenance, security, information, janitorial and other services as may be necessary to assure the preservation and operation of the site; and

(3)   prescribe reasonable terms and conditions by which the Committee can fulfill its responsibilities under sections 470 to 470a, 470b, and 470c to 470w-6 of this title.

Matching grants-in-aid to Committee; limitation on amounts

(c)   The Secretary is authorized and directed to provide matching grants-in-aid to the Committee referred to in subsection (a) of this section for its programs related to historic preservation. The Committee shall match such grants-in-aid in a manner and with such funds and services as shall be satisfactory to the Secretary, except that no more than $500,000 may be provided to the Committee in any one fiscal year.

Renovation of site

(d)   The renovation of the site shall be carried out by the Administrator with the advice of the Secretary. Such renovation shall, as far as practicable—
    (1)   be commenced immediately,
    (2)   preserve, enhance, and restore the distinctive and historically authentic architectural character of the site consistent with the needs of a national museum of the building arts and other compatible use, and
    (3)   retain the availability of the central court of the building, or portions thereof, for appropriate public activities.

Annual Committee report to Secretary and Administrator

(e)   The Committee shall submit an annual report to the Secretary and the Administrator concerning its activities under this section and shall provide the Secretary and the Administrator with such other information as the Secretary may, from time to time, deem necessary or advisable.

Definition of "building arts"

(f)   For purposes of this section, the term "building arts" includes, but shall not be limited to, all practical and scholarly aspects of prehistoric, historic, and contemporary architecture, archaeology, construction, building technology and skills, landscape architecture, preservation and conservation, building and construction, engineering, urban and community design and renewal, city and regional planning, and related professions, skills, and trades, and crafts.

§470w-6. Effective date of regulations—Copy to Congress prior to publication in Federal Register; effective date of final regulations

(a)   At least thirty days prior to publishing in the Federal Register any proposed regulation required by sections 470 to 470a, 470b, and 470c to 470w-6 of this title, the Secretary shall transmit a copy of the regulation to the Committee on Interior and Insular Affairs of the House of Representatives and the Committee on Energy and Natural Resources of the Senate. The Secretary also shall transmit to such committees a copy of any final regulation prior to its publication in the Federal Register. Except as provided in subsection (b) of this section, no final regulation of the Secretary shall become effective prior to the expiration of thirty calendar days after it is published in the Federal Register during which either or both Houses of Congress are in session.

Effective date of final regulation in case of emergency

(b)   In the case of an emergency, a final regulation of the Secretary may become effective without regard to the last sentence of subsection (a) of this section if the

Secretary notified in writing the Committee on Interior and Insular Affairs of the United States House of Representatives and the Committee on Energy and Natural Resources of the United States Senate setting forth the reasons why it is necessary to make the regulation effective prior to the expiration of the thirty-day period.

Disapproval of regulation by resolution of Congress

(c)   Except as provided in subsection (b) of this section, the regulation shall not become effective if, within ninety calendar days of continuous session of Congress after the date of promulgation, both Houses of Congress adopt a concurrent resolution, the matter after the resolving clause of which is as follows: "That Congress disapproves the regulation promulgated by the Secretary dealing with the matter of   , which regulation was transmitted to Congress on   ," the blank spaces therein being appropriately filled.

Failure of Congress to adopt resolution of disapproval of regulation

(d)   If at the end of sixty calendar days of continuous session of Congress after the date of promulgation of a regulation, no committee of either House of Congress has reported or been discharged from further consideration of a concurrent resolution disapproving the regulation, and neither House has adopted such a resolution, the regulation may go into effect immediately. If, within such sixty calendar days, such a committee has reported or been discharged from further consideration of such a resolution, the regulation may go into effect not sooner than ninety calendar days of continuous session of Congress after its promulgation unless disapproved as provided for.

Sessions of Congress

(e)   For the purposes of this section –
    (1)   continuity of session is broken only by an adjournment sine die; and
    (2)   the days on which either House is not in session because of an adjournment of more than three days to a day certain are excluded in the computation of sixty and ninety calendar days of continuous session of Congress.

Congressional inaction or rejection of resolution of disapproval
not deemed approval of regulation

(f)   Congressional inaction on or rejection of a resolution of disapproval shall not be deemed an expression of approval of such regulation.

Appendix B

National Park Service, "Historic Preservation Certification," Final Rule, 36 C.F.R. Part 67

Part 67—Historic Preservation Certifications Pursuant to the Tax Reform Act of 1976, the Revenue Act of 1978, the Tax Treatment Extension Act of 1980, and the Economic Recovery Tax Act of 1981

Authority: Sec. 101(a) (1), National Historic Preservation Act of 1966 U.S.C. 470a-1(a) (170 ed.), as amended: sec. 2124. Tax Reform Act of 1976, 90 Stat. 1519: secs. 701(f) and 315. Revenue Act of 1978, 92 Stat. 2828; sec. 6. Tax Treatment Extension Act of 1980, 94 Stat. 3204: and secs. 212 and 214. Economic Recovery Tax Act of 1981, 95 Stat. 172.

§ 67.1. The Tax Reform Act of 1976, the Revenue Act of 1978, the Tax Treatment Extension Act of 1980, and the Economic Recovery Tax Act of 1981.

(a)   The Tax Reform Act of 1976, 90 Stat. 1519, the Revenue Act of 1978, 92 Stat. 2828, the Tax Treatment Extension Act of 1980, 94 Stat. 3204, and the Economic Recovery Tax Act of 1981, 95 Stat. 172, require the Secretary to make certifications of historic district statutes and of State and local districts, certifications of significance, and certifications of rehabilitation in connection with certain tax incentives involving historic preservation. These certification responsibilities have been delegated to the National Park Service ("NPS"); the following five regional offices issue certifications for the States listed below them:

Alaska Regional Office, National Park Service, 2525 Gambell Street, Room 107, Anchorage, Alaska 99503:
Alaska

Mid-Atlantic Regional Office, National Park Service, 143 South Third Street, Philadelphia, Pennsylvania 19106:

Connecticut	New Jersey
Delaware	New York
District of Columbia	Ohio
Indiana	Pennsylvania
Maine	Rhode Island
Maryland	Vermont
Massachusetts	Virginia
Michigan	West Virginia
New Hampshire	

Rocky Mountain Regional Office, National Park Service, 655 Parfet Street, P.O. Box 25287, Denver, Colorado 80225:

Colorado	New Mexico
Illinois	North Dakota
Iowa	Oklahoma
Kansas	South Dakota
Minnesota	Texas
Missouri	Utah
Montana	Wisconsin
Nebraska	Wyoming

Southeast Regional Office, National Park Service, 75 Spring Street, S.W. Atlanta, Georgia 30303:

Alabama	Mississippi
Arkansas	North Carolina
Florida	Puerto Rico
Georgia	South Carolina
Kentucky	Tennessee
Louisiana	Virgin Islands

Western Regional Office, National Park Service, 450 Golden Gate Avenue, San Francisco, California 94102:

Arizona	Nevada
California	Oregon
Hawaii	Washington
Idaho	

(b)   The Washington office of the National Park Service establishes program direction and considers appeals of certification denials. The procedures for obtaining certifications are set forth below. It is the responsibility of owners wishing certifications to provide sufficient documentation to the Secretary to make certification decisions. These procedures, upon their effective date, are applicable to future and pending certification requests, except as otherwise noted herein.

(c)   Most States participate in the review of requests for certification, through recommendations to the Secretary, although this participation is voluntary and by law all certification decisions are made by the Secretary. Three levels of participation are available to all States:

(1)   Regular Participation. States wishing to participate in the review process are given a 30-day opportunity to comment on all certification requests upon receipt of a complete, adequately documented application. In these situations, requests for certification and approvals of proposed rehabilitation work are sent first to the appropriate State official. State comments are carefully considered by the Secretary before a certification decision is made. Certification requests channeled through "regular participation" States are normally processed within 30 days by the Secretary.

(2)   Expedited Review Participation. States wishing to participate in the review of Part 1 and Part 2 certification requests, and which meet qualifications in sec. 67.11, are also given a 30-day opportunity to comment on these requests. Like the "regular participation" explained above, certification requests are first sent to the appropriate State official. Because qualified States assume greater responsibility for making certification recommendations, certification requests channeled through "expedited review" States are normally processed within 15 days by the Secretary. The recommendations of qualified States are generally followed, but by law, all certification decisions are made by the Secretary, based upon his review of the application and related information. Expedited review does not apply to the review of State or local statutes or districts.

(3)   No Participation. A State may choose not to participate in the review and comment of certification requests. States not wishing to participate in the commenting process are requested to notify the Secretary in writing of this fact. Owners requesting certification from these States may send their applications directly to the appropriate NPS office listed above. In all other situations certification requests are sent first to the appropriate State official.

(d)   The Internal Revenue Service is responsible for all procedures, legal determinations and rules and regulations concerning the tax consequences of the historic preservation provisions described above. Any certifications made by the Secretary pursuant to this part shall not be considered as binding upon the Internal Revenue Service or the Secretary of the Treasury with respect to tax consequences under the Internal Revenue Code. For example, certifications made by the Secretary do not constitute determinations that a structure is of the type subject to the allowance for depreciation under Section 167 of the Code.

§ 67.2. Definitions.

As used in these regulations:

"Certified Historic Structure" means a building (and its structural components) which is of a character subject to the allowance for depreciation provided in Section 167 of the Internal Revenue Code of 1954 which is either (a) individually listed in the National Register; or (b) located in a registered historic district and certified by the Secretary as being of historic significance to the district. Portions of larger buildings, such as single condominium apartment units, are not independently considered certified historic structures. Rowhouses, even with abutting or party walls, are considered as separate buildings.

For purposes of the charitable contribution provisions only, a certified historic structure need not be depreciable to qualify, may be a structure other than a building and may also be a remnant of a building such as a facade, if that is all that remains, and may include the land area on which it is located. For purposes of the demolition expense provisions and the 15 percent and 20 percent tax investment credits under the Economic Recovery Tax Act of 1981, any building located in a registered historic district is considered a certified historic structure; exemption from this provision can generally occur only if the Secretary has determined, prior to the demolition or rehabilitation of the building, that it is not of historic significance to the district.

"Certified Rehabilitation" means any rehabilitation of a certified historic structure which the Secretary has certified to the Secretary of the Treasury as being consistent with the historic character of such structure and, where applicable, with the district in which such structure is located.

"Duly Authorized Representative" means a State or locality's Chief Elected Official or his or her representative who is authorized to apply for certification of State/local statutes and historic districts.

"Historic District" means a geographically definable area, urban or rural, that possesses a significant concentration, linkage or continuity of sites, buildings, structures or objects united by past events or aesthetically by plan or physical development. A district may also comprise individual elements separated geographically but linked by association or history.

"Inspection" means a visit by an authorized representative of the Secretary to a certified historic structure for the purposes of reviewing and evaluating the significance of the structure and the ongoing or completed rehabilitation work.

"National Register of Historic Places" means the National Register of districts, sites, buildings, structures, and objects significant in American history: architecture, archeology, engineering, and culture that the Secretary is authorized to expand and maintain pursuant to Section 101(a)(1) of the National Historic Preservation Act of 1966, as amended.

"National Register Program" means the survey, planning, and registration program that is administered by the Secretary pursuant to 101(a)(1) of the National Historic Preservation Act of 1966, as amended. The procedures of the National Register program appear in 36 CFR part 60, *et seq.*

"Owner" means a person, partnership, corporation, or public agency holding a fee-simple interest in a building or any other person or entity recognized by the Internal Revenue Code for purposes of the applicable tax benefits.

"Qualified State" means a State which has agreed to participate in the certification program and which the Secretary has determined to meet established professional and review standards.

"Registered Historic District" means any district listed in the National Register or any district (a) which is designated under a State or local statute which has been certified by the Secretary as containing criteria which will substantially achieve the purpose of preserving and rehabilitating buildings of significance to the district; and, (b) which is certified by the Secretary as meeting substantially all of the requirements for the listing of districts in the National Register.

"Rehabilitation" means the process of returning a building or buildings to a state of utility, through repair or alteration, which makes possible an efficient contemporary use while preserving those portions and features of the building(s) which are significant to its historic, architectural and cultural values.

"Secretary" means the Secretary of the Interior or the designee authorized to carry out his responsibilities.

"Standards for Rehabilitation" mean the Secretary's "Standards for Rehabilitation" set forth in § 67.7 hereof.

"State or Local Statute" means a law of a State or local government designating, or providing a method for the designation of, a historic district or districts.

"State official" means an official within each State, designated by the Governor or by State statute, to act as liaison for purposes of reviewing and commenting upon historic preservation certification applications. In most cases this will be the State Historic Preservation Officer (SHPO). In the event the Governor or a State statute has not designated such an official, the term "State official" shall refer to the Governor.

§ 67.3. Introduction to certifications of significance and rehabilitation and information collection.

(a)   Who may apply:

(1)   Ordinarily, only the fee simple owner of the building in question may apply for the certification described in § § 67.4 and 67.6 hereof. If an application for an evaluation of significance or rehabilitation project is made by someone other than the fee simple owner, however, the application must be accompanied by a written statement from the fee simple owner indicating that he or she is aware of the application and has no objection to the request for certification.

(2)   Upon request of a State official the Secretary may determine whether or not a particular building located within a registered historic district qualifies as a certified historic structure. The Secretary shall do so, however, only after notifying the fee simple owner of record of the request, informing such owner of the possible tax consequences of such a decision, and permitting the property owner a 30 day time period to submit written comments to the Secretary prior to decision. Such time period for comment may be waived by the fee simple owner.

(3)   The Secretary may undertake the certifications described in § § 67.4 and 67.6 on his own initiative after notifying the fee simple owner and the appropriate State official and allowing a comment period as specified in § 67.3 (a) (2).

(4)   Owners of buildings which appear to meet National Register criteria but are not yet listed in the National Register or which are located within potential historic districts may request preliminary determinations from the Secretary as to whether such buildings may qualify as certified historic structures when and if the buildings or the potential historic districts in which they are located are listed in the

National Register. Preliminary determinations may also be requested for buildings outside the period or area of significance of registered historic districts as specified in § 67.5 (c). Procedures for obtaining these determinations shall be the same as those described in § 67.4. Such determinations are preliminary only and are not binding upon the Secretary. Preliminary determinations will be made final as of the date of the listing of the individual building or district in the National Register. For buildings outside the period or area of significance of a registered historic district, preliminary determinations will be made final, except as provided below, when the district documentation on file with NPS is formally amended. If during review of a Request for Certification of Rehabilitation, it is determined that the building does not contribute to the significance of the district because of changes made after the preliminary determination was made, certified historic structure designation will be denied.

(5) Owners of buildings not yet designated certified historic structures may obtain determinations from the Secretary on whether or not rehabilitation proposals meet the Secretary of the Interior's "Standards for Rehabilitation." Such determinations will be made only when the owner has requested a preliminary determination of the significance of the building as described in paragraph (a) (4) of this section and such determination has been acted upon by the NPS. Final certifications of rehabilitation will be issued only to owners of certified historic structures. Procedures for obtaining these determinations shall be the same as those described in § 67.6.

(b) How to apply:

(1) Requests for certifications of historic significance and/or of rehabilitation shall be made on "Historic Preservation Certification Applications" (approved OMB form No. 1024-0009). The information collection requirements contained in the application and in this part have been approved by the Office of Management and Budget under 44 U.S.C. 3507 and assigned clearance number 1024-0009. Part 1 of the application shall be used in requesting a certification of historic significance or non-significance and preliminary determinations, while Part 2 shall be used in requesting an evaluation of a proposed rehabilitation project or a certification of a completed rehabilitation project. Information contained in the application is required to obtain a benefit.

(2) Application forms are available from National Park Service regional offices or the appropriate State official.

(3) Requests for certifications, preliminary determinations, and approvals of proposed rehabilitation projects shall be sent to the appropriate State official in participating States. Requests in nonparticipating States should be sent to the appropriate NPS regional office.

(4) Generally review of certification requests shall be concluded within 60 days of receipt of a complete, adequately documented application, as defined in § § 67.4 and 67.6 (30 days at the State level and 30 days at the Federal level). Where certification requests come from qualified States, review shall be concluded within 45 days (30 days at the State level and 15 days at the Federal level; see § 67.11 for procedures regarding qualified States). Where a State has chosen not to participate in the review process, review by the NPS shall be concluded within 60 days of receipt of a complete, adequately documented application. Where adequate documentation is not provided, the owner will be notified of the additional information needed to undertake or complete review. The time periods in this

part are based on the receipt of a complete application: they will be adhered to as closely as possible and are defined as calendar days. They are not, however, considered to be mandatory, and the failure to complete review within the designated periods does not waive or alter any certification requirement.

(5) State comments received within the time period will be considered by the Secretary in the review process. Reviews of complete certification requests, taking longer than 30 days at the State level may be brought to the attention of the Secretary. The Secretary in turn will consult with the appropriate State to ensure that a review is completed in a timely manner. If necessary the Secretary may process a certification request without the recommendations of the State. The recommendations of qualified States are generally followed, but by law, all certification decisions are made by the Secretary based on his review of the application and related information.

(6) Although certifications of significance and rehabilitation are discussed separately below, owners are encouraged to submit Part 1 of the "Historic Preservation Certification Application" prior to, or with, Part 2. Part 2 of the application will not be processed until an adequately documented Part 1 is on file and acted upon unless the building is already a certified historic structure. Reviews of rehabilitation projects will also not be undertaken if the owner has objected to the listing of the building in the National Register.

§ 67.4. Certifications of historic significance.

(a) Requests for certifications of historic significance should be made by the owner to determine—

(1) That a building within a registered historic district is of historic significance to such district; or

(2) That a building located within a registered historic district is not of historic significance to such district; or

(3) That a building not yet on the National Register appears to meet National Register criteria; or

(4) That a building located within a potential historic district appears to contribute to the significance of such district.

(b) If the building is individually listed in the National Reigster it automatically is considered a certified historic structure, except as provided below.

(1) To determine whether or not a building is individually listed or is part of a district in the National Register, the owner may consult the listing of National Register properties in the "Federal Register" (found in most large libraries), or contact the appropriate SHPO for current information.

(2) If the building is individually listed in the National Register and the owner believes it has lost the characteristics which caused it to be nominated and therefore wishes it delisted, the owner should refer to the delisting procedures outlined in 36 CFR Part 60.

(3) Many individual listings in National Register include more than one building. In such cases, the Secretary will utilize the Standards for Evaluating Significance within Registered Historic Districts for the purpose of determining which of the buildings included within the listing are of historic significance to the property. An individual listing containing more than one building where the buildings

are judged by the Secretary to have been functionally related to serve an overall purpose, such as a mill complex or an industrial plant, will be treated as a single certified historic structure, when rehabilitated as part of an overall project. For questions concerning demolition of separate structures as part of an overall rehabilitation project, see § 67.6.

(4)   If it is proposed that a building individually listed in the National Register be moved as a part of a request for certification of rehabilitation, the owner must follow the procedures outlined in 36 CRF 60. When a building is moved, every effort should be made to reestablish its historic orientation, immediate setting, and general environment.

(c)   If the building is located within the boundaries of a registered historic district and the owner wishes the Secretary to certify as to whether the building contributes or does not contribute to the historic significance of the district or if the owner is requesting a preliminary determination in accordance with subsection 67.3(a)(4), the owner must complete Part 1 of the "Historic Preservation Certification Application" according to instructions accompanying the application. Such documentation includes but is not limited to:

(1)   Name and mailing address of owner;

(2)   Name and address of building;

(3)   Name of historic district;

(4)   Current photographs of building; photographs of the building prior to alteration if rehabilitation has been completed; photograph(s) showing the building along with adjacent buildings and structures on the street; and photographs of interior features and/or spaces adequate to document significance;

(5)   Brief description of appearance including alterations, distinctive features and spaces, and date(s) of construction;

(6)   Brief statement of significance summarizing how the building reflects the values that give the district its distinctive historical and visual character, and explaining any significance attached to the building itself (i.e., unusual building techniques, important events that took place there, etc.);

(7)   Sketch map clearly delineating building's location within the district; and

(8)   Signature of fee simple owner requesting or concurring in a request for evaluation.

(d)   Applications for preliminary determinations for individual listing must show how the building individually meets the National Register Criteria for Evaluation. An application for a building located in a potential historic district must document how the district meets the criteria and how the building contributes to the significance of that district. An application for a preliminary determination for a building in a registered historic district which is outside the period or area of significance in the district documentation on file with the NPS must document and justify the expanded significance of the district and how the building contributes to the significance of the district or document the individual significance of the building. Applications must contain substantially the same level of documentation as National Register nominations, as specified in 36 CFR Part 60 and "How to Complete National Register Forms." Applications must also include written assurance from the State that the district nomination is being revised to expand its significance or, for certified districts, written assurance from the duly authorized representative that the district documentation is being revised to expand its significance. Owners should understand that confirmation of intent to nominate by a State does not constitute listing in the National Register, nor does it constitute a certification of significance as required by law for Federal tax incentives. Owners should further understand that they are proceeding at their own risk. In the event that: (1) the building or district is not listed in the National Register for procedural, substantive or other reasons, (2) district documentation is not formally amended, or (3) the significance of the building has been lost as a result of alterations, final certifications will not be issued. The SHPO for National Register districts and the duly authorized representative for certified districts must submit documentation and have it approved by the NPS to amend the National Register nomination or certified district before the preliminary certification can become final.

(e)   For purposes of the 15 and 20 percent tax credits under the Economic Recovery Tax Act of 1981, buildings within registered historic districts are presumed to contribute to the significance of such districts unless certified as non-significant by the Secretary. Owners of non-historic buildings within registered historic districts, therefore, must obtain certification of non-significance in order to qualify for either the 15 percent (buildings 30-39 years old) or the 20 percent (buildings 40 years or older) investment tax credit. If an owner begins or completes demolition or substantial alteration (within the meaning of Section 167(n) of the Internal Revenue Code) of a building in a registered historic district without knowledge of requirements for certification of non-significance, he or she may request certification that the building was not of historic significance to the district prior to substantial alteration or demolition in the same manner as stated in (c). The owner should be aware, however, of the requirements under section 701(f)(2)(B)(iii) and 701(f)(5)(b) of the Revenue Act of 1978 and section 212(b) of the Economic Recovery Tax Act of 1981 that the taxpayer must certify to the Secretary of the Treasury that, at the beginning of such demolition or substantial alteration, he or she in good faith was not aware of the certification requirement by the Secretary of the Interior.

(f)   If an owner wishes to obtain certification of a building which has been moved (or is proposed to be moved) into a registered historic district or which is within a registered historic district and which has been moved (or will be moved) elsewhere in the district, he must complete Part 1 of the Historic Preservation Certification Application and should submit additional documentation which demonstrates:

(1)   The effect of the move on the building's appearance (and proposed demolition, proposed changes in foundations, etc.);

(2)   The new setting and general environment of the proposed site;

(3)   The effect of the move on the distinctive historical and visual character of the district;

(4)   The method to be used for moving the building.

Photographs showing the proposed location must be sent with the documentation. When a building is moved, every effort should be made to reestablish its historic orientation, immediate setting, and general environment.

(g)   Buildings within registered historic districts will be evaluated to determine if they contribute to the significance of the district by application of the Secretary's "Standards for Evaluating Significance within Registered Historic Districts" as set forth in § 67.5.

(h)   Once the significance of a building located within a registered historic district or a potential historic district has been determined by the Secretary, written notification will be sent to the owner and the State official in the form of a certification of significance or non-significance.

§ 67.5. Standards for evaluating significance within registered historic districts.

(a)   Buildings located within registered historic districts are reviewed by the Secretary to determine if they contribute to the significance of the district by applying the following "Standards for Evaluating Significance within Registered Historic Districts."

(1)   A building contributing to the historic significance of a district is one which by location, design, setting, materials, workmanship, feeling and association adds to the district's sense of time and place and historical development.

(2)   A building not contributing to the historic significance of a district is one which does not add to the district's sense of time and place and historical development; or one where the location, design, setting, materials, workmanship, feeling and association have been so altered or have so deteriorated that the overall integrity of the building has been irretrievably lost.

(3)   Ordinarily buildings that have been built within the past 50 years shall not be considered to contribute to the significance of a district unless a strong justification concerning their historical or architectural merit is given or the historical attributes of the district are considered to be less than 50 years old.

(b)   A condemnation order may be presented as evidence of physical deterioration of a building but will not of itself be considered sufficient evidence to warrant certification of non-significance for loss of integrity. In certain cases it may be necessary for the owner to submit a structural engineer's report to help substantiate physical deterioration and/or structural damage.

(c)   Certifications of significance and non-significance must be consistent with documentation on official file for registered historic districts and individually listed properties. In the event that a certification request is received for a building which is outside a district's established period or area of significance, a preliminary determination of significance will only be issued if the request includes adequate documentation to support the revision and if there is written assurance from the State that the district nomination in question is being revised to expand its significance or for certified districts, written assurance from the duly authorized representative that the district documentation is being revised to expand the significance. Final certifications will be issued when the district documentation is officially amended unless the significance of the building has been lost as a result of alterations. For procedures on amending listings to the National Register, consult the appropriate SHPO or NPS regional office.

(d)   Where rehabilitation credits are sought, certifications of significance will be made on the appearance and condition of the building before rehabilitation was begun.

(e)   In cases where a nonhistoric surface material obscures a facade so that it is impossible to discern whether the building contributes to the significance of the historic district, it may be necessary for the owner to remove a portion of the surface material prior to requesting certification so that a determination of significance can be made.

(f)   Additional guidance on certifications of historic significance is available from State officials and NPS regional offices.

§ 67.6. Certifications of rehabilitation.

(a)   Owners wanting rehabilitation projects for certified historic structures to be certified by the Secretary as being consistent with the historic character of the structure, and, where applicable, the district in which the structure is located, thus qualifying as "certified rehabilitations," shall comply with the procedures listed below. A fee, as described in § 67.12, for reviewing all proposed, ongoing or completed rehabilitation work is charged by the Secretary. Final action will not be taken on any application until the appropriate remittance is received.

(1)   To initiate review of a rehabilitation project for certification purposes, an owner must complete Part 2 of the Historic Preservation Certification Application according to instructions accompanying the application. These instructions explain in detail the documentation required for certification of a rehabilitation project. The application may describe a proposed rehabilitation project, a project in progress, or a completed project. In all cases, documentation, including photographs adequate to document the appearance of the building(s) prior to rehabilitation, both on the exterior and on the interior, must accompany the application as well as the social security or taxpayer identification number(s) of the owner(s). Other documentation may be required by reviewing officials to evaluate certain rehabilitation projects. Plans for any attached or adjacent new construction also must accompany the application. Where such documentation is not provided, review and evaluation may not be completed. Owners are encouraged to submit Part 2 of the application prior to undertaking any rehabilitation work. Owners who undertake rehabilitation projects without prior approval from the Secretary do so at their own risk.

(2)   If requesting certification of a completed rehabilitation project, the owner shall also provide the project completion date and a signed statement indicating that, in the owner's opinion, the completed rehabilitation project meets the Secretary's "Standards for Rehabilitation" and is consistent with the work described in Part 2 of the Historic Preservation Certification Application. Also required in requesting certification of a completed rehabilitation project are: costs attributed to the rehabilitation and photographs adequate to document the completed rehabilitation. A determination that the completed rehabilitation of a building not yet designated a certified historic structure meets the Secretary's "Standards for Rehabilitation" does not constitute a certification of rehabilitation.

(b)   A rehabilitation project for certification purposes encompasses all work on the significant interior and exterior features of the certified historic structure(s) and its setting and environment, as determined by the Secretary, and, related demolition, construction or rehabilitation work which may affect the historic qualities, integrity or setting of the certified historic structure(s). More specific considerations in this regard are as follows:

(1)   All elements of the rehabilitation project must meet the Secretary's ten "Standards for Rehabilitation" (§ 67.7); portions of the rehabilitation project

not in conformance with the Standards may not be exempted. In general, an owner undertaking a rehabilitation project will not be held responsible for rehabilitation work undertaken by previous owners or third parties. However, if the Secretary considers or has reason to consider that a project submitted for certification does not include the entire rehabilitation project undertaken by the owner, or beneficial owner, the Secretary may choose to deny a rehabilitation certification or to withhold a decision on such a certification until such time as the Internal Revenue Service, through a private letter ruling, has determined, pursuant to these regulations and applicable provisions of the Internal Revenue Code and income tax regulations, the proper scope of the rehabilitation project to be reviewed by the Secretary. Factors to be taken into account by the Secretary and the Internal Revenue Service in this regard include, but are not limited to, the facts and circumstances of each application and (i) whether previous demolition, construction or rehabilitation work irrespective of ownership or control at the time was in fact undertaken as part of the rehabilitation project for which certification is sought, and (ii) whether property conveyances, reconfigurations, ostensible ownership transfers or other transactions were transactions which purportedly limit the scope of a rehabilitation project for the purpose of review by the Secretary without substantially altering beneficial ownership or control of the property. The fact that a building may still qualify as a certified historic structure after having undergone inappropriate rehabilitation, construction or demolition work does not preclude the Secretary or the Internal Revenue Service from determining that such inappropriate work is part of the rehabilitation project to be reviewed by the Secretary.

(2)   Conformance to the Standards will be determined by evaluating the building as it existed prior to the commencement of the rehabilitation project, regardless of when the building becomes or became a certified historic structure.

(3)   For rehabilitation projects involving more than one certified historic structure where the structures are judged by the Secretary to have been functionally related historically to serve an overall purpose, such as a mill complex or an industrial plant, rehabilitation certification will be issued on the merits of the overall project rather than on individual components.

(4)   In situations involving rehabilitation of a certified historic structure in a historic district, the Secretary will review the rehabilitation project both as it affects the certified historic structure and its district and make a certification decision accordingly.

(5)   In the event that an owner of a portion of a certified historic structure requests certification for a rehabilitation project related only to that portion, but there is or was a larger related rehabilitation project(s) occurring with respect to the certified historic structure, the Secretary's decision on the requested certification will be based on review of the overall rehabilitation project(s) for the certified historic structure.

(c)   Upon receipt of the complete application describing the rehabilitation project, the Secretary shall determine if the project is consistent with the "Standards for Rehabilitation." If the project does not meet the "Standards for Rehabilitation," the owner shall be advised of that fact in writing and, where possible, will be advised of necessary revisions to meet such standards. For additional procedures regarding rehabilitation projects determined not to meet the "Standards for Rehabilitation," see § 67.6(f).

(d)   Once a proposed or ongoing project has been approved, substantive changes

in the work as described in the application shall be promptly brought to the attention of the Secretary by written statement, with a copy to the appropriate State
official, to ensure continued conformance to the Standards; such changes should
be made using a Historic Preservation Certification Application Continuation Sheet.
The Secretary will notify the owner and the State official in writing whether the
revised project continues to meet the Standards. Oral approvals of revisions are
not authorized or valid.

(e)   Completed projects may be inspected by an authorized representative of
the Secretary to determine if the work meets the "Standards for Rehabilitation."
The Secretary reserves the right to make inspections at any time up to five years
after completion of the rehabilitation and to revoke a certification, after giving
the owner 30 days to comment on the matter, if it is determined that the rehabilitation project was not undertaken as presented by the owner in his or her application and supporting documentation, or the owner, upon obtaining certification,
undertook unapproved further alterations as part of the rehabilitation project
inconsistent with the Secretary's "Standards for Rehabilitation." The tax consequences of a revocation of certification will be determined by the Secretary of the
Treasury.

(f)   In the event that a proposed, ongoing, or completed rehabilitation project
does not meet the "Standards for Rehabilitation," an explanatory letter will be
sent to the owner with a copy to the State official. A rehabilitated building not
in conformance with the "Standards for Rehabilitation" and which is determined
to have lost those qualities which caused it to be nominated to the National Register, will be removed from the National Register in accord with Department of the
Interior regulations 36 CFR Part 60. Similarly, if a building has lost those qualities
which caused it to be designated a certified historic structure, it will be certified as
non-contributing (see § § 67.4 and 67.5). In either case, the delisting or certification of non-significance is considered effective as of the date of issue and is not
considered to be retroactive. In these situations, the Internal Revenue Service will
be notified of the substantial alteration. The tax consequences of a denial of certification will be determined by the Secretary of the Treasury.

§ 67.7. Standards for rehabilitation.

(a)   The following "Standards for Rehabilitation," a section of the Secretary's
"Standard for Historic Preservation Projects" (see 36 CFR Part 68), are the guidelines used to determine if a rehabilitation project of a certified historic structure
qualifies as a certified rehabilitation. The Standards shall be applied taking into
consideration the economic and technical feasibility of each project; in the final
analysis, however, to be certified, the rehabilitation project must be consistent
with the historic character of the structure(s) and, where applicable, the district
in which it is located.

(1)   Every reasonable effort shall be made to provide a compatible use
for a property which requires minimal alteration of the building, structure, or site
and its environment, or to use a property for its originally intended purpose.

(2)   The distinguishing original qualities or character of a building, structure, or site and its environment shall not be destroyed. The removal or alteration
of any historic material or distinctive architectural features should be avoided when
possible.

(3)   All buildings, structures, and sites shall be recognized as products of their own time. Alterations that have no historical basis and which seek to create an earlier appearance shall be discouraged.

(4)   Changes which may have taken place in the course of time are evidence of the history and development of a building, structure, or site and its environment. These changes may have acquired significance in their own right, and this significance shall be recognized and respected.

(5)   Distinctive stylistic features or examples of skilled craftsmanship which characterize a building, structure, or site shall be treated with sensitivity.

(6)   Deteriorated architecture features shall be repaired rather than replaced, wherever possible. In the event replacement is necessary, the new material should match the material being replaced in composition, design, color, texture, and other visual qualities. Repair or replacement of missing architectural features should be based on accurate duplications of features, substantiated by historic, physical, or pictorial evidence rather than on conjectural designs or the availability of different architectural elements from other buildings or structures.

(7)   The surface cleaning of structures shall be undertaken with the gentlest means possible. Sandblasting and other cleaning methods that will damage the historic building materials shall not be undertaken.

(8)   Every reasonable effort shall be made to protect and preserve archeological resources affected by, or adjacent to any project.

(9)   Contemporary design for alterations and additions to existing properties shall not be discouraged when such alterations and additions do not destroy significant historical, architectural, or cultural material, and such design is compatible with the size, scale, color, material, and character of the property, neighborhood or environment.

(10)   Wherever possible, new additions or alterations to structures shall be done in such a manner that if such additions or alterations were to be removed in the future, the essential form and integrity of the structure would be unimpaired.

(b)   Certain treatments, if improperly applied, or certain materials by their physical properties, may cause or accelerate physical deterioration of historic buildings. Inappropriate physical treatments include, but are not limited to: improper repointing techniques and improper exterior masonry cleaning methods; and the introduction of insulation into cavity walls of historic woodframe buildings where damage to historic fabric would result. In almost all situations, use of these materials and treatments will result in certification denial. Similarly, exterior additions that duplicate the form, material, style and detailing of the structure to the extent that they compromise the historic character of the structure will result in certification denial. For specific information on appropriate and inappropriate rehabilitation treatments, owners should consult the "Preservation Briefs" series published by the National Park Service. Additional guidelines and other technical information to help property owners formulate plans for the rehabilitation, preservation, and continued use of historic properties consistent with the intent of the Secretary's "Standards for Rehabilitation" are available from the appropriate SHPO or NPS regional office.

(c)   In certain limited cases, it may be necessary to dismantle and rebuild portions of a certified historic structure to stabilize and repair weakened structural members and systems. In such cases, the Secretary will consider such extreme interventions as part of a certified rehabilitation if (1) the necessity for dismantling is justified in supporting documentation; (2) significant architectural features and overall design are retained; and (3) adequate historic materials are retained to maintain the architectural and historic integrity of the overall structure. Substantial alterations undertaken between June 30, 1976, and December 31, 1981, may be subject to the provisions of section 167(n) of the Internal Revenue Code. The Economic Recovery Tax Act of 1981 requires that 75 percent or more of the existing external walls remain as external walls in the rehabilitation process to qualify for the investment tax credit.

(d)   Prior approval of a project by Federal, State and local agencies and organizations does not ensure certification by the Secretary for Federal tax purposes. The Secretary's Standards take precedence over other regulations and codes in determining whether the rehabilitation project is consistent with the historic character of the building and/or the district in which it is located.

§ 67.8. Certification of statutes.

(a)   State or local statutes which will be certified by the Secretary. For the purpose of this regulation, a State or local statute is a law of the State or local government designating, or providing a method for the designation of, a historic district or districts. This includes any by-laws or ordinances that contain information necessary for the certification of the statute. A statute must contain criteria which will substantially achieve the purpose of preserving and rehabilitating buildings of historic significance to the district. To be certified by the Secretary, the statute generally must provide for a duly designated review body, such as a review board or commission, with power to review proposed alterations to structures of historic significance within the boundaries of the district or districts designated under the statute except those owned by the State.

(b)   When the certification of State statutes will have an impact on districts in specific localities, the Secretary encourages State governments to notify and consult with appropriate local officials prior to submitting a request for certification of the statue.

(c)   State enabling legislation which authorizes local governments to designate, or provides local governments with a method to designate, a historic district or districts will not be certified unless accompanied by local statutes that implement the purpose of the State law. Adequate State statutes which designate specific historic districts and do not require specific implementing local statutes will be certified. If the State enabling legislation contains provisions which do not meet the intent of the law, local statutes designated under the authority of the enabling legislation will not be certified. When State enabling legislation exists, it must be certified before any local statutes enacted under its authority can be certified.

(d)   *Who may apply.* Requests for certification of State or local statutes may be made only by the Chief Elected Official of the government which enacted the statute or his or her duly authorized representative. The applicant shall certify in writing that he or she is authorized by the appropriate State or local governing body to apply for certification.

(e)   *Statute certification process.* Requests for certification of State or local statutes shall be made as follows:

(1)   The request shall be made in writing from the duly authorized representative certifying that he or she is authorized to apply for certification. The request should include the name or title of a person to contact for further information and his or her address and telephone number. The authorized representative is responsible for providing historic district documentation for review and certification prior to the first certification of significance in a district unless another responsible person is indicated including his or her address and telephone number. The request shall also include a copy of the statute(s) for which certification is requested, including the bylaws or ordinances that contain information necessary for the certification of the statute. Local governments shall also submit a copy of the State enabling legislation, if any, authorizing the designation of historic districts.

(2)   The address to which requests should be sent may be obtained by contacting the appropriate NPS regional office or State official. These requests shall be sent to the appropriate State official in participating States. Requests from owners in non-participating States should be sent to the appropriate NPS regional office.

(3)   The Secretary shall review the statute(s) and assess whether the statute(s) and any bylaws or ordinances that contain information necessary for the certification of the statute contain criteria which will substantially achieve the purposes of preserving the rehabilitating building of historic significance to the district(s) based upon the standards set out above in § 67.3(a). The State shall be given a 30-day opportunity to comment upon the request. State comments received within this time period will be considered by the Secretary in the review process. If the statute(s) contain such provisions and if this and other provisions in the statute will substantially achieve the purpose of preserving and rehabilitating buildings of historic significance to the district, the Secretary will certify the statute(s).

(4)   The Secretary shall provide written notification within 60 days to the duly authorized representative and the State official when certification of the statute is given or denied. If certification is denied, the notification will provide an explanation of the reason(s) for such denial.

(f)   Amendment or Repeal of statute(s). State or local governments, as appropriate, must notify the Secretary in the event that certified statutes are repealed, whereupon the certification of the statute (and any districts designated thereunder) will be withdrawn by the Secretary. If a certified statute is amended, the duly authorized representative shall submit the amendment(s) to the Secretary, with a copy to the State, for review in accordance with procedures outlined above. Written notification of the Secretary's decision as to whether the amended statute continues to meet these criteria will be sent to the duly authorized representative and the State official within 60 days of receipt.

(g)   The Secretary may withdraw certification of a statute (and any districts designated thereunder) on his own initiative if it is repealed or amended to be inconsistent with certification requirements after providing the duly authorized representative and the State official 30 days in which to comment prior to the withdrawal of certification.

§ 67.9. Certifications of State or local historical districts.

(a)   The particular State or local historic district must also be certified by the Secretary as substantially meeting National Register criteria, thereby qualifying it as a "registered historic district," before the Secretary will process requests for certification of individual buildings within a district or districts established under a certified statute.

(b)   The provisions described herein will not apply to buildings within a State or local district until the district has been certified, even if the statute creating the district has been certified by the Secretary.

(c)   The Secretary considers the duly authorized representative requesting certification of a statute to be the official responsible for submitting district documentation for certification. If another person is to assume responsibility for the district documentation, the letter requesting statute certification shall indicate that person's name, address and telephone number. The Secretary considers the authorizing statement of the duly authorized representative to indicate that the jurisdiction involved wishes not only that the statute in question be certified but also wishes all historic districts designated by the statute to be certified unless otherwise indicated.

(d)   The address to which requests should be sent may be obtained by contacting the appropriate NPS regional office or State official. These requests shall be sent to the appropriate State official in participating States. Requests from duly authorized representatives in non-participating States should be sent to the appropriate NPS regional office. The State shall be given a 30 day opportunity to comment upon an adequately documented request. State comments received within this time period will be considered by the Secretary in the review process. Each request should include the following documentation:

(1)   A concise description of the general physical or historical qualities which make this a district; an explanation for the choice of boundaries for the district; descriptions of typical architectural styles and types of buildings in the district.

(2)   A concise statement of why the district has significance and why it substantially meets National Register criteria for listing (see 36 CFR 60); the relevant criteria should be identified (A, B, C, or D).

(3)   A definition of what types of buildings do not contribute to the significance of the district as well as an estimate of the percentage of buildings within the district that do not contribute to its significance.

(4)   A map showing all district buildings with, if possible, identification of contributing and non-contributing buildings; the map should clearly show the district's boundaries.

(5)   Photographs of typical areas in the district as well as major types of contributing and non-contributing buildings (all photos should be keyed to the map).

(e)   Districts designated by certified State or local statutes shall be evaluated using the National Register criteria (36 CFR 60) within 60 days of the receipt of the required documentation. Written notification of the Secretary's decision will be sent to the duly authorized representative or to the person designated as responsible for the district documentation.

(f)   Certification of statutes and districts does not constitute certification of significance of individual buildings within the district or of rehabilitation projects by the Secretary.

(g)   Districts certified by the Secretary as substantially meeting the requirements for listing will be determined eligible for listing in the National Register at the time of certification and will be published as such in the **Federal Register.**

(h)   Documentation on additional districts designated under a State or local statute that has been certified by the Secretary should be submitted to the Secretary for certification following the same procedure and including the same information outlined in the section above.

(i)   State or local governments, as appropriate, shall notify the Secretary if a certified district designation is amended (including boundary changes) or repealed. If a certified district designation is amended, the duly authorized representative shall submit documentation describing the change(s) and, if the district has been increased in size, information on the new areas as outlined in § 67.9. A revised statement of significance for the district as a whole shall also be included to reflect any changes in overall significance as a result of the addition or deletion of areas. Review procedures shall follow those outlined in § 67.9 (d) and (e). The Secretary will withdraw certification of repealed or inappropriately amended certified district designations, thereby disqualifying them as registered historic districts.

(j)   The Secretary may withdraw certification of a district on his own initiative if it ceases to meet the National Register Criteria for Evaluation after providing the duly authorized representative and the State official 30 days in which to comment prior to withdrawal of certification.

(k)   The Secretary urges State and local review boards or commissions to become familiar with the standards used by the Secretary of the Interior for certifying the rehabilitation of historic buildings and to consider their adoption for local design review.

§ 67.10. Appeals.

(a)   An appeal by the owner or duly authorized representative as appropriate may be made from any of the certifications or denials of certification made pursuant to this part or any decisions made pursuant to § 67.6(e). Such appeals must be in writing and received by the Chief Appeals Officer, Cultural Resources, National Park Service, U.S. Department of the Interior, Washington, D.C. 20240, within 30 days of receipt of the decision which is the subject of the appeal. The appellant may request an opportunity for a meeting to discuss the appeal. The State official will be notified that an appeal is pending. The Chief Appeals Officer will review such appeals and the written record of the decision in question, and shall notify the appellant of his decision within 30 days of its receipt if circumstances permit.

(b)   The denial of a preliminary determination of significance for an individual building may not be appealed by the owner because the denial itself does not exhaust the administrative remedy that is available. The owner instead must seek recourse by undertaking the usual nomination process (36 CFR Part 60). Similarly, the denial of preliminary certification for a rehabilitation project for a building that is not a certified historic structure may not be appealed. The owner must seek a final certification of significance as the next step, rather than appealing the denial of rehabilitation certification. Administrative reviews in these circumstances may be performed at the discretion of the Chief Appeals Officer. The decision to undertake an administrative review will be made on a case-by-case basis, depending on particular facts and circumstances and the Chief Appeals Officer's schedule, the expected date for nomination, and the nature of the rehabilitation project (proposed, ongoing or completed). Administrative reviews of rehabilitation projects will not be undertaken if the owner has objected to the listing of the building in the National Register.

(c)   In such appeals or administrative reviews:

(1)   The Chief Appeals Officer shall consider: (i) alleged errors in professional judgment; (ii) alleged prejudicial procedural errors; and (iii) any additional information provided.

(2)   The Chief Appeals Officer's decision may: (i) reverse the appealed decision; (ii) affirm the appealed decision, (iii) resubmit the matter to the appropriate Regional Director for further consideration; or (iv) where appropriate, withhold a decision until issuance of a ruling from the Internal Revenue Service pursuant to Section 67.6(b)(1). The Chief Appeals Officer is authorized to issue the certifications discussed in this part only if he considers that the requested certification meets the applicable statutory standard upon application of the guidelines set forth herein or prejudicial procedural error legally compels issuance of the request certification.

(d)   The decision of the Chief Appeals Officer shall be the final administrative decision on the appeal. No person shall be considered to have exhausted his or her administrative remedies with respect to the certifications or decisions described in this part until the Chief Appeals Officer has issued a final administrative decision pursuant to this section.

§ 67.11. Expedited review system for qualified States.

(a)   Expedited review of certification requests is an objective of the Secretary. Qualified States wishing to participate in the review and processing of Part 1 and Part 2 certification requests can greatly assist in expediting the process. The procedures detailed below will eliminate duplication of effort and enable qualified States to assume greater responsibilty for making certification recommendations. The procedures will enable the Secretary to make certification decisions in 15 days, shortening the total review time for applications from 60 to 45 days. This system does not apply to the review of State or local statutes or districts.

(b)   States wishing to obtain "qualified State" status will be evaluated by the Secretary prior to receiving such status to ensure that:

(1)   Each certification request, for evaluations of significance and rehabilitation, is reviewed by professionally qualified staff in accordance with procedures set forth herein.

(2)   The State is able to document that it has reviewed, and is reviewing, certification requests and has made, and is making certification recommendations consistent with established standards, within the specified 30-day time frame, and other guidelines established by the Secretary.

Guidelines for evaluating whether or not a State is qualified shall be established by the Secretary.

(c)   A request for "qualified State" status may be made in writing at any time by the State official to the appropriate National Park Service regional office. The Secretary shall evaluate each request and notify the State in writing of his determination.

(d)   The performance of a qualified State in administering the certification program shall be reviewed on an ongoing basis. "Qualified" status, however, may be revoked at any time, with 30 days notice, if it is determined by the Secretary that the State is not meeting the guidelines for qualified State status.

(e)   All certification requests from qualified States will generally be processed within 15 days by the Secretary. These time frames, however, are not binding upon the Secretary.

(f)   Qualified States are encouraged to provide local governments certified in accordance with the National Historic Preservation Act and Department of the Interior guidelines an opportunity to participate in the certification program, within the time periods established in § 67.3.

(g)   States not wishing "qualified State" status may continue to comment on any or all certification requests within their jurisdiction, as described in § 67.1(c).

§ 67.12. Fees for processing rehabilitation certification requests.

(a)   Fees are charged for reviewing rehabilitation certification requests in accordance with the schedule below. The fee schedule described in this part shall apply to all requests for certification of rehabilitation received by the State official or the NPS regional office after the effective date of this regulation.

(b)   Payment shall not be made until requested by the NPS Regional Office according to instructions accompanying the Historic Preservation Certification Application. All checks shall be made payable to: NATIONAL PARK SERVICE. Final action will not be taken on an application until the appropriate remittance is received. Fees are nonrefundable.

(c)   The fee for review of proposed or ongoing rehabilitation projects for projects over $20,000 is $250. The fees for review of completed rehabilitation projects are based on the dollar amount of the costs attributed solely to the rehabilitation of the certified historic structure provided by the owner in the Historic Preservation Certification Application, Request for Certification of Completed Work, as follows:

Fee	Size of rehabilitation
$500	$20,000 to $99,000
$800	$100,000 to $499,999
$1,500	$500,000 to $999,999
$2,500	$1,000,000 or more

If review of a proposed or ongoing rehabilitation project had been undertaken by the Secretary prior to submission of a Request for Certification of Completed Work, the initial fee of $250 will be deducted from these fees. No fee will be charged for rehabilitations under $20,000.

(d)   In general, each rehabilitation of a separate certified historic structure will be considered a separate project for purposes of computing the size of the fee.

(1)   In the case of a rehabilitation project which includes more than one certified historic structure where the structures are judged by the Secretary to have been functionally related historically to serve an overall purpose, the fee for preliminary review is $250 and the fee for final review is computed on the basis of the total rehabilitation costs.

(2)   In the case of multiple building projects which are under the same ownership; are located in the same historic district; are adjacent or contiguous; are of the same architectural type (e.g., rowhouses, loft buildings, etc.), and are submitted for review at the same time, the maximum total fee is $2,500. In this situation, the fee for preliminary review is $250 per building to a maximum of $2,500 and the fee for final review is computed on the basis of the total rehabilitation costs of the entire multiple building project to a maximum of $2,500. If the $2,500 maximum fee was paid at the time of review of the proposed or ongoing rehabilitation project, no further fee will be charged for a Request for Certification of Completed Work.

[FR Doc. 84-8452 Filed 3-11-84: 8.45 am]

Appendix C

National Park Service, Historic Preservation Certification Application Form

Form 10-168
Rev. 3/84

UNITED STATES DEPARTMENT OF THE INTERIOR

NATIONAL PARK SERVICE

OMB Approved
No. 1024-0009
Expires 8/31/86

HISTORIC PRESERVATION CERTIFICATION APPLICATION
PART 1 — EVALUATION OF SIGNIFICANCE

NPS Office Use Only

Project Number:

Instructions: Read the instructions carefully before completing application. No certification will be made unless a completed application form has been received. Use typewriter or print clearly in black ink. If additional space is needed, use continuation sheets or attach blank sheets.

---

1. Name of property: _____

   Address of property: _____

   City _____ County _____ State _____ Zip Code _____

   Name of historic district: _____

   ☐ National Register district   ☐ certified state or local district   ☐ potential historic district

2. Check nature of request:
   ☐ certification that the building contributes to the significance of the above-named historic district for the purpose of rehabilitation.
   ☐ certification that the structure or building and, where appropriate, the land area on which such a structure or building is located contributes to the significance of the above-named historic district for a charitable contribution for conservation purposes.
   ☐ certification that the building does not contribute to the significance of the above-named district.
   ☐ preliminary determination for individual listing in the National Register.
   ☐ preliminary determination that a building located within a potential historic district contributes to the significance of the district.
   ☐ preliminary determination that a building outside the period or area of significance contributes to the significance of the district.

3. Authorized project contact:

   Name _____ Title _____

   Street _____ City _____

   State _____ Zip _____ Telephone Number (during day): _____

4. Owner:

   Name _____

   Street _____ City _____

   State _____ Zip _____ Telephone Number (during day): _____

   I hereby attest that the information I have provided is, to the best of my knowledge, correct, and that I own the above-named property.

   Owner's Signature _____ Date _____

   Social Security Number or Taxpayer Identification Number _____

NPS Office Use Only

The National Park Service has reviewed the "Historic Preservation Certification Application — Part 1" for the above-named property and hereby determines that the property:

☐ contributes to the significance of the above-named district and is a "certified historic structure" for the purpose of rehabilitation.
☐ contributes to the significance of the above-named district and is a "certified historic structure" for a charitable contribution for conservation purposes in accordance with the Tax Treatment Extension Act of 1980.
☐ does not contribute to the significance of the above-named district.

Preliminary Determinations:

☐ appears to meet the National Register Criteria for Evaluation and will likely be listed in the National Register of Historic Places if nominated by the State Historic Preservation Officer according to the procedures set forth in 36 CFR Part 60.
☐ does not appear to meet the National Register Criteria for Evaluation and will likely not be listed in the National Register.
☐ appears to contribute to the significance of a potential historic district, which will likely be listed in the National Register of Historic Places if nominated by the State Historic Preservation Officer.
☐ appears to contribute to the significance of a registered historic district but is outside the period or area of significance as documented in the National Register nomination or district documentation on file with the NPS.
☐ does not appear to qualify as a certified historic structure.

Date _____   National Park Service Authorized Signature _____   National Park Service Office _____

HISTORIC PRESERVATION CERTIFICATION APPLICATION— PART 1

Property Name _____

Project Number: _____

Property Address _____

Owner Name/Social Security or Taxpayer ID Number _____

5. Description of physical appearance:

Date of Construction: _____ Source of Date: _____

Date(s) of Alteration(s): _____

Has building been moved? ☐ yes ☐ no. If so, when? _____

6. Statement of significance:

7. Photographs and maps.

   Attach photographs and maps to application.

Continuation sheets attached: ☐ yes ☐ no

Form 10-168a
Rev. 3/84

UNITED STATES DEPARTMENT OF THE INTERIOR

NATIONAL PARK SERVICE

OMB Approved
No. 1024-0009
Expires 8/31/86

HISTORIC PRESERVATION CERTIFICATION APPLICATION
PART 2 — DESCRIPTION OF REHABILITATION

NPS Office Use Only

Project Number:

Instructions: Read the instructions carefully before completing application. No certification will be made unless a completed application form has been received. Use typewriter or print clearly in black ink. If additional space is needed, use continuation sheets or attach blank sheets. A copy of this form may be provided to the Internal Revenue Service.

1. Name of property: _____

   Address of property: Street _____

   City _____ County _____ State _____ Zip Code _____

   If located in a Registered Historic District, specify: _____

   If listed individually in the National Register of Historic Places, give date of listing: _____

   Has a Part 1 Application (Evaluation of Significance) been submitted for this project? ☐ yes ☐ no

   If yes, date Part 1 submitted: _____ Date of certification: _____ NPS Project Number: _____

2. Data on building:

   Date of construction: _____ Number of housing units before rehabilitation: _____

   Use before rehabilitation: _____ Floor area before rehabilitation: _____

   Type of construction: _____

3. Data on rehabilitation project:

   Project starting date (est.): _____ Project completion date (est.): _____

   Estimated cost of rehabilitation: _____ Proposed use: _____

   Number of housing units after project completion: _____ Floor area after rehabilitation: _____

   Has the property received Federal or State financial assistance? ☐ yes ☐ no

   If yes, specify source by program title: _____

4. Authorized project contact:

   Name _____ Title _____

   Street _____ City _____

   State _____ Zip _____ Telephone Number (during day): _____

5. Owner:

   Name _____

   Street _____ City _____

   State _____ Zip _____ Telephone Number (during day): _____

   I hereby attest that the information I have provided is, to the best of my knowledge, correct, and that I am the owner of the property described above.

   Owner's Signature _____ Date _____

   Social Security Number or Taxpayer Identification Number _____

NPS Office Use Only

The National Park Service has reviewed the "Historic Preservation Certification Application — Part 2" for the above-named property and has determined:

☐ that the rehabilitation described herein is consistent with the historic character of the property or the district in which it is located and that the project meets the Secretary of the Interior's "Standards for Rehabilitation."

☐ that the rehabilitation or proposed rehabilitation will meet the Secretary of the Interior's "Standards for Rehabilitation" if the attached conditions are met.

☐ that the rehabilitation or proposed rehabilitation is not consistent with the historic character of the property or the district in which it is located and that the project does not meet the Secretary of the Interior's "Standards for Rehabilitation."

Date _____ National Park Service Authorized Signature _____ National Park Service Office _____

HISTORIC PRESERVATION
CERTIFICATION APPLICATION—
PART 2

NPS Office Use Only

Property Name _____

Project Number: _____

Property Address _____

Owner Name/Social Security or Taxpayer ID Number _____

6. DETAILED DESCRIPTION OF REHABILITATION/PRESERVATION WORK—Includes site work, new construction, alterations, etc. Complete blocks below.

| NUMBER 1 | Architectural feature _____ | Describe work and impact on existing feature: |

Approximate date of feature _____

Describe existing feature and its condition:

Photo no. _____ Drawing no. _____

| NUMBER 2 | Architectural feature _____ | Describe work and impact on existing feature: |

Approximate date of feature _____

Describe existing feature and its condition:

Photo no. _____ Drawing no. _____

| NUMBER 3 | Architectural feature _____ | Describe work and impact on existing feature: |

Approximate date of feature _____

Describe existing feature and its condition:

Photo no. _____ Drawing no. _____

| NUMBER 4 | Architectural feature _____ | Describe work and impact on existing feature: |

Approximate date of feature _____

Describe existing feature and its condition:

Photo no. _____ Drawing no. _____

Form 10-168c
Rev. 3/84

UNITED STATES DEPARTMENT OF THE INTERIOR

NATIONAL PARK SERVICE

OMB Approved
No. 1024-0009
Expires 8/31/86

HISTORIC PRESERVATION CERTIFICATION APPLICATION
REQUEST FOR CERTIFICATION OF COMPLETED WORK

Instructions: Upon completion of the rehabilitation, return this form with representative photographs of the completed work (both exterior and interior views) to the appropriate reviewing office. If a Part 2 application has not been submitted in advance of project completion, it must accompany this Request for Certification of Completed Work. A copy of this form will be provided to the Internal Revenue Service. Use typewriter or print clearly in black ink.

1. **Name of property:** _____

   Address of property: Street _____

   City _____ County _____ State _____ Zip Code _____

   Is property a certified historic structure? ☐ yes ☐ no If yes, date of certification by NPS: _____

   or date of listing in the National Register: _____

2. **Data on rehabilitation project:**

   National Park Service assigned rehabilitation project number: _____

   Rehabilitation work on this property was completed and the building placed in service as of _____ (date)

   Estimated costs attributed solely to the rehabilitation of the historic structure: $ _____

   Estimated costs attributed to new construction associated with the
   rehabilitation, including additions, site work, parking lots, landscaping: $ _____

3. **Owner:** (space on reverse for additional owners)

   Name _____

   Street _____ City _____

   State _____ Zip _____ Telephone Number (during day): _____

   I hereby apply for certification of rehabilitation work described above for purposes of the Federal tax incentives. I hereby attest that the information I have provided is, to the best of my knowledge, correct, and that in my opinion the completed rehabilitation meets the Secretary's "Standards for Rehabilitation" and is consistent with the work described in Part 2 of the Historic Preservation Certification Application. I also attest that I am the owner of the property described above.

   Signature _____ Date _____

   Social Security or Taxpayer Identification Number: _____

NPS Office Use Only

The National Park Service has reviewed the "Historic Preservation Certification Application — Part 2" for the above-listed "certified historic structure" and has determined:

☐ that the completed rehabilitation meets the Secretary of the Interior's "Standards for Rehabilitation" and is consistent with the historic character of the property or the district in which it is located. Effective the date indicated below, the rehabilitation of the "certified historic structure" is hereby designated a "certified rehabilitation." A copy of this certification has been provided to the Department of the Treasury in accordance with Federal law. This letter of certification is to be used in conjunction with appropriate Internal Revenue Service regulations. Questions concerning specific tax consequences or interpretations of the Internal Revenue Code of 1954 should be addressed to the appropriate local Internal Revenue Service office.

☐ that the rehabilitation is not consistent with the historic character of the property or the district in which it is located and that the project does not meet the Secretary of the Interior's "Standards for Rehabilitation."

_____   _____   _____
Date   National Park Service Authorized Signature   National Park Service Office

REQUEST FOR CERTIFICATION OF COMPLETED WORK, *continued*

NPS Project Number _____

Additional Owners:

Name _____

Street _____

City _____ State _____ Zip Code _____

Social Security or Taxpayer Identification Number: _____

Name _____

Street _____

City _____ State _____ Zip Code _____

Social Security or Taxpayer Identification Number: _____

Name _____

Street _____

City _____ State _____ Zip Code _____

Social Security or Taxpayer Identification Number: _____

Name _____

Street _____

City _____ State _____ Zip Code _____

Social Security or Taxpayer Identification Number: _____

Name _____

Street _____

City _____ State _____ Zip Code _____

Social Security or Taxpayer Identification Number: _____

For State Office Use Only

This office _____ recommends _____ does not recommend the above-listed "certified historic structure" be designated as a "certified rehabilitation."

_____ Additional comments attached

_____   _____
Date   State Official Signature

Appendix D

Interpreting the Secretary of the Interior's Standards for Rehabilitation

(selected issues)

[*Note:* As distributed by the National Park Service, these publications contain copies of photographs of the buildings to which they relate, illustrating the particular features being discussed. These photographs have not been reproduced here.]

Number: 84-058

Applicable Standards

2. Retention of Distinguishing Architectural Character (nonconformance)
5. Sensitive Treatment of Distinctive Features and Craftsmanship (nonconformance)
9. Compatible Contemporary Design for New Alterations/Additions (nonconformance)

Subject: Inappropriate Size and Scale of New Exterior Additions: Loss of Historic Character

Issue: In the Secretary of the Interior's "Standards for Rehabilitation," the Department of the Interior acknowledges that a new exterior addition to a historic building (such as a fire stair, service wing, or additional story) may be essential to return the property to a state of utility for an efficient contemporary use; however, at the same time, the cumulative effect of the design and installation process of a new addition must not radically change, damage, destroy, or obscure those "portions and features of the property which are significant to its historic, architectural, or cultural values" (36 CFR 67.2).

Therefore, in evaluating the appropriateness of a new addition, it is critical that the important character-defining materials, form, features, and detailing of the historic building be properly identified so that they may be protected and preserved. This identification process will also make clear those "portions and features" of the historic property that are *not* important in defining the historic character and may thus be reasonably altered or added to in the course of rehabilitating for the new use.

Because of the difficulty in designing sensitive new additions and to clarify what constitutes a compatible new addition, the NPS has expanded its guidance in this area (see pp. 56-57, "New Exterior Additions to Historic Buildings" in the Revised Guidelines to the Standards for Rehabilitation, 1983). The advice listed first in the guidelines is to avoid constructing a new exterior addition altogether because of the potential for altering and expanding the historic form and thereby diminishing the historic character. Rather, it is recommended that services and functions required by the new use be located in non-character-defining *interior* spaces. Only after it is determined that interior spaces cannot be utilized, should a new exterior addition be considered at all. Then, the new addition should be designed so that its size and scale are limited in proportion in relationship to the historic building— and located on an inconspicuous side of a historic building to further assure that there will be no radical changes to the historic form and appearance.

The failure to recognize those qualities that comprise a building's historic character (its materials, form, features, and detailing as well as relationship to the site and the district) prior to designing and attaching a new exterior addition can result in overall changes that are inconsistent with the historic character. In consequence, Standard 2, 5, or 9 may be violated, thus jeopardizing project certification.

Application: A small late 1920s Mission Revival building of brick construction with stucco finish is primarily distinguished on the main facade by a waved parapet cap and symmetrically placed openings (see illus. 1). In rehabilitating the building for use as law offices, interior and exterior work was undertaken, including replacement of damaged plastered walls, re-stuccoing of the brick, cleaning and painting of windows, and the construction of two new exterior additions.

The first new addition consisted of enclosing existing stairs at one end of the facade for the clients' main entrance, as well as serving as handicapped access to a ground floor elevator. The second new addition was a non-functional matching wing wall at the other end of the facade which the developer felt would preserve the sense of symmetry which was so strong in the historic building (see illus. 2 and 3).

After reviewing the Part II application, the State office recommended denial of the project, citing violation of Standards 5 and 9; the regional office, completing its review, concurred with the State's assessment. In a denial letter to the owner, the regional office stated:

> The new additions, consisting of the exterior stairs enclosure at one end of
> the facade and the wing wall at the other end, increase the length of the
> facade by at least one-third, thereby altering significantly its overall mass,
> scale, and proportional relationships. Further, these additions extend and
> expand on the symmetrical historic design of the facade in a way that lends
> to it a degree of expansiveness . . . not present in the simple design character
> of the structure's original design features. It is apparent that the attempt
> to match the color, texture, and detail of the original design and to con-
> tinue its symmetry by extending the facade wall was motivated by a desire
> to preserve the historic character of the building. In effect, however, this
> matching new design is incompatible: it compounds the additions' negative

visual impacts on the original design by making contemporary and historic portions of the building indistinguishable from one another.

When the project was subsequently appealed, the Chief Appeals Officer sustained the regional office's decision that the new additions violated Standards 5 and 9, adding that "they also give the building a monumentality that, historically, it never possessed, thus changing its historic character." In consequence, the project also failed to conform to Standard 2. As part of the appeals process, the architect forwarded three drawings (schemes A, B, and C; see illus. 4, 5, and 6) for possible changes to the new additions to bring the project into conformance with the Standards and thus qualify for Federal historic preservation tax incentives. After reviewing all of the drawings, the Chief Appeals Officer concluded in his final letter to the owner:

> The only remedial action that can now be taken . . . would be to follow scheme "C": insert a wide expansion joint between the historic building facade and the new stair enclosure, demolish the new wing wall, lower the parapet on the stair tower by at least one foot, and paint the new addition a different color than the original facade. These actions would make the distinction between the old and new construction clear; and would restore to the building its aspect of a modest, simplified Spanish Colonial Revival commercial structure. Demolition of the wing wall would allow one to view the continuous wavy cornice as it carries around the corner. If the final revised project fails to meet any of the above conditions, it will not meet the Standards and cannot be certified.

Prepared by: Kay D. Weeks, TPS

These bulletins are issued to explain preservation project decisions made by the U.S. Department of the Interior. The resulting determinations, based on the Secretary of the Interior's Standards for Rehabilitation, are not necessarily applicable beyond the unique facts and circumstances of each particular case.

Number: 84-059

Applicable Standards:

2. Retention of Distinguishing Architectural Character (nonconformance)
5. Sensitive Treatment of Distinctive Features and Craftsmanship (non-conformance)
6. Repair/Replacement of Deteriorated or Missing Features Based on Historical Evidence (nonconformance)

Subject: Replacing a Significant Interior Feature to Meet Health and Safety Code Requirements

Issue: To comply with health and safety codes in rehabilitation projects, the Revised Guidelines to the Secretary of the Interior's "Standards for Rehabilitation" first recommendation to owners and developers is to work with local code officials to investigate variances available under some codes or to devise creative and safe alternatives so that alterations and additions to historic buildings can be avoided completely, if possible. Because such variances or alternatives may not always be feasible, owners and developers are next advised to identify significant spaces, features, and finishes, so that they can be preserved in the process of successfully meeting code requirements (such as providing barrier-free access, upgrading historic stairways or elevators, or installing fire suppression systems).

While it is understood that owners must often undertake work necessary to meet health and safety codes, the Department of the Interior—by law—cannot approve rehabilitation projects if significant interior spaces, features, or finishes are lost as a result of such code-required work and, in consequence, the rehabilitation is not consistent with the historic character of the building. In reviewing an overall project, it is thus critical that administrators evaluate work proposals to assure that significant interior features are properly identified so that they may be protected and preserved in the process of meeting health and safety code requirements. Where a conflict exists between code requirements and the Secretary of the Interior's "Standards for Rehabilitation," it should be noted that ". . . The Secretary of the Interior's Standards take precedence over other regulations and the codes in determining whether the historic character of the building is preserved in the process of rehabilitation and should be certified." 36 CFR 67.7(d).

Application: An early 20th century commercial building was being rehabilitated for use as medical offices (see illus. 1). As a result of an inspection by a structural engineer to assure compliance with State health and safety codes, proposed rehabilitation work involved removal of a historic ornamental iron cage-type elevator that was manually operated (see illus. 2) and replacement with a modern elevator (see illus. 3) featuring automatic pushbutton operation. (The ANSI building code specifically requires an enclosed cab and hollow metal shaft doors.) Additional proposed work included removal of the ground floor elevator doors; removal of one set of the existing west-side elevator doors on floors 3 through 7; and the subsequent blocking of access to the elevator on that side due to limited passenger use after rehabilitation (see illus. 4 and 5).

When the project was initially reviewed by the SHPO, recommendation for certification was made because it was felt that loss of the elevator—although unfortunate—did not constitute a radical change to the building's interior. However, when the National Park Service evaluated the proposed work that principally involved removal of the historic elevator and replacement with a modern elevator to meet code, a final determination was made that such removal of a significant interior feature violated Standards 2, 5, and 6. The denial letter to the owner stated:

> The elevator with its highly elaborate iron grillwork and the decoratively molded elevator doors in the lobby is a significant historic feature which contributes to the historic character of this early twentieth century commercial building. The features of the elevator, particularly the decorative cab and the lobby doors are historically significant elements which should be preserved. Your rehabilitation . . . will lead to the loss of a significant feature of the building, in violation of the Standards of Rehabilitation, and the rehabilitation will not be consistent with the historic character of the building. For purposes of the historic preservation tax incentives, the Standards for Rehabilitation take precedence over other regulations and codes in determining whether the historic character of the building is preserved in the process of rehabilitation and should be certified (36 CFR 67.7(d)).

The denial was subsequently appealed and, in spite of the owner's referral to ANSI codes requiring enclosure of the elevator, the NPS decision was sustained by the Chief Appeals Officer, who reiterated in the letter to the owner,". . . since a rehabilitation must preserve the historic character of a property to be certified, I have determined that this project is not consistent with the historic character of the building and does not meet the "Standards for Rehabilitation." In the same letter —in order to achieve a certifiable project—the owner was encouraged to pursue alternative means of preserving the elevator by enclosing the cab itself with fire-rated glass or by constructing a fire-rated enclosure for the elevator shaft.

Prepared by: Kay D. Weeks, TPS

These bulletins are issued to explain preservation project decisions made by the U.S. Department of the Interior. The resulting determinations, based on the Secretary of the Interior's Standards for Rehabilitation, are not necessarily applicable beyond the unique facts and circumstances of each particular case.

Number: 84-061

Applicable Standards:

2. Retention of Distinguishing Architectural Character (nonconformance)
3. Recognition of Historic Period (nonconformance)
4. Retention of Significant Later Alterations/Additions (nonconformance)
6. Repair/Replacement of Deteriorated or Missing Architectural Features Based on Historical Evidence (nonconformance)

Subject: Alterations to Non-Original 20th Century Storefronts

Issue: Storefronts on many 19th and early 20th century buildings were changed in the 1920s and 1930s, incorporating new materials and designs popular at that time and introducing trademarks of the increasing number of commercial chains. Some of these later storefronts today have no intrinsic value while others merit preservation as part of the historic structure.

As guidance in evaluating non-original storefronts, those that meet one or more of the following categories usually are worthy of retention:

1. Exhibit high quality workmanship;
2. Show evidence of being architect-designed;
3. Incorporate materials not commonly used today but are characteristic of a particular period (e.g., curved glass, Carrara glass, bronze frames);
4. Are representative of a particular architectural style;
5. Are compatible with the rest of the building in terms of design and scale and date to a historically significant period of the building and/or district.

Application: A two-story commercial building located in a historic district in the Southwest was operated until recently as part of the S. H. Kress Company store chain (see illus. 1). While the building dates to the early teens, the storefront had been altered in the late 1930s, incorporating a distinctive design which was a trademark of many Kress Company buildings. The band of transom windows, recessed entries, metal framing and large glass display windows sections created the visual image characteristic of, and historically associated with, the Kress Company chain and its buildings constructed or renovated in the 1920s and 1930s (see illus. 2 and

3). Thus, while the 50-year age criteria of the National Register was minimally met, greater significance was attached to the storefront because it was part of the nation-wide Kress Company effort in storefront design. While the new owners of the building originally had intended to maintain the existing storefront, breakage of one of the large curved glass sections posed an unforseen rehabilitation problem since such glass was not readily available locally in the required safety glass. With the overall rehabilitation progressing quickly, the decision was made to replace the entire store-front with a composite design referencing features from other buildings in the his-toric district (see illus. 4). Regretfully, little physical or pictorial evidence of the original appearance of the building had survived. The completed rehabilitation was denied certification and the decision sustained on appeal primarily because of the loss of the intact 1930s storefront (Standard 4), but also because the new storefront was a conjectural historic design and contained inappropriate detailing (Standards 2, 3, and 6). Regarding the problem of availability of materials—curved glass sections—cost was not the major factor but rather time. Given time, com-panies could have been located which make such custom shapes in safety glass. Unfortunately, expedience and perhaps only mild apprecation of the historic importance of the 1930s storefront did not facilitate the careful investigation of such alternatives.

Prepared by: Charles E. Fisher, TPS

These bulletins are issued to explain preservation project decisions made by the U.S. Depart-ment of the Interior. The resulting determinations, based on the Secretary of the Interior's Standards for Rehabilitation, are not necessarily applicable beyond the unique facts and circumstances of each particular case.

Number: 84-062

Applicable Standards:

2. Retention of Distinguishing Architectural Character (conformance)
4. Retention of Significant Later Alterations /Additions (conformance)
5. Sensitive Treatment of Distinctive Features and Craftsmanship (conformance)

Subject: Replacing Altered Features of a Historic Storefront: Compatible Contemporary Design

Issue: Standards 2, 4, and 5 call for the retention of distinctive architectural features—whether original or changes that reflect the history and development of the building or the craftsmanship of its builders—and Standard 6 states that such distinctive features should be repaired rather than replaced, wherever possible. However, there may be cases where, over time, there has been a cumulative loss of historic material comprising these features and introduction of new material that neither exhibits a distinctive style nor special craftsmanship. (Examples of material loss may include decorative portions of a building such as a storefront cornice; more functional portions, such as its display windows, entrance doors, metal kick plates, or transoms; larger portions that combine structural and design roles within the overall storefront such as masonry, wood, or cast-iron pilasters between bays; or even the individual storefront bays themselves.)

If individual features of a storefront have been altered and the alterations are not "changes that have acquired significance in their own right," then the preservation and repair requirements of Standard 6 do not apply. In these cases, the nonsignificant later features may be removed and compatible replacement features designed and installed as long as the new work preserves any remaining historic material, the storefront character is preserved, and the overall rehabilitation is consistent with the historic character of the building. The option of replacing features, such as storefront doors or windows would, however, never extend to later, distinctive features that help define the storefront character.

In summary, it is cautioned that a thorough professional evaluation be made prior to removal to ascertain both the significance of individual storefront features as well as their potential for repair. Demolition of distinctive architectural features and craftsmanship can be the basis for denying an entire rehabilitation project.

Application: A 6-story brownstone and terra-cotta structure built in the 1890s and located in a historic district in a southeastern city was being rehabilitated for retail and office use. Proposed exterior work included removal of nonoriginal 20th century storefront infill features—transoms, double doors, glass display windows, and concrete block kick panels (see illus. 1, 2). A contemporary replacement storefront would then be installed within the original cast-iron columns, pilasters, and framing, thus retaining the three-bay division of the historic storefront. The owner's primary reason for removing much of the later storefront—those nonoriginal portions—was to integrate an additional code-required fire exit into an overall design scheme that he felt would successfully reflect the building's new use as an art gallery.

In its initial review, the SHPO recommended approval of the project work, but expressed concern over whether or not the 20th storefront infill features had acquired significance over time. In the regional review, the project was denied certification. In a letter to the owner, the reasons for denial were explained:

> We have reviewed your proposal to replace the existing storefront with a new entrance of contemporary design that would meet the code requirement of providing a second fire exit. Though not original to the building, the storefront appears to be of sufficient age and design quality to have gained significance in its own right; we feel that its removal would violate Standards 4 and 5. Although we recognize the need to install a fire exit through one of the side display windows, alternative methods were suggested to the architect by this office that would avoid damaging the significant portions of the storefront (i.e., the gridded transom windows and double doors) and which would not require replacement of the entire storefront ... In the absence of documentation demonstrating that the existing storefront is not significant in terms of its age, period, style, materials, or condition, we cannot approve its removal for the purpose of installing a modern entrance to the building.

Because the owner felt that the existing storefront needed to be altered to accommodate code; that the altered portions were not important historically; and that the contemporary storefront met Standard 9, the region's decision was appealed. Prior to appeal, the SHPO offered a final recommendation on the storefront replacement issue in a letter to the Chief Appeals Officer, supporting the owner's contention that new evidence seemed to indicate that most of the later alterations to the storefront had post-dated the 1930s:

> In our initial review of the project, much discussion occurred concerning the significance of the existing storefront. While the existing storefront, which is obviously not original, is of nice design, it is not of sufficient quality to say that the storefront has acquired special significance in its own right or that it is important to retain the storefront to show the evolution of the building through history. In addition, I have personally inspected the building and believe that the storefront is not representative of any particular stylistic period and is not an example of skilled craftsmanship or a good example of design and use of material.

On appeal, the regional decision was overturned and the project subsequently certified for preservation tax benefits. In a final letter to the owner, the Chief Appeals Officer stated:

After carefully considering information submitted by your architect concerning the construction detail and dating of the existing storefront and comments provided by the State Historic Preservation Officer, I have determined that the proposed project meets the Secretary's Standards. I share, however, some of the concerns of the regional office regarding proposed storefront design. While I have concluded that the existing storefront has not acquired special significance over time nor exhibits significant stylistic features or craftsmanship, I would encourage you to consider a contemporary design that provides greater visual distinction between the transom and the display windows. I would also encourage you to revise your design to provide for solid base panels beneath the windows and doors. These alterations would, I feel, be more in keeping with the historic character of the building and district yet would clearly "read" as new construction.

After removal of the altered, nonhistoric portions of the storefront, the compatible new infill was installed, thus retaining and preserving those original portions identified as historically significant (see illus. 3, 4).

Prepared by: Kay D. Weeks and Charles E. Fisher, TPS

These bulletins are issued to explain preservation project decisions made by the U.S. Department of the Interior. The resulting determinations, based on the Secretary of the Interior's Standards for Rehabilitation, are not necessarily applicable beyond the unique facts and circumstances of each particular case.

Number: 84-063

Applicable Standard:

7. Cleaning with Gentlest Method Possible (nonconformance)

**Subject: Inappropriate Chemical Cleaning of Historic Masonry
Buildings**

Standard 7 of the Secretary of the Interior's Standards for Rehabilitation states
that "the surface cleaning of structures shall be undertaken with the gentlest means
possible. Sandblasting and other cleaning methods that will damage the historic
building materials shall not be undertaken." While "the gentlest means possible"
is usually interpreted to mean chemical cleaning, water, or water with the addition
of detergents, it is important to realize that these methods, too, can be damaging
to historic building fabric. Cleaning techniques involving water or chemicals are
not infallible, and must always be tested. If carried out improperly—for instance,
if the chemical mixture is too strong, if chemicals are not adequately rinsed out of
the masonry, if wet cleaning methods are undertaken during cold weather or if
there is still a possibility of freezing temperatures—such cleaning methods can physically abrade or otherwise visually damage historic masonry. In short, chemical
cleaning may not be "the gentlest means possible." Historic masonry buildings
(and brick buildings in particular) which have been chemically cleaned in a way that
has resulted in damage to the visual or aesthetic qualities of the masonry, may be
denied certification for tax benefits.

Application No. 1: A 1912 bank and office building constructed of brick with
stone and terra cotta trim was rehabilitated for contemporary office use after being
vacant for several years (see illus. 1). Located at a major downtown intersection,
this nine-story building is a prominent and highly visible landmark throughout the
city, towering as it does above the more modestly scaled two or three story neighboring buildings. The proposed project which was given preliminary approval by
the National Park Service, and was carried out in 1982, included refurbishing of
office suites on the interior, chemical cleaning of the exterior masonry, and replacement of the later 1940s storefront infill with more appropriately scaled window
glass.

When the completed project was submitted to the National Park Service for final review, however, it was denied certification on the basis of the cleaning techniques which had resulted in "severe discoloration and splotching of the brick surfaces" (see illus. 2). The region's denial letter went on to say: "The brick was apparently cleaned with an inappropriate chemical cleaner which was not adequately tested before its use, contrary to the recommendations contained in the Secretary's Guidelines for Rehabilitating Historic Buildings. Although the physical damage to the brick was not documented, the region felt that the *visual* change to the brick surface was sufficient to deny the project, citing violation of Standards 7 and 2.

When the owner appealed the denial he explained that the exterior of the building had actually been cleaned *and* treated with a water repellent *two* times. Unsatisfied with the result after the first chemical cleaning, the owner required the cleaning contractor to reclean the building in what turned out to be a futile attempt to improve the appearance of the brick. During the appeal, the owner was unable to identify the type of chemicals or the methods used in the cleaning, nor did he provide any close-up photographs of the discolored brick. Consequently, it remained unknown whether the chemical cleaning had also caused physical damage to the brick.

After careful review of the project, the Chief Appeals Officer sustained the region's decision, stating that: "I concur with the regional office's finding that this treatment (cleaning of the exterior brickwork) 'has so altered the appearance of the building as to detract from its historic character.' Standard 7 permits only the gentlest means of surface cleaning. . . . Close-up photographs showing the conditions of the brick before and after this process (the second cleaning) were not submitted, nor were technical details of the cleaning methods and substances made available. Nevertheless, it is convincingly evident from the extent and degree of the persistent discoloration that the brickwork was subjected to unacceptably harsh cleaning. Accordingly, I find a violation of Standard 7."

Application No. 2: In a second case, a mid-nineteenth century brick rowhouse was rehabilitated for rental residential use (see illus. 3). A major aspect of the rehabilitation of the exterior was the removal of paint covering the brick facade. The project application stated that the building was to be chemically cleaned, generally an acceptable paint removal technique in accordance with the Secretary of the Interior's "Standards for Rehabilitation," and the proposal was given preliminary approval by the National Park Service. However, when the request for final certification was submitted, photographs showed that the "cleaned" brick appeared to have been damaged by the cleaning method (see illus. 4). When questioned, the owner revealed that the paint had been removed with sodium hydroxide, more commonly called caustic soda or lye. With the knowledge that some types of chemical cleaning may be just as damaging to historic brick as sandblasting, it was decided that an on-site inspection of the property by the National Park Service was necessary in order to determine if, indeed, the brick really had been damaged by this method of paint removal. At the project site, comparison of the cleaned brick with the painted brick of an identical row house on the same block provided evidence (see illus. 4 and 5) that the surface of the rather soft brick had been "etched" by lye.

On that basis, the project was denied certification by the National Park Service Regional office. The denial letter sent to the owner stated: "The National Park

Service has been cautioning property owners for some time about the dangers of paint removal and cleaning of soft masonry. The (State Historic Preservation Office) has been advising property owners concerning the early practice of painting many . . . rowhouses for aesthetic reasons and as a protective treatment for inherently poor quality brick. We strongly urge you to be more cautious in future projects when you consider removing paint from historic masonry, we would encourage you not to remove paint where historically such surface treatment has acquired significance over time. Where paint removal is an appropriate treatment, only the gentlest means possible, determined by careful testing, should be used. *If no method can be found which does not damage the brick or change its original visual appearance, the paint should not be removed.*"

When the owner appealed this decision, the Chief Appeals Officer upheld the denial of the regional office, explaining that "as a result of the cleaning, the surface of the brick has been eroded, exposing additional folds and irregularities in the clay and creating a rougher texture to the brick. These visual and physical changes to the brick have altered the character of the masonry facade."

Prepared by: Anne E. Grimmer, TPS

These bulletins are issued to explain preservation project decisions made by the U.S. Department of the Interior. The resulting determinations, based on the Secretary of the Interior's Standards for Rehabilitation, are not necessarily applicable beyond the unique facts and circumstances of each particular case.

Number: 84-064

Standards for Evaluating Significance Within Registered Historic
Districts (36 CFR 67.5 (a) (2))

Subject: Extensive Replacement of Historic Materials/Features:
Loss of Integrity

Issue: In planning any rehabilitation project, it is assumed that some historic materials (masonry, wood, and metal) will be deteriorated or damaged and need repair or replacement in preparation for the new use. While a reasonable level of replacement of such deteriorated or damaged exterior and interior material is acceptable, at the same time the preservation requirements outlined in 36 CFR 67 must always be met. To receive Part 1 certification, the building, *prior* to rehabilitation, must convey historic significance through its intact features, i.e., display integrity of design, materials, and workmanship, location, feeling, and association according to the Secretary of the Interior's "Standards for Evaluating Historic Significance Within Registered Historic Districts"; and to receive Part 2 certification, the building, *after* rehabilitation, must retain those portions and features of the building that have been identified as significant prior to work, in accordance with the Secretary of the Interior's "Standards for Rehabilitation."

If, after close inspection, it becomes clear that the significant portions and features of the building cannot be retained and preserved because of the extent of physical deterioration or damage, then the building will generally not possess sufficient integrity of design, materials, and workmanship to be designated as a "certified historic structure" and, in consequence, Part 1 certification should be denied. In unusual cases where Part 1 certification has already been issued and, during the course of rehabilitation, it is discovered that the structure does not possess sufficient integrity, the Part 1 certification should be rescinded and the Part 2 application returned to the owner, unprocessed, with a letter explaining the action.

Application: A deteriorated, three-story, three-bay wide brick structure built in 1843 was certified in the Part 1 application as contributing to the significance of the registered historic district—a 13-block area of 19th century Federal and Greek Revival structures (see illus. 1, 2, 3, 4).

A Part 2 application was submitted at the same time as the Part 1 application, but a determination on Part 2 could not be given due to a lack of information concerning the below-grade storefront which the owner proposed removing as part of the work to return the building to a residential appearance. The letter from NPS, WASO requesting additional information, stated:

> Although the application material indicates that the structure was originally residential, the photographs suggest that the storefront, including the projecting bay with side entrance door and cornice, may have acquired historic significance over time. For this office to make a Part 2 assessment, however, you will have to provide information concerning the building's conversion on the lower floor to commercial use and the approximate date of the existing storefront. Photographs of the storefront showing in more detail what had survived should be submitted. When additional information and photographic documentation is received, a determination can be made as to whether the project meets the Standards for Rehabilitation.

In response, the owner submitted the requested information on the storefront in order to process the Part 2 application; this particular work component was reviewed and found to be in conformance with the Standards.

The amended application also included new photographic documentation that revealed the severely deteriorated condition of previously blocked-up portions of the rear of the building and the extent of damage and loss of both exterior and interior features. This portion of the building had not been assessed in the initial application, but was assumed to be substantially intact when Part 1 certification was issued. The newly submitted photographic documentation called into question the integrity of design, materials, and workmanship of the building, and it was decided to re-evaluate the Part 1 certification (see illus. 5, 6, 7). Following re-assessment, a second letter was sent to the owner, explaining the region's findings:

> Based on the information submitted in the original application, the National Park Service determined that the property contributed to the significance of the registered historic district in which it was located, and thus qualified (for tax benefits) as a "certified historic structure." This certification was based on the assumption that a majority of the structure was still standing and that character-giving features such as interior trim, moldings, and fireplace details would be retained.

> The new photographic documentation that you submitted shows that barely one-third of the building was standing at the time rehabilitation work commenced. As a result of the building's extremely deteriorated condition, significant architectural features are too deteriorated to be preserved on the remaining portion of the building. In addition, nearly all interior finishes are to be replaced and rebuilt using new materials. As a result of the new information, we have determined that No. 2 of the "Standards for Evaluating Significance Within Registered Historic Districts" has been met (e.g., the structure does *not* contribute to the significance of the district) and, therefore, the building cannot qualify as a "certified historic structure." This decision supersedes the earlier decision . . . since the building does not qualify as a "certified historic structure," in accordance with Department of Interior regulations, the project is not eligible for certification of rehabilitation.

Because the owner felt preservation tax incentives should be made available and the Part 2 processed, the project was appealed. On appeal, the region's denial of Part 1 was affirmed by the Chief Appeals Officer, who reiterated: "Similarly, I have determined that it is not a certified historic structure because the integrity of the original design, individual architectural features and spaces have been irretrievably lost through physical deterioration and structural damage . . ."

Prepared by: Kay D. Weeks, TPS

These bulletins are issued to explain preservation project decisions made by the U.S. Department of the Interior. The resulting determinations, based on the Secretary of the Interior's Standards for Rehabilitation, are not necessarily applicable beyond the unique facts and circumstances of each particular case.

Number: 85-065

Applicable Standards:

1. Compatible New Use (conformance)
2. Retention of Distinguishing Architectural Character (nonconformance, conformance)
5. Sensitive Treatment of Distinctive Features and Craftsmanship (nonconformance, conformance)
6. Repair/Replacement of Deteriorated or Missing Features (conformance)
9. Compatible Design for New Alterations/Additions (nonconformance, conformance)

Subject: Alterations to Historic Auditorium Spaces

Issue: Changing the use of historic auditorium spaces, such as those in theaters, churches and schools, poses difficult design problems. Some new uses cannot be accommodated in such auditoriums without destroying character-defining spaces or features. Dividing the space, or altering or destroying its features will result in a denial of certification for noncompliance with Standards 2 and 5. However, there are cases where earlier insensitive alteration to, or extensive deterioration of, the materials comprising significant features and spaces has already resulted in loss of the historic character. In such cases, further alterations to accommodate a new use will generally not result in denial of rehabilitation certification. It is particularly important, however, that a careful professional evaluation be made of altered spaces and deteriorated features to assure that repair is, indeed, infeasible.

Applications: A small church built in 1875 in the Gothic style and located in a historic district had been purchased by a neighboring church in 1923 for use as an educational facility. During the 1960s it had been used as a theater and recreational center (see illus. 1 and 2). A proposal was made to rehabilitate the structure into residential condominiums (see illus. 3). In order to accomplish this conversion, the owner proposed to subdivide the interior space and to insert three new floor levels into the sanctuary.

The regional office denied the project preliminary certification on the basis that the "austere interior is of major importance" in defining the "ecclesiastical character of the structure." It found that inserting seven residential units into the interior would seriously impair that character. While the concept of inserting residences into the church was not ruled out, the plans as submitted were deemed unsatisfactory because they involved the "total loss of the original volume and space of the sanctuary."

Upon appeal the owner stressed the alterations made to the interior during the previous 20 years. The "austere" appearance resulted, he stated, from the gutting of the interior to provide a basketball court. The interior did not, therefore, contribute to the overall character of the building. He further stated that "the sense of volume and the ecclesiastical character of the former church will be retained in the individual apartment units. After the rehabilitation, this building will look like a church, as it does now."

In his decision upholding the denial of certification, the Chief Appeals Officer determined that changes made to the interior over the years had not seriously diminished the historic character of that space. The alterations, he said, "appear to amount to little more than removal of church furnishings." He noted that the church retains such features as the regularly spaced windows, the conspicuous roof structure and exposed scissor trusses, and that the extent and form of the space remain. Overall, he concluded, the interior still conveys a sense of the purpose for which it was designed—assembly. The interior space, therefore, was determined to be integral to the historic character of the building. Because that space would be destroyed by the insertion of apartments as planned, certification was denied.

A second case involved an 1890's brownstone, Romanesque Revival church with an octagon plan sanctuary, individually listed in the National Register, and located in a residential section of a major northeastern city. A rehabilitation was proposed to convert the building, which had been empty for fourteen years, to medical offices. The new use necessitated insertion of three floors and office partitions into the sanctuary (see illus. 4). The interior had ornate, clustered, engaged colonettes; acanthus leaf entablatures; a wooden chair rail; four arched tripartite windows; an egg-and-dart ceiling cornice; and a shallow dished ceiling. Plans called for enclosing most of the deteriorated plaster detailing on the walls with furred-out walls, and removal of the lath and plaster of the dished ceiling (see illus. 5).

The church had been converted to a synagogue in 1948, at which time the organ; organ chamber; choir, choir gate, and railing; pulpit; stained glass windows; and pendant lighting fixtures had been removed. Shortly afterward (early 1950s), an acoustical tile ceiling and recessed lighting were installed. During fourteen years of disuse, the building's attic and tower had become infested with pigeons, little maintenance had been done, the building was without heat, and had been vandalized.

The NPS regional office denied the proposal preliminary certification, citing Standards 1, 2, 5, 6, and 9. The decision was predicated on an evaluation of the sanctuary space and its elaborate ornamentation as essential to the historic character of the building. The region determined that, "although parts of the historic fabric were water-damaged and although alterations had occurred, the sanctuary had not lost its ability to convey historical associations and the damaged features were repairable." The denial letter stated that the installation of new floors and partitions that "leave no area for perception of even part of the original, grand, open plan" violates Standards 1, 2, and 9. The removal of the ceiling, enclosure of decorative detailing, and replacement of (1948) windows violates Standards 2, 5, and 6.

In appealing the regional denial, the owners stated that the dished ceiling plaster and lath (as well as the applied acoustical tile) would have to be removed, as they were soaked with water from the numerous roof leaks, and had a thick layer (as much as one foot) of pigeon excrement above. Further, due to water penetration and freeze-thaw cycles, the decorative plaster on the sanctuary walls was severely damaged and so unstable as to be unable to withstand even the slightest impact.

At the appeal meeting, close-up photographs of deteriorated plaster details were shown (see illus. 6), and the condition of the plasterwork was fully discussed. The Appeals Officer overturned the regional office denial and determined that the project was consistent with the *existing* historic character of the church. In certifying the project, he said:

> The information and photographs (as well as the physical evidence) you provided clarified for me the condition of the building . . . I am convinced that the plaster-work has deteriorated to such an extent that it cannot now be repaired, and that the interior wall and ceiling finishes have lost their physical integrity and their historic character.

Church sanctuaries are often character-defining features of historic churches. The importance of these spaces, however, is not dependent on the ornateness of detailing. The first space discussed here was plain; the second was elaborate. In neither case did evaluation of the proposed project depend on the level of ornamentation. Minor changes had been made to the first church interior, but the materials and the sanctuary space had remained intact. In the second case, the sanctuary had lost its character due to extreme deterioration. Regardless of the original level of detail, if a character-defining historic interior remains largely intact, it must be retained in a rehabilitation. Subdivision or other alteration that destroys the form or features of a significant space will result in denial of certification.

Prepared by: Michael Auer and Susan Dynes, TPS

These bulletins are issued to explain preservation project decisions made by the U.S. Department of the Interior. The resulting determinations, based on the Secretary of the Interior's Standards for Rehabilitation, are not necessarily applicable beyond the unique facts and circumstances of each particular case.

Appendix E

Internal Revenue Service – Proposed Rules Relating to Qualified Conservation Contributions

Federal Register Volume 48, May 23, 1983
pages 22940-22949 to be codified at
26 C.F.R. Parts 1, 20 and 25

Department of the Treasury

Internal Revenue Service

26 CFR Parts 1, 20 and 25

[LR-200-76]

Qualified Conservation Contribution;
Proposed Rulemaking

Agency: Internal Revenue Service, Treasury.

Action: Notice of proposed rulemaking.

Summary: This document contains proposed regulations relating to contributions of partial interests in property for conservation purposes. Changes to the applicable tax law were made by section 6 of the Tax Treatment Extension Act of 1980. This document is intended to clarify the statutory rules in effect under that Act.

Date: Written comments and requests for a public hearing must be delivered or mailed by July 22, 1983. The amendments are proposed to be applicable for contributions made on or after December 18, 1980, and are proposed to be effective the date final regulations are published in the **Federal Register.**

Address: Internal Revenue Service, 1111 Constitution Avenue, N.W., Washington, D.C. 20224, Attention: CC:LR:T (LR-200-76).

For further information contact: John R. Harman of the Legislation & Regulations Division, Office of the Chief Counsel, Internal Revenue Service, 1111 Constitution Avneue, N.W., Washington, D.C. 20224. Attention: CC:LR:T, 202-566-3287, not a toll-free call.

Supplementary Information:

Background

This document contains amendments proposed to conform the Income Tax Regulations (26 CFR Part 1) under section 170 of the Internal Revenue Code of 1954 (Code) relating to contributions not in trust of partial interests in property to section 6 of the Tax Treatment Extension Act of 1980.

The Tax Reform Act of 1969 and the regulations promulgated thereunder limited the deductibility of the donation of easements generally to charitable contributions of perpetual open space easements in gross, (section 170(f) (3) (B) (ii) of the Code and § 1.170A-7(b) (1) (ii) of the Income Tax Regulations). Although subsequent revenue rulings held that a variety of easements were deductible under the limitation of the 1969 Act (See, *e.g.,* Rev. Rul. 75-358, 1975-2 C.B. 76), Congress in 1976 added further legislative authority for the deductibility of easement donations.

Section 2124 (e) of the Tax Reform Act of 1976 (Pub. L. 94-455, 90 Stat. 1919) provided, for the first time, specific statutory authority under section 170 of the Code for the deductibility of the donation to a qualified organization of easements, remainder interests, and certain other partial interests in property. The 1976 Act allowed the deduction for partial interests donated for a term of 30 years or more, but required that the donation be made "for conservation purposes." Conservation purposes was defined in section 170(f) (3) (C).

Section 309 of the Tax Reduction and Simplification Act of 1977 (Pub. L. 95-30, 91 Stat. 154) made two changes in the statutory language codified by the 1976 Act. The first change eliminated the deductibility of term easements for conservation purposes and required that such easements be perpetual in order to qualify for a deduction under section 170. The second change set the expiration date of these provisions at June 14, 1981.

Section 6 of the Tax Treatment Extension Act of 1980 made extensive changes in the existing statute, eliminated the expiration date, and incorporated the relevant language into a new section 170(h). The House and Senate Committee reports accompanying the legislation also provided, for the first time, an in-depth statement of congressional intent concerning the donation of partial interests for conservation purposes (H.R. Rep. No. 96-1278, S. Rep. No. 96-1007). The regulations reflect the major policy decisions made by the Congress and expressed in these committee reports.

Additional Information

Generally, the donation of an easement to preserve open space is deductible under section 170(h) (4) (A) (iii) if such preservation will yield a significant public benefit and is either for the scenic enjoyment of the general public or its pursuant to a clearly delineated governmental policy. The most difficult problem posed in this regulation was how to provide a workable framework for donors, donees, and the Internal Revenue Service to judge the deductibility of open space easements.

Defining "Significant public benefit" with any degree of precision is impossible. Any attempt to reduce the test to a mathematical formula would be arbitrary. The factors included at § 1.170A-13(d) (4) (iv) are not intended to be exclusive; however, a longer list of factors would always fall short of being all-inclusive. The same statements can be made concerning the list of factors proposed under § 1.170A-13 (d) (4) (ii) with respect to "scenic enjoyment."

It is believed, however, that the "sliding scale" approach proposed in § 1.170A-13(d) (4) (vi) that establishes a relationship between the requirements of "significant public benefit" and "clearly delineated governmental policy" will eliminate much of the uncertainty that surrounds this part of the statute. Additionally, by including prior state and local governmental determinations of specific resources to be protected as a criteria for meeting the "significant public benefit" and "scenic enjoyment" tests, a degree of certainty will be available to taxpayers in jurisdictions that have carefully articulated preservation policies. In the end, of course, some exercise of judgment and of responsibility is ultimately required by both donors and donees.

Comments and Requests for a Public Hearing

Before adopting these proposed regulations, consideration will be given to any written comments that are submitted (preferably seven copies) to the Commissioner of Internal Revenue. All comments will be available for public inspection and copying. A public hearing will be held upon written request to the Commissioner by any person who has submitted written comments. If a public hearing is held, notice of the time and place will be published in the **Federal Register.**

Executive Order 12291 and Regulatory Flexibility Act

The Commissioner of Internal Revenue has determined that this proposed rule is not a major rule as defined in Executive Order 12291. Accordingly, a Regulatory Impact Analysis is not required. The Internal Revenue Service has concluded that although this document is a notice of proposed rulemaking that solicits public comments, the regulations proposed herein are interpretative and the notice and public procedure requirements of 5 U.S.C. 553 do not apply. Accordingly, no Regulatory Flexibility Anaylsis is required for this rule.

Drafting Information

The principal authors of this regulation are John R. Harman and Stephen J. Small of the Legislation and Regulations Division of the Office of Chief Counsel, Internal Revenue Service. However, personnel from other offices of the Internal Revenue Service and Treasury Department participated in developing the regulation, both on matters of substance and style.

List of Subjects

26 CFR 1.61-1-1.281-4

Income taxes, Taxable income, Deductions, Exemptions.

26 CFR Part 20

Estate taxes.

26 CFR Part 25

Gift taxes.

Proposed Amendments to the Regulations

The proposed amendments to 26 CFR Parts 1, 20, and 25 are as follows:

Part 1—[Amended]

§ 1.167(a)-5 [Amended]

Paragraph 1. Section 1.167(a)-5 is amended by adding at the end thereof the following new sentences: "For the adjustment to the basis of a structure in the case of a donation of a qualified conservation contribution under section 170(h), see § 1.170A-13(h) (3) (iii)."

Par. 2. Paragraph (a) (2) (ii) of § 1.170A-1 is amended by adding at the end thereof the following new paragraph:

§ 1.170A-1 Charitable, etc., contributions and gifts; allowance of deduction.

(a)   In general. * * *

(2)   Information required in support of deductions. * * *

(ii)   Contribution by individual of property other than money. * * *

(j)   In the case of a "qualified conservation contribution" under section 170(h), see § 1.170A-13(i).

Par. 3. Section 1.170A-7 is amended as follows:

a. The first sentence of paragraph (b) (1) (ii) is revised to begin with the phrase "With respect to contributions made on or before December 17, 1980."

b. Paragraph (b) (1) (ii) is revised by adding at the end the following new sentence: "For the deductibility of a qualified conservation contribution, see § 1.170A-13."

c. Paragraph (b) (3) is revised by adding at the end the following new sentence: "For the deductibility of the donation of a remainder interest in real property exclusively for conservation purposes, see § 1.170A-13."

d. Paragraph (b) (4) is revised by adding at the end the following new sentence: "For the deductibility of the donation of a remainder interest in real property exclusively for conservation purposes, see § 1.170A-13."

e. A new paragraph (b) (5) is added immediately after paragraph (b) (4), as set forth below.

f. The first sentence of paragraph (c) is revised to begin with the phrase "Except as provided in § 1.170A-13."

g. Paragraph (e) is revised as set forth below.

§ 1.170A-7. Contributions not in trust of partial interest in property.

* * * *

(b)   Contributions of certain partial interests in property for which a deduction is allowed. * * *

(5)   *Qualified conservation contribution.* A deduction is allowed under section 170 for the value of a qualified conservation contribution. For the definition of a qualified conservation contribution, see § 1.170A-13.

* * * *

(e)   *Effective date.* This section applies only to contributions made after July 31, 1969. The deduction allowable under § 1.170A-7(b) (1) (ii) shall be available only for contributions made on or before December 17, 1980. The deduction allowable under § 1.170A-7(b) (5) shall be available for contributions made on or after December 18, 1980.

Par. 4. A new § 1.170A-13 is added after § 1.170A-12 to read as set forth below.

§ 1.170A-13. Qualified conservation contributions.

(a)   *Qualified conservation contributions.* A deduction under section 170 is generally not allowed for a charitable contribution of any interest in property that consists of less than the donor's entire interest in the property other than certain transfers in trust (see § 1.170A-6 relating to charitable contributions in trust and § 1.170A-7 relating to contributions not in trust of partial interests in property). However, a deduction may be allowed under section 170(f) (3) (B) (iii) for the value of a qualified conservation contribution if the requirements of this section are met. A qualified conservation contribution is the contribution of a qualified real property interest to a qualified organization exclusively for conservation purposes. To be eligible for a deduction under this section, the conservation purpose must be protected in perpetuity.

(b)   *Qualified real property interest*

(1)   *Entire interest of donor other than qualified mineral interest.* The entire interest of the donor other than a qualified mineral interest is a qualified real property interest. A qualified mineral interest is the taxpayer's interest in subsurface oil, gas, or other minerals and the right of access to such minerals. A property interest shall not be treated as a qualified real property interest by reason of section 170(h) (2) (A) or this paragraph (b) (1), if any time over the entire term of the taxpayer's interest in such property the taxpayer transferred any portion of that interest (except in the case of a donation of a perpetual conservation restriction under paragraph (b) (3) of this section) to any other person (except for minor interests, such as rights-of-way, that will not interfere with the conservation purposes of the donation).

(2)   *Remainder interest in real property.* A remainder interest in real property is a qualified real property interest. A property interest shall not be treated as a qualified real property interest by reason of section 170(h) (2) (B) or this paragraph (b) (2), if at any time over the entire term of the taxpayer's interest in such property the taxpayer transferred any portion of that interest (except in the case of a donation under paragraph (b) (3) of this section) to any other person (except for minor interests, such as rights-of-way, that will not interfere with the conservation purposes of the donation).

(3)   *Perpetual conservation restriction.* A perpetual conservation restriction is a qualified real property interest. A "perpetual conservation restriction" is a restriction granted in perpetuity on the use which may be made of real property—including, an easement or other interest in real property that under state law has attributes similar to an easement (*e.g.*, a restrictive covenant or equitable servitude). For purposes of this section, the terms "easement'" "conservation restriction," and "perpetual conservation restriction" have the same meaning. The definition of "perpetual conservation restriction" under this paragraph (b) (3) is not intended to preclude the deductibility of a donation of affirmative rights to use a land or water area under § 1.170A-13(d) (2). Any rights reserved by the donor in the donation of a perpetual conservation restriction must conform to the requirements of this section. See *e.g.,* paragraphs (d) (4) (ii), (d) (5) (i), (e) (3), and (g) (4) of this section.

(c)   *Qualified organization—*

(1)   *Eligible donee.* To be considered an eligible donee under this section, an organization must have the resources to enforce the restrictions and must be able to demonstrate a commitment to protect the conservation purposes of the donation. An established group organized exclusively for conservation purposes, for example, would meet this test. A qualified organization need not set aside funds, however, to enforce the restrictions that are the subject of this contribution. For purposes of this section, the term "qualified organization" means:

(i)   A governmental unit described in section 170(b) (1) (A) (v);

(ii)   An organization described in section 170(b) (1) (A) (vi);

(iii)   A charitable organization described in section 501(c) (3) that meets the public support test of section 509(a) (2);

(iv)   A charitable organization described in section 501(c) (3) that meets the requirements of section 509(a) (3) and is controlled by an organization described in paragraphs (c) (1) (i), (ii), or (iii) of this section.

(2)   *Transfers by donee.* A deduction shall be allowed for a contribution under this section only if in the instrument of conveyance the donor prohibits the donee from subsequently transferring the easement (or, in the case of a remainder interest or the reservation of a qualified mineral interest, the property), whether or not for consideration, unless the donee organization, as a condition of the subsequent transfer, requires that the conservation purposes which the contribution was originally intended to advance continue to be carried out. Moreover, subsequent transfers must be restricted to organizations qualifying, at the time of the subsequent transfer, as an eligible donee under paragraph (c) (1) of this section. When a later unexpected change in the conditions surrounding the property that is the subject of a donation under paragraph (b) (1), (2), or (3) of this section makes impossible or impractical the continued use of the property for conservation purposes, the requirement of this paragraph will be met if the property is sold or exchanged and any proceeds are used by the donee organization in a manner consistent with the conservation purposes of the original contribution. In the case of a donation under paragraph (b) (3) of this section to which the preceding sentence applies, see also paragraph (g) (5) (ii) of this section.

(d)   *Conservation purposes—*

(1)   *In general.* For purposes of section 170(h) and this section, the term "conservation purposes" means—

(i)   The preservation of land areas for outdoor recreation by, or the education of, the general public, within the meaning of paragraph (d) (2) of this section.

(ii)   The protection of a relatively natural habitat of fish, wildlife, or plants, or similar ecosystem, within the meaning of paragraph (d) (3) of this section.

(iii)   The preservation of certain open space (including farmland and forest land) as described in paragraph (d) (4) of this section, or

(iv)   The preservation of a historically important land area or a certified historic structure, within the meaning of paragraph (d) (5) of this section.

(2)   *Recreation or education—*

(i)   *In general.* The donation of a qualified real property interest to preserve land areas for the outdoor recreation of the general public or for the education of the general public will meet the conservation purposes test of this section. Thus, conservation purposes would include, for example, the preservation of a water area for the use of the public for boating or fishing, or a nature or hiking trail for the use of the public.

(ii)   *Public use.* The preservation of land areas for recreation or education will not meet the test of this section unless the recreation or education is for the substantial and regular use of the general public or the community.

(3)   *Protection of environmental system—*

(i)   *In general.* The donation of a qualified real property interest to protect a significant relatively natural habitat in which a fish, wildlife, or plant community, or similar ecosystem, normally lives will meet the conservation purposes test of this section. The fact that the habitat or environment has been altered to some extent by human activity will not result in a deduction being denied under this section if the fish, wildlife, or plants continue to exist there in a relatively natural state. For example, the preservation of a lake formed by a man-made dam or a salt pond formed by a man-made dike would meet the conservation purposes test if the lake or pond were a nature feeding area for a wildlife community that included rare, endangered, or threatened native species.

(ii) *Significant habitat or ecosystem.* Significant habitats and ecosystems include, but are not limited to, habitats for rare, endangered, or threatened species of animal, fish, or plants; natural areas that represent high quality examples of a terrestrial community or aquatic community, such as islands that are undeveloped or not intensely developed where the coastal ecosystem is relatively intact; and natural areas which are included in, or which contribute to, the ecological viability of a local, state, or national park, nature preserve, wildlife refuge, wilderness area, or other similar conservation area.

(4) *Preservation of open space—*

(i) *In general.* The donation of a qualified real property interest to preserve open space (including farmland and forestland) will meet the conservation purposes test of this section if such preservation is—

(A) Pursuant to a clearly delineated Federal, state, or local governmental policy and will yield a significant public benefit, or

(B) For the scenic enjoyment of the general public, and will yield a significant public benefit.

An open space easement donated on or after December 18, 1980, must meet the requirements of this section in order to be deductible under section 170. See § 1.170A-7(b) (1) (ii).

(ii) *Scenic enjoyment—*

(A) *Factors.* A contribution made for the preservation of open space may be for the scenic enjoyment of the general public. "Scenic enjoyment" will be evaluated by considering all pertinent facts and circumstances germane to the contribution. Regional variations in topography, geology, biology, and cultural and economic conditions require flexibility in the application of this test, but do not lessen the burden on the taxpayer to demonstrate the scenic characteristics of a donation under this paragraph. The application of a particular objective factor to help define a view as "scenic" in one setting may in fact be entirely inappropriate in another setting. Among the factors to be considered are:

(*1*) The compatibility of the land use with other land in the vicinity;

(*2*) The degree of contrast and variety provided by the visual scene;

(*3*) The openness of the land (which would be a more significant factor in an urban or densely populated setting or in a heavily wooded area);

(*4*) Relief from urban closeness;

(*5*) The harmonious variety of shapes and textures;

(*6*) The degree to which the land use maintains the scale

and character of the urban landscape to preserve open space, visual enjoyment, and sunlight for the surrounding area;

(7) The consistency of the proposed scenic view with a methodical state scenic identification program, such as a state landscape inventory; and

(8) The consistency of the proposed scenic view with a regional or local landscape inventory made pursuant to a sufficiently rigorous review process, especially if the donation is endorsed by an appropriate state agency.

(B) *Preservation of a view.* To satisfy the requirements of scenic enjoyment by the general public, visual (rather than physical) access to or across the property by the general public is sufficient. This, preservation of land may be for the scenic enjoyment of the general public if development of the property would impair the scenic character of the local rural or urban landscape or would interfere with a scenic panorama that can be enjoyed from a park, nature preserve, road, waterbody, trail, or historic structure or land area, and such area or transportation way is open to, or utilized by, the public.

(C) *Visible to public.* Under the terms of an open space easement on scenic property, the entire property need not be visible to the public for a donation to qualify under this section, although the public benefit from the donation may be insufficient if only a small portion of the property is visible to the public.

(iii) *Governmental conservation policy* —

(A) *In general.* The requirement that the preservation of open space be pursuant to a clearly delineated Federal, state, or local governmental policy is intended to protect the types of property identified by representatives of the general public as worthy of preservation or conservation. A general declaration of conservation goals by a single official or legislative body is not sufficient. This requirement will be met by donations that further a specific, identified conservation project, such as the preservation of land within a state or local landmark district that is locally recognized as being significant to that district; the preservation of a wild or scenic river; the preservation of farmland pursuant to a state program for flood prevention and control; or the protection of the scenic, ecological, or historic character of land that is contiguous to, or an integral part of, the surroundings of existing recreation or conservation sites. For example, the donation of a perpetual conservation restriction to a qualified organization pursuant to a formal declaration (in the form of, for example, a resolution or certification) by a local governmental agency established under state law specifically identifying the subject property as worthy of protection for conservation purposes will meet the requirement of this paragraph. A program need not be funded to satisfy this requirement, but the program must involve a significant commitment by the government with respect to the conservation project.

(B) *Effect of acceptance by governmental agency.* Acceptance of an easement by an agency of the Federal Government or by an agency of a state or local government for by a commission, authority, or similar body duly constituted by the state or local government and acting on behalf of the state or local government) tends to establish the requisite clearly delineated governmental policy, although such acceptance, without more, is not sufficient. The more rigorous the review process by the governmental agency, the more that acceptance of the easement tends to establish the requisite clearly delineated governmental policy.

(iv) *Significant public benefit—*

(A) *Factors.* All contributions made for the preservation of open space must yield a significant public benefit. Public benefit will be evaluated by considering all pertinent facts and circumstances germane to the contribution. Factors germane to the evaluation of public benefit from one contribution may be irrelevant in determining public benefit from another contribution. No single factor will necessarily be determinative. Among the factors to be considered are:

(*1*) The uniqueness of the property to the area;

(*2*) The intensity of land development in the vicinity of the property (both existing development and foreseeable trends of development);

(*3*) The consistency of the proposed open space used with public programs (whether Federal, state or local) for conservation in the region, including programs for outdoor recreation, irrigation or water supply protection, water quality maintenance or enhancement, flood prevention and control, erosion control, shoreline protection, and protection of land areas included in, or related to, a government approved master plan or land management area;

(*4*) The consistency of the proposed open space use with existing private conservation programs in the area, as evidenced by other land, protected by easement or fee ownership by organizations referred to in § 1.170A-13(c) (1), in close proximity to the property;

(*5*) The likelihood that development of the property would lead to or contribute to degradation of the scenic, natural, or historic character of the area;

(*6*) The opportunity for the general public to use the property or to appreciate its scenic values.

(*7*) The importance of the property in preserving a local or regional landscape or resource that attracts tourism or commerce to the area;

(*8*) The likelihood that the donee will acquire equally desirable and valuable substitute property or property rights;

(*9*) The cost to the donee of enforcing the terms of the conservation restriction;

(*10*) The population density in the area of the property; and

(*11*) The consistency of the proposed open space use with a legislatively mandated program identifying particular parcels of land for future protection.

(B) *Illustrations.* The preservation of an ordinary tract of land would not in and of itself yield a significant public benefit, but the preservation of ordinary land areas in conjunction with other factors that demonstrate significant public benefit or the preservation of a unique land area for public enjoyment would yield a significant public benefit. For example, the preservation of a vacant downtown lot would not by itself yield a significant public benefit, but the preservation of the downtown lot as a public garden would, absent countervailing factors, yield a significant public benefit. The following are other examples of contributions which would, absent countervailing factors, yield a significant public benefit; the preservation of farmland pursuant to a state program for flood prevention and control; the preservation of a unique natural land formation for the enjoyment of the general public; the preservation of woodland along a Federal highway pursuant to a government program to preserve the appearance of the area so as to maintain the scenic view from the highway; and the preservation of a stretch of undeveloped property located between a public highway and the ocean in order to maintain the scenic ocean view from the highway.

(v)   *Limitation.* A deduction will not be allowed for the preservation of open space under section 170(h) (4) (A) (iii), if the terms of the easement permit a degree of intrusion or future development that would interfere with the essential scenic quality of the land or with the governmental conservation policy that is being furthered by the donation.

(vi)   *Relationship of requirements—*

(A)   *Clearly delineated governmental policy and significant public benefit.* Although the requirements of "clearly delineated governmental policy" and "significant public benefit" must be met independently, for purposes of this section the two requirements may also be related. The more specific the governmental policy with respect to the particular site to be protected, the more likely the governmental decision, by itself, will tend to establish the significant public benefit associated with the donation. For example, while a statute in State X permitting preferential assessment for farmland is, by definition, governmental policy, it is distinguishable from a state statute, accompanied by appropriations, naming the X River as a valuable resource and articulating the legislative policy that the X River and the relatively natural quality of its surrounding be protected. On these facts, an open space easement on farmland in State X would have to demonstrate additional factors to establish "significant public benefit." The specificity of the legislative mandate to protect the X River, however, would by itself tend to establish the significant public benefit associated with an open space easement on land fronting the X River.

(B)   *Scenic enjoyment and significant public benefit.* With respect to the relationship between the requirements of "scenic enjoyment" and "significant public benefit," since the degrees of scenic enjoyment offered by a variety of open space easements are subjective and not as easily delineated as are increasingly specific levels of governmental policy, the significant public benefit of preserving a scenic view must be independently established in all cases.

(C)   *Donations may satisfy more than one test.* In some cases, open space easements may be both for scenic enjoyment and pursuant to a clearly delineated governmental policy. For example, the preservation of a particular scenic view identified as part of a scenic landscape inventory by a rigorous governmental review process will meet the tests of both paragraphs (d) (4) (i) (A) and (d) (4) (i) (B) of this section.

(5)   *Historic preservation—*

(i)   *In general.* The donation of a qualified real property interest to preserve an historically important land area or a certified historic structure will meet the conservation purposes test of this section. When restrictions to preserve a building or land area within a registered historic district permit future development on the site, a deduction will be allowed under this section only if the terms of the restrictions require that such development conform with appropriate local, state, or Federal standards for construction or rehabilitation within the district. See also, § 1.170A-13(h) (3) (ii).

(ii)   *Historically important land area.* The term "historically important land area" includes:

(A)   An independently significant land area (for example, an archaeological site or a Civil War bettlefield) that substantially meets the National Register Criteria for Evaluation in 36 CFR 60.4 (Pub. L. 89-665, 80 Stat. 915);

(B)   Any building or land area within a registered historic district (except buildings that cannot reasonably be considered as contributing to the significance of the district); and

(C)   Any land area adjacent to a property listed individually in the National Register of Historic Places (but not within a registered historic district) in a case where the physical or environmental features of the land area contribute to the historic or cultural integrity of the structure.

(iii)   *Certified Historic structure—*

(A)   *Definition.* The term "certified historic structure," for purposes of this section, generally has the same meaning as in section 191(d) (1) (as it existed prior to the Economic Recovery Tax Act of 1981, relating to 5-year amortization of expenditures incurred in the rehabilitation of certified historic structures). However, a "structure" for purposes of this section means any structure, whether or not it is depreciable. Accordingly, easements on private residences may qualify under this section. In addition, a structure would be considered to be a certified historic structure if it were certified either at the time the transfer was made or at the due date (including extensions) for filing the donor's return for the taxable year in which the contribution was made.

(B)   *Interior and exterior easements.* A deduction under this section will not be allowed for the donation of an interior or exterior easement prohibiting destruction or alteration of architectural characteristics inside or on the outside of a certified historic structure unless there is substantial and regular opportunity for the general public to view the architectural characteristics that are the subject of the easement.

(c)   *Exclusively for conservation purposes—*

(1)   *In general.* to meet the requirements of this section, a donation

must be exclusively for conservation purposes. See paragraphs (c) (1) and (g) (1) through (g) (5) (ii) of this section. A deduction will not be denied under this section when incidental benefit inures to the donor merely as a result of conservation restrictions limiting the uses to which the donor's property may be put.

(2)   *Access.* Any limitation on public access to property that is the subject of a donation under this section shall not render the donation nondeductible if such limitation is necessary for protection of the conservation interests that are the basis of the deduction. For example, a restriction on all public access to the habitat of a threatened native animal species protected by a donation under paragraph (d) (3) of this section would be appropriate if such restriction were necessary for the survival of the species.

(3)   *Inconsistent use.* Except as provided in paragraph (e) (4) of this section, a deduction will not be allowed if the contribution would accomplish one of the enumerated conservation purposes but would permit destruction of other significant conservation interests. For example, the preservation of farmland pursuant to a State program for flood prevention and control would not qualify under paragraph (d) (4) of this section if under the terms of the contribution a significant naturally occurring ecosystem could be injured or destroyed by the use of pesticides in the operation of the farm. A donor is not required to demonstrate that all possible conservation interests associated with the property will be protected; rather, the terms of the donation must not permit destruction of significant conservation interests.

(4)   *Inconsistent use permitted.* A use that is destructive of conservation interests will be permitted only if such use is necessary for the protection of the conservation interests that are the subject of the contribution. For example, a deduction for the donation of an easement to preserve an archaeological site that is listed on the National Register of Historic Places will not be disallowed if site excavation consistent with sound archaeological practices may impair a scenic view of which the land is a part. A donor may continue a pre-existing use of the property that does not conflict with the conservation purposes of the gift.

(f)   *Examples.* The provisions of this section relating to conservation purposes may be illustrated by the following examples.

Example (1). State S contains many large track forests that are desirable recreation and scenic areas for the general public. The forests' scenic values attract millions of people to the State. However, due to the increasing intensity of land development in State S, the continued existence of forestland parcels greater than 45 acres is threatened. J grants a perpetual easement on a 100-acre parcel of forestland that is part of one of the State's scenic areas to a qualifying organization. The easement imposes restrictions on the use of the parcel for the purpose of maintaining its scenic values. The restrictions include a requirement that the parcel be maintained forever as open space devoted exclusively to conservation purposes and wildlife protection, and that there be no commercial, industrial, residential, or other development use of such parcel. The law of State S recognizes a limited public right to enter private land, particularly for recreational pursuits, unless such land is posted or the landowner objects. The easement specifically restricts the landowner from posting the parcel, thereby maintaining public access to the parcel according to the custom of the State. J's parcel is regarded by the local community as providing the opportunity for the public to enjoy the use of the property and appreciate its scenic values. Accordingly, J's donation qualifies for a deduction under this section.

Example (2). A qualified conservation organization owns Greenacre in fee as a nature preserve. Greenacre contains a high quality example of a tall grass prairie ecosystem. Farmacre, an operating farm, adjoins Greenacre and is a compatible buffer to the nature preserve. Conversion of Farmacre to a more intense use, such as a housing development, would adversely affect the continued use of Greenacre as a nature preserve because of human traffic generated by the development. The owner of Farmacre donates an easement preventing any future development on Farmacre to the qualified conservation organization for conservation purposes. Normal agricultural uses will be allowed on Farmacre. Accordingly, the donation qualifies for a deduction under this section.

Example (3). H owns Greenacre, a 900-acre parcel of woodland, rolling pasture, and orchards on the crest of a mountain. All of Greenacre is clearly visible from a nearby national park. Because of the strict enforcement of an applicable zoning plan, the highest and best use of Greenacre is as a subdivision of 40-acre tracts. H wishes to donate a scenic easement on Greenacre to a qualifying conservation organization, but H would like to reserve the right to subdivide Greenacre into 90-acre parcels with no more than one single-family home allowable on each parcel. Random building on the property, even as little as one home for each 90 acres, would destroy the scenic character of the view. Accordingly, no deduction would be allowable under this section.

Example (4). Assume the same facts as in *example (3),* except that not all of Greenacre is visible from the park and the deed of easement allows for limited cluster development of no more than five nine-acre clusters (with four houses on each cluster) located in areas generally not visible from the national park and subject to site and building plan approval by the donee organization in order to preserve the scenic view from the park. The donor and the donee have already identified sites where limited cluster development would not be visible from the park or would not measurably impair the view. Owners of homes in the clusters will not have any rights with respect to the surrounding Greenacre property that are not also available to the general public. Accordingly, the donation qualifies for a deduction under this section.

Example (5). State S has experienced a marked decline in open acreage well suited for agricultural use. In the parts of the State where land is highly productive in agricultural use, substantially all active farms are small, family-owned enterprises under increasing development pressures. In response to those pressures, the legislature of State S passed a statute authorizing the purchase of "agricultural land development rights" from farm owners and the placement of "agricultural preservation restrictions" on their land, in order to preserve the State's open space and farm resources. Agricultural preservation restrictions prohibit or limit construction or placement of buildings except those used for agricultural purposes or dwellings used for family living by the farmer and his family and employees; removal of mineral substances in any manner that adversely affects the land's agricultural potential; or other uses detrimental to retention of the land for agricultural use. Money has been appropriated for this program and some landowners have in fact sold their "agricultural land development rights" to State S. K owns and operates a small dairy farm in State S. K desires to preserve his farm for agricultural purposes in perpetuity. Rather than selling the development rights to State S, K grants to a qualifying conservation organization an agricultural preservation restriction on his property in the form of a conservation easement. K reserves to himself, his heirs and assigns the right to manage the farm consistent with sound agricultural and management practices. K's farm is located in one of the more agriculturally productive areas within State S. Accordingly, a deduction is allowed under this section.

(g) *Enforceable in perpetuity* —

(1) *In general.* In the case of any donation under this section, the interest in the property retained by the donor (and the donor's successors in interest)

must be subject to legally enforceable restrictions that will prevent uses of the retained interest inconsistent with the conservation purposes of the donation. In the case of a contribution of a remainder interest, the contribution will not qualify if the tenants, whether they are tenants for life or a term of years, can use the property in a manner that diminishes the conservation values which are intended to be protected by the contributions.

(2)   *Remote future event.* A deduction shall not be disallowed under section 170(f) (3) (B) (iii) and this section merely because the interest which passes to, or is vested in, the donee organization may be defeated by the performance of some act or the happening of some event, if on the date of the gift it appears that the possibility that such act or event will occur is so remote as to be negligible. See paragraph (e) of § 1.170A-1. For example, a state's statutory requirement that use restrictions must be rerecorded every 30 years to remain enforceable shall not, by itself, render an easement nonperpetual.

(3)   *Retention of qualified mineral interest*—

(i)   *In general.* The requirements of this section are not met and no deduction shall be allowed in the case of a contribution of any interest when there is a retention of a qualified mineral interest if at any time there may be extractions or removal of minerals by any surface mining method. Moreover, in the case of a qualified mineral interest gift, the requirement that the conservation purposes be protected in perpetuity is not satisfied if any method of mining that is inconsistent with the particular conservation purposes of a contribution is permitted at any time. See also § 1.170A-13(e) (3). However, a deduction under this section will not be denied in the case of certain methods of mining that may have limited, localized impact on the real property but that are not irremediably destructive of significant conservation interests. For example, a deduction will not be denied in a case where production facilities are concealed or compatible with existing topography and landscape and when surface alteration is to be restored to its original state.

(ii)   *Examples.* The provisions of paragraph (g) (3) (i) of this section may be illustrated by the following examples:

Example (1). K owns 5,000 acres of bottomland hardwood property along a major watershed system in the southern part of the United States. Agencies within the Department of the Interior have determined that southern bottomland hardwoods are a rapidly diminishing resource and a critical ecosystem in the south because of the intense pressure to cut the trees and convert the land to agricultural use. These agencies have further determined (and have indicated in correspondence with K) that bottomland hardwoods provide a superb habitat for numerous species and play an important role in controlling floods and in purifying rivers. K donates to a qualifying conservation organization all his interest in this property other than his interest in the gas and oil deposits that have been identified under K's property. K convenants and can ensure that, although drilling for gas and oil on the property may have some temporary localized impact on the real property, the drilling will not interfere with the overall conservation purpose of the gift, which is to protect the unique bottomland hardwood ecosystem. Accordingly, the donation qualifies for a deduction under this section.

Example (2). Assume the same facts as in *example (1),* except that K does not own the mineral rights (or the right of access to those minerals) on the 5,000 acres and cannot ensure that the mining and drilling will not interfere with the overall conservation purpose. Accordingly, a deduction for the donation of the easement would not be allowable under this section. The same rule would apply to disallow a deduction by K for the donation of a remainder interest in the land for conservation purposes. A different result would follow if under applicable State law the qualifying organization had sufficient rights to protect the conservation purpose of the gift. Additionally, a donation of K's entire interest in the 5,000 acres to an eligible organization would qualify for a deduction under section 170(f) (3) (A) without regard to this section.

Example (3). Assume the same facts as in *example (1)*, except that K sell the mineral rights to an unrelated party in an arm's length transaction, subject to a recorded prohibition on the removal of any minerals by any surface mining method and a recorded prohibition against any mining technique that will harm the bottomland hardwood ecosystem. After the sale, K donates an easement for conservation purposes to a qualifying organization to protect the bottomland hardwood ecosystem. Since K can ensure in the easement that the mining of minerals on the property will not interfere with the conservation purposes of the gift, the donation qualifies for a deduction under this section.

(4)   *Protection of conservation purpose where taxpayer reserves certain rights.*

(i)   *Documentation.* In the case of a donation made after (the date final regulations are published in the **Federal Register**) of any qualified real property interest when the donor reserves rights the exercise of which may have an adverse impact on the conservation interests associated with the property, for a deduction to be allowable under this section the donor must make available to the donee, prior to the time the donation is made, documentation sufficient to establish the condition of the property at the time of the gift. Such documentation may include:

(A)   The appropriate survey maps from the United States Geological Survey, showing the property line and other contiguous or nearby protected areas;

(B)   A map of the area drawn to scale showing all existing man-made improvements or incursions (such as roads, buildings, fences, or gravel pits), vegetation and identification of flora and fauna (including, for example, rare species locations, animal breeding and roosting areas, and migration routes), land use history (including present uses and recent past disturbances), and distinct natural features (such as large trees and aquatic areas);

(C)   An aerial photograph of the property at an appropriate scale taken as close as possible to the date the donation is made; and

(D)   On-site photographs taken at appropriate locations on the property. If the terms of the donation contain restrictions with regard to a particular natural resource to be protected, such as water quality or air quality, the condition of the resource at or near the time of the gift must be established. The documentation, including the maps and photographs, must be accompanies by a statement signed by the donor and a representative of the donee clearly referencing the documentation and in substance saying "This natural resources inventory is an accurate representation of [the protected property] at the time of the transfer."

(ii)   *Donee's right to inspection and legal remedies.* In the case of any donation referred to in paragraph (g) (4) (i) of this section, the donor must agree to notify the donee, in writing, before exercising any reserved right, *e.g.,* the right to extract certain minerals which may have an adverse impact on the conservation interests associated with the property. The terms of the donation must provide a right of the donee to enter the property at reasonable times for the purpose of inspecting the property to determine if there is compliance with the terms of the donation. Additionally, the terms of the donation must provide a right of the donee to enforce the conservation restrictions by appropriate legal proceedings, including but not limited to, the right to require the restoration of the property to its condition at the time of the donation.

(5)   *Extinguishment.*

(i)   *In general.* If a subsequent unexpected change in the conditions surrounding the property that is the subject of a donation under this paragraph can make impossible or impractical the continued use of the property for conservation purposes, the conservation purpose can nonetheless be treated as protected in perpetuity if the restrictions are extinguished by judicial proceeding and all of the donee's proceeds (determined under paragraph (g) (5) (ii) of this section) from a subsequent sale or exchange of the property are used by the donee organization in a manner consistent with the conservation purposes of the original contribution.

(ii)   *Proceeds.* In case of a donation made after (the date final regulations are published in the **Federal Register**), for a deduction to be allowed under this section, at the time of the gift the donor must agree that the donation of the perpetual conservation restriction gives rise to a property right, immediately vested in the donee organization, with a fair market value that is a minimum ascertainable proportion of the fair market value to the entire property. See § 1.170A-13(h) (3) (iii). For purposes of this paragraph (g) (5) (ii), that original minimum proportionate value of the donee's property rights shall remain constant. Accordingly, when a change in conditions give rise to the extinguishment of a perpetual conservation restriction under paragraph (g) (5) (i) of this section, the donee organization, on a subsequent sale, exchange, or involuntary conversion of the subject property, must be entitled to a portion of the proceeds at least equal to the original proportionate value of the perpetual conservation restriction, unless state law provides that the donor is entitled to the full proceeds from the conversion without regard to the terms of the prior perpetual conservation restriction.

(h)   *Valuation—*

(1)   *Entire interest of donor other than qualified mineral interest.* The value of the contribution under section 170 in the case of a contribution of a taxpayer's entire interest in property other than a qualified mineral interest is the fair market value of the surface rights in the property contributed. The value of the contribution shall be computed without regard to the mineral rights. See paragraph (h) (4), *example (1),* of this section.

(2)   *Remainder interest in real property.* In the case of a contribution of any remainder interest in real property, section 170(f) (4) provides that in determin-

ing the value of such interest for purposes of section 170, depreciation and deple-
tion of such property shall be taken into account. See § 170.1A-12. In the case
of the contribution of a remainder interest for conservation purposes, the current
fair market value of the property (against which the limitations of § 1.170A-12
are applied) must take into account any pre-existing or contemporaneously
recorded rights limiting, for conservation purposes, the use to which the subject
property may be put.

    (3)   *Perpetual conservation restriction–*

      (i)   *In general.* The value of the contribution under section 170
in the case of a charitable contribution of a perpetual conservation
restriction is the fair market value of the perpetual conservation
restriction at the time of the contribution. See § 1.170A-7(c). If no
substantial record of market-place sales is available to use as a meaningful or valid comparison, as a general rule (but not necessarily in all
cases) the fair market value of a perpetual conservation restriction is
equal to the difference between the fair market value of the land it
encumbers before the granting of the restriction and the fair market
value of the encumbered land after the granting of the restriction. The
amount of the deduction in the case of a charitable contribution of a
perpetual conservation restriction covering a portion of the contiguous
land owned by a taxpayer and the taxpayer's family (see section 267(c)
(4)) is the difference between the fair market value of the entire contiguous tract before and after the granting of the restriction. Accordingly, in the case of the donation of a perpetual conservation restriction
in all or a portion of the contiguous land owned by a taxpayer and the
taxpayer's family (see section 267(c) (4)), if the donor or the donor's
family receives, or can reasonably expect to receive, financial or economic benefits that are greater than those that will inure to the general
public from the transfer, no deduction is allowable under this section.
However, if the transferor receives, or can reasonably expect to receive,
a financial or economic benefit that is substantial, but it is clearly
shown that the benefit is less than the amount of the transfer, then a
deduction under this section is allowable for the excess of the amount
transferred over the amount of the financial or economic benefit received or reasonably expected to be received by the transferor. (See
example (11) of paragraph (h) (4) of this section.)

      (ii)   *Fair market value of property before and after restriction.* If
before and after valuation is used, the fair market value of the property
before contribution of the conservation restriction must take into account not only the current use of the property but also an objective
assessment of how immediate or remote the likelihood is that the property, absent the restriction, would in fact be developed, as well as any
effect from zoning, conservation, or historic preservation laws that already restrict the property's potential highest and best use. Further,
there may be instances where the grant of a conservation restriction
may have no material effect on the value of the property or may in fact
serve to enhance, rather than reduce, the value of property. In such
instances no deduction would be allowable. In the case of a conservation restriction that allows for any development, however, limited, on

the property to be protected, the fair market value of the property after contribution of the restriction must take into account the effect of the development. Additionally, if before and after valuation is used, an appraisal of the property after contribution of the restriction must take into account the effect of restrictions that will result in a reduction of the potential fair market value represented by highest and best use but will, nevertheless, permit uses of the property that will increase its fair market value above that represented by the property's current use. The value of a perpetual conservation restriction shall not be reduced by reason of the existence of restrictions on transfer designed solely to ensure that the conservation restriction will be dedicated to conservation purposes. See § 1.170A-13(c) (2).

(iii) *Allocation of basis.* In the case of the donation of a qualified real property interest for conservation purposes, the basis of the property retained by the donor must be adjusted by the elimination of that part of the total basis of the property that is properly allocable to the qualified real property interest granted. The amount of the basis that is allocable to the qualified real property interest shall bear the same ratio to the total basis of the property as the fair market value of the qualified real property interest bears to the fair market value of the property before the granting of the qualified real property interest. When a taxpayer donates to a qualifying conservation organization an easement on a structure with respect to which deductions are taken for depreciation, the reduction required by this paragraph (h) (3) (ii) in the basis of the property retained by the taxpayer must be allocated between the structure and the underlying land.

(4) *Examples.* The provisions of this section may be illustrated by the following examples. In examples illustrating the value or deductibility of donations, the applicable restrictions and limitations of § 1.170A-4, with respect to reduction in amount of charitable contributions of certain appreciated property, and § 1.170A-8, with respect to limitations on charitable deductions by individuals, must also be taken into account.

Example (1). A owns Goldacre, a property adjacent to a state park. A wants to donate Goldacre to the state to be used as part of the park, but A wants to reserve a qualified mineral interest in the property, to exploit currently and to devise at death. The fair market value of the surface rights in Goldacre is $200,000 and the fair market value of the mineral rights is $100,000. In order to ensure that the quality of the park will not be degraded, restrictions must be imposed on the right to extract the minerals that reduce the fair market value of the mineral rights to $80,000. Under this section, the value of the contribution is $200,000 (the value of the surface rights).

Example (2). Assume the same facts as in *example (1),* except that A would like to retain a life estate in Goldacre. A donates a remainder interest in Goldacre to the county government to use Goldacre as a park after A's death, but reserves the mineral rights in Goldacre (with restrictions on extraction similar to those in *example (1)).* A's gift does not meet the requirements of § 1.170A-13(b) (2), with respect to remainder interests in real property, Accordingly, no income tax deduction is allowable under this section.

Example (3). In 1982, B, who is 62, donates a remainder interest in Greenacre to a qualifying organization for conservation purposes. Greenacre is a tract of 200 acres of undeveloped woodland that is valued at $200,000 at its highest and best use. Under § 1.170A-12(b), the value of a remainder interest in real property following one life is determined under § 25.2512-9 of the Gift Tax Regulations. Accordingly, the value of the remainder interest, and thus the amount eligible for an income tax deduction under section 170(f), is $95,358 ($200,000 x .47679).

Example (4). Assume the same facts as in *example (3),* except that Greenacre is B's 200-acre estate with a home built during the colonial period. Some of the acreage around the home is cleared; the balance of Greenacre, except for access roads, is wooded and undeveloped. See § 170(f) (3) (E) (i). However, B would like Greenacre to be maintained in its current state after his death, so he donates a remainder interest in Greenacre to a qualifying organization for conservation purposes pursuant to sections 170(f) (3) (B) (iii) and (h) (2) (B). At the time of the gift the land has a value of $200,000 and the house has a value of $100,000. The value of the remainder interest, and thus the amount eligible for an income tax deduction under section 170(f), is computed pursuant to § 1.170A-12. See § 1.170A-12(b) (3).

Example (5). Assume the same facts as in *example (3),* except that at age 62 instead of donating a remainder interest B donates an easement in Greenacre to a qualifying organization for conservation purposes. The fair market value of Greenacre after the donation is reduced to $110,000. Accordingly, the value of the easement, and thus the amount eligible for a deduction under section 170(f), is $90,000 ($200,000 less $110,000).

Example (6). Assume the same facts as in *example (5),* and assume that three years later, at age 65, B decides to donate a remainder interest in Greenacre to a qualifying organization for conservation purposes. Increasing real estate values in the area have raised the fair market value of Greenacre (subject to the easement) to $130,000. Accordingly, the value of the remainder interest, and thus the amount eligible for a deduction under section 170(f), is $67,324 ($130,000 x .51788).

Example (7). Assume the same facts as in *example (3)*, except that at the time of the donation of a remainder interest in Greenacre. B also donates an easement to a different qualifying organization for conservation purposes. Based on all the facts and circumstances, the value of the easement is determined to be $100,000. Therefore, the value of the property after the easement is $100,000 and the value of the remainder interest, and thus the amount eligible for deduction under section 170(f), is $47,679 ($100,000) x .47679).

Example (8). C owns Greenacre, a 200-acre estate containing a house built during the colonial period. At its highest and best use, for home development, the fair market value of Greenacre is $300,000. C donates an easement (to maintain the house and Greenacre in their current state) to a qualifying organization for conservation purposes. The fair market value of Greenacre after the donation is reduced to $125,000. Accordingly, the value of the easement and the amount eligible for a deduction under section 170(f) is $175,000 ($300,000 less $125,000).

Example (9). Assume the same facts as in *example (8)* and assume that three years later, C decides to donate a remainder interest in Greenacre to a qualifying organization for conservation purposes. Increasing real estate values in the area have raised the fair market value of Greenacre to $180,000. Assume that because of the perpetual easement prohibiting any development of the land, the value of the house is $120,000 and the value of the land is $60,000. The value of the remainder interest, and thus the amount eligible for an income tax deduction under section 170(f), is computed pursuant to § 1.170A-12. See § 1.170A-12(b)(3).

Example (10). D owns property with a basis of $20,000 and a fair market value of $80,000. D donates to a qualifying organization an easement for conservation purposes that is determined under this section to have a fair market value of $60,000. The amount of basis allocable to the easement is $15,000 ($60,000/$80,000 = $15,000/$20,000). Accordingly, the basis of the property is reduced to $5,000 ($20,000 minus $15,000).

Example (11). E owns 10 one-acre lots that are currently woods and parkland. The fair market value of each of E's lots is $15,000 and the basis of each lot is $3,000. E grants to the county a perpetual easement for conservation purposes to use and maintain eight of the acres as a public park and to restrict any future development on those eight acres. As a result of the restrictions, the value of the eight acres is reduced to $1,000 an acre. However, by perpetually restricting development on this portion of the land, E has ensured that the two remaining acres will always be bordered by parkland, thus increasing their fair market value to $22,500 each. If the eight acres represented all of E's land, the fair market value of the easement would be $112,000, an amount equal to the fair market value of the land before the granting of the easement (8 x $15,000 = $120,000) minus the fair market value of the encumbered land after (8 x $1,000 = $8,000). However, because the easement only covered a portion of the taxpayer's contiguous land, the amount of the deduction under section 170 is reduced to $97,000 ($150,000 − $53,000), that is, the difference between the fair market value of the entire tract of land before ($150,000) and after ((8 x $1,000) + (2 x $22,500)) the granting of the easement.

Example (12). Assume the same facts as in *example (11).* Since the easement covers a portion of E's land, only the basis of that portion is adjusted. Therefore, the amount of basis allocable to the easement is $22,400 ((8 x $3,000) x ($112,000/$120,000)). Accordingly, the basis of the eight acres encumbered by the easement is reduced to $1,600 ($24,000 − $22,400), or $200 for each acre. The basis of the two remaining acres is not affected by the donation.

Example (13). F owns and uses as professional offices a two-story building that lies within a registered historic district. F's building is an outstanding example of period architecture with a fair market value of $125,000. Restricted to its current use, which is the highest and best use of the property without making changes to the facade, the building and lot would have a fair market value of $100,000, of which $80,000 would be allocable to the building and $20,000 would be allocable to the lot. F's basis in the property is $50,000, of which $40,000 is allocable to the building and $10,000 is allocable to the lot. F's neighborhood is a mix of residential and commercial uses, and it is possible that F (or another owner) could enlarge the building for more extensive commercial use, which is its highest and best use. However, this would require changes to the facade. F would like to donate to a qualifying preservation organization an easement restricting any changes to the facade and promising to maintain the facade in perpetuity. The donation would qualify for a deduction under this section. The fair market value of the easement is $25,000 (the fair market value of the property before the easement, $125,000, minus the fair market value of the property after the easement, $100,000). Pursuant to § 1.170A-13(h) (3) (iii), the basis allocable to the easement is $10,000 and the basis of the underlying property (building and lot) is reduced to $40,000.

(i)   *Substantiation requirement.* If a taxpayer makes a qualified conservation contribution and claims a deduction, the taxpayer must maintain written records of the fair market value of the underlying property before and after the donation and the conservation purpose furthered by the donation and such information shall be stated in the taxpayer's income tax return if required by the return or its instructions.

(j)   *Effective date.* This section applies only to contributions made on or after December 18, 1980.

Part 20—[Amended]

Par. 5. Paragraph (e) (2) of § 20.2055-2 is amended as follows:

a. The sixth sentence of paragraph (e) (2) (i) is revised to read: "However, except as provided in paragraphs (e) (2) (ii), (iii), and (iv) of this section, for purposes of this subdivision a charitable contribution of an interest in property not in trust where the decedent transfers some specific rights to one party and transfers other substantial rights to another party will not be considered a contribution of an undivided portion of the decedent's entire interest in property."

b. The eighth sentence of paragraph (e) (2) (i) is revised to read: "A bequest to charity made on or before December 17, 1980, of an open space easement in gross in perpetuity shall be considered the transfer to charity of an undivided portion of the decedent's entire interest in the property."

c. Paragraphs (e) (2) (iv), (e) (2) (v), a (e) (2) (vi) are redesignated (e) (2) (v), (e) (2) (vi), and (e) (2) (vii), respectively.

d. A new paragraph (e) (2) (iv) is inserted after paragraph (e) (2) (iii) to read as set forth below.

§ 20.2055-2. Transfers not exclusively for charitable purposes

* * * *

(e)   *Limitations applicable to decedents dying after December 31, 1969.* * * *

(2)   *Deductible interests.* * * *

(iv)   *Qualified conservation contribution.* The charitable interest is a qualified conservation contribution. For the definition of a qualified conservation contribution, see § 1.170A-13.

Part 25—[Amended]

Par. 6. Paragraph (c) (2) of § 25.2522(c)-3 is amended as follows:

a. The sixth sentence of paragraph (c) (2) (i) is revised to read: "However, except as provided in paragraphs (e) (2) (ii), (iii), and (iv) of this section, for purposes of this subdivision a charitable contribution of an interest in property not in trust where the decedent transfers some specific rights to one party and transfers other substantial rights to another party will not be considered a contribution of an undivided portion of the decedent's entire interest in property."

b. The eighth sentence of paragraph (c) (2) (i) is revised to read: "A bequest to charity made on or before December 17, 1980, of an open space easement in gross in perpetuity shall be considered the transfer to charity of an undivided portion of the decedent's entire interest in property."

c. Paragraphs (c) (2) (iv), (c) (2) (v), and (c) (2) (vi) are redesignated (c) (2) (v), (c) (2) (vi), and (c) (2) (vii), respectively.

§ 25.2522(c)-3. Transfers not exclusively for charitable, etc., purposes in the case of gifts made after July 31, 1969.

* * * *

(c)   *Transfers of partial interest in property.*   * * *

(2)   *Deductible interest.*   * * *

(iv)   *Qualified Conservation Contribution.* The charitable interest is a qualified conservation contribution. For the definition of a qualified conservation contribution, see § 1.170A-13.

Roscoe L. Egger, Jr.,

Commissioner of Internal Revenue.

[FR Doc. 8; 13693 Filed 5-20-83; 8:45 am]

Appendix F

Agreement of AAA Associates Limited Partnership Formed to Conduct Real Property Rehabilitation

Dated: As of January   , 1985

TABLE OF CONTENTS

AGREEMENT OF LIMITED PARTNERSHIP OF AAA Associates, dated as of January , 1985, among GP Associates, a partnership consisting of [Names of Individuals], as general partner (the "General Partner") and the other persons executing this Agreement as limited partners and Class B limited partners (individually, a "Limited Partner" and collectively, the "Limited Partners"; the General Partner and the Limited Partners as constituted from time to time are hereinafter sometimes referred to individually as "Partner" or collectively as the "Partners").

WITNESSETH:

WHEREAS, the Partners wish to form a limited partnership pursuant to the Partnership Law of the State of [   ] (the "Uniform Act") for the purposes and term and upon the conditions set forth herein;

NOW, THEREFORE, in consideration of the premises and the agreements herein contained, the Partners hereby agree as follows:

I. Formation

1.1. *Formation—Name—Office.* The Partners hereby form a [State in which partnership is formed] limited partnership (the "Partnership"), under and pursuant to the Uniform Act, which partnership shall be conducted under the name of AAA Associates. The office of the Partnership shall be [Address of Office], or at such other place or places as the General Partner may from time to time designate. Each Partner shall be notified by the General Partner of any change in the office of the Partnership.

1.2. *Purposes.* The purposes for which the Partnership has been formed are—

(i)   to acquire the fee estate to certain real property and the buildings (the "Buildings") and improvements situate thereon located in [Location of property] (the "Property"),

(ii)   to renovate the Buildings,

(iii)   to manage and operate the Property and hold the Property as the owner thereof, and

(iv)   to engage in such other activities as shall be necessary or desirable in connection with or incidental to the foregoing.

1.3. *Term.* The term of the Partnership shall begin as of the date hereof, and shall end on the date which is the earlier of

(i)   December 31, 2021;

(ii)   the date of the disposition of all, or substantially all, of the Property and other assets of the Partnership, or

(iii)   a date determined by all of the Partners.

II. Capital Contributions

2.1. *Capital Contributions.* As of the date of execution hereof, each Partner has made an initial cash contribution to the Partnership in the amount set forth opposite his name on the signature page of this agreement.

2.2. *Withdrawal of Capital.* No Partner shall have the right to withdraw any part of his capital contributions prior to the liquidation and termination of the Partnership pursuant to Article IX hereof, unless such withdrawal is provided for in this Agreement.

2.3. *Interest.* No Partner shall receive any interest on his capital contributions to the Partnership.

2.4. *Use of Capital Contributions.* The funds received by the Partnership pursuant to Section 2.1 hereof shall be held by the Partnership in trust in a special bank account or bank accounts for the benefit of the Partners, until disbursed as provided in this Agreement.

2.5. *Limited Partner's Liability.* The liability of each Limited Partner, as such, shall be limited to the amount of the capital contribution which he has made pursuant to this Article II. No Limited Partner shall have any further obligation to contribute money to or in respect of, nor shall any Limited Partner be personally liable for, any liability or other obligation to the Partnership.

III. Title to the Property of the Partnership

3.1. Title to the Property and to any other property, real or personal, owned by or leased to the Partnership shall be held in the name of the Partnership or in the name of any nominee the General Partner, in its sold discretion, may designate.

IV. The General Partner; Conduct of Business; Powers; Other Activities

4.1. *Management*

(a)   The business and affairs of the Partnership shall be conducted and managed solely by the General Partner in accordance with the Uniform Act. Subject to the other provisions of this Agreement, the General Partner shall be authorized, among other things, in furtherance of the objects of the business of the Partnership to—

(i)   sell, or grant options relating to a sale of, all or any portion of the Property,

(ii)   incur indebtedness, secured or unsecured,

(iii)   mortgage, lease or otherwise encumber the Property, whether or not the term of any such mortgage or lease shall extend beyond the date of termination of the Partnership,

(iv)   enter into and perform contracts of any kind in connection with or necessary or incidental to the management and operation of the Partnership's business, including employing such persons, firms or corporations as the General Partner, in his sole discretion, deems necessary or advisable, and open accounts and maintain funds in the Partnership's name in any bank or savings and loan association, and

   (v)   prepare or cause to be prepared reports and other relevant information for distribution to the Partners.

(b)   The Limited Partners consent to the exercise by the General Partner of the powers conferred on him by this Agréement. No Limited Partner (except one who may also be a general partner and then only in his capacity as General Partner) shall participate in or have any control over the Partnership's business or have any right or authority to act for or bind the Partnership. No grantee, lessee or mortgagee shall be required to investigate the General Partner's authority to sell, mortgage or lease the Property or any part thereof, and any deed, lease, mortgage or option executed by the General Partner shall be binding upon the Partnership. The power of the General Partner to sell the Partnership Property shall be deemed to include full authority and power in his discretion to sell or exchange the Property, or any part thereof, for all cash or for part cash and part deferred payment whether or not secured by mortgage, or wholly or partly for other property or for stock or other securities of any corporation, trust, or other entity, as the General Partner in his sole discretion, may deem advisable.

4.2. *Compensation*

(a)   Except as otherwise provided in this Agreement, no Partner shall be entitled to receive any salary or other remuneration from the Partnership.

(b)   Should the General Partner assume responsibility for management of the Buildings, for reconstruction or renovation of the Buildings, or for the performance of any other service to the Partnership or relating to the Buildings or the Property, the General Partner shall be entitled to compensation at a rate no more favorable than would be charged for comparable service by an unaffiliated party of comparable stature in the [Name of Locality] area.

4.3. *Limitations on Powers*

(a)   The General Partner, in its conduct and management of the business and affairs of the Partnership, shall have all the powers conferred by the Uniform Act on a general partner of a limited partnership.

(b)   The General Partner, with the consent of 51% in interest of the Limited Partners, may cause the Partnership to:

   (i)   make loans to other persons,

   (ii)   invest in the equity securities of other entities,

   (iii)   repurchase, redeem or otherwise acquire Limited Partner's interests, or

   (iv)   issue senior or other securities of the Partnership. However, such consent of 51% in interest of the Limited Partners shall not be required in connection with the refinancing of any of the Partnership's indebtedness.

(c)   The General Partner is hereby authorized on behalf of the Partnership to execute, deliver or accept any document, agreement or instrument which shall be required in connection with and any transaction related to the acquisition of the Property and the activities referred to in Section 1.2 hereof. The execution, delivery or acceptance by the General Partner of any document, agreement or instrument related to the foregoing shall be deemed to be the act of and shall be binding upon the Partnership.

4.4. *Duties.* The General Partner shall devote only so much of its time and attention to the business and affairs of the Partnership as it in his sole discretion may deem reasonably necessary for the purposes of the Partnership.

4.5. *Exculpation.* The General Partner shall not be liable to the Partnership or to the Limited Partners for any act performed by it in good faith on behalf of the Partnership or any failure to so act, if in the sole discretion of the General Partner such action or inaction was in the best interests of the Partnership and did not constitute bad faith or willful misconduct.

4.6. *Outside Interests.* The General Partner and the two individual partners of the General Partner may engage in, invest in, participate in or otherwise enter into such other bsuiness ventures of any kind, nature and description, individually and with others, including, without limitation, the ownership of, or investment in or the operation and management of real estate, whether or not any such business venture competes with the Partnership, and neither the Partnership nor any other Partner shall have any right in or to any such activities or the income or profits derived therefrom.

4.7. *Conflicts of Interest*

(a)   The General Partner may employ, on behalf of the Partnership, such persons, firms or corporations as he, in his sole judgment, shall deem advisable for the operation and management of the business of the Partnership on such terms and for such compensation, as he in his sole judgment, shall determine. Any such entity also may be employed or retained by the General Partner in connection with other business ventures of the General Partner.

(b)   The fact that any partner or a member of his family is directly or indirectly interested in or connected with any person, firm or corporation employed by the Partnership or from whom the Partnership may buy merchandise, services or other property shall not prohibit the General Partner from employing, or from dealing with, such person, firm or corporation on behalf of the Partnership.

V. Accounting Provisions

5.1. *Fiscal Year.* The General Partner, in its sole discretion, shall elect, and from time to time may change, the fiscal year of the Partnership.

5.2. *Books and Accounts*

(a)   Complete and accurate books and accounts shall be kept and maintained for the Partnership at the principal place of business of the Partnership. Each Partner or his duly authorized representative, at his own expense, shall at all reasonable times have access to, and may inspect such books and accounts and any other records of the Partnership.

(b)   All funds received by the Partnership shall be deposited in such bank account or accounts as the General Partner may designate from time to time, and withdrawals therefrom shall be made upon the signature of the General Partner and/ or upon such other signature or signatures on behalf of the Partnership as the General Partner may designate from time to time. All deposits (including security deposits) and other funds not needed in the operation of the Partnership's business may, in the sole discretion of the General Partner, be deposited in interest-bearing bank accounts or invested in United States Government or municipal obligations, money market investments or banking certificates of deposit, or such other investments as the General Partner, in its sole discretion, may determine.

5.3. *Financial and Other Reports*

(a)   The Partnership will report its operations on an accrual or a cash basis, as the General Partner in its sole discretion shall from time to time determine.

Each year, each Partner will be provided with financial statements of the Partnership. The books of account of the Partnership shall be audited on a regular basis at least annually by certified public accountants selected from time to time by the General Partner.

(b)   The General Partner shall cause to be prepared and filed after the end of each fiscal year of the Partnership, all Federal and State income tax returns of the Partnership for such fiscal year and shall take all action as may be necessary to permit the Partnership to prepare timely and thereafter to file promptly such returns. A copy of each such tax return and Form 1065 (Schedule K-1) shall be transmitted to each Partner after the end of each fiscal year relating to the Partner's *pro rata* share of income, credit and deductions for such fiscal year.

5.4. *Tax Elections.* The Partnership, in the sole discretion of the General Partner, may make any election permitted under the Internal Revenue Code of 1954, as amended (the "Code"), including, without limitation, the election permitted under Section 754 of the Code.

5.5. *Expenses.* To the extent practicable, all expenses of the Partnership shall be billed directly to and be paid by the Partnership.

VI. Admission of Successor Limited Partners; Transfer of a Limited Partner's Interest; Disability of a Limited Partner

6.1. *Admission of Successor Limited Partners.* The General Partners shall have the right to admit successor limited partners to the Partnership, pursuant to the provisions of Sections 6.2 and 6.3 hereof. Upon the admission of a successor limited partner, an amendment to the Certificate of Partnership of the Partnership reflecting such admission shall be filed by the General Partner. The admission of any successor limited partner pursuant to this Section 6.1 shall not be cause for dissolution of the Partnership.

6.2. *Transfer of a Limited Partner's Interest.*

(a)   Subject to the provisions of Sections 6.2(b) and 6.4 hereof, a Limited Partner may not, without the consent of the General Partner, transfer, assign, or otherwise dispose of all or any portion of his interest in the Partnership (other than an economic interest) (hereinafter, a "Transfer"). Any purported Transfer without such consent shall be void *ab initio* and shall not bind the Partnership. However, if

(i)   the General Partner consents thereto,

(ii)   the provisions of Section 6.3 hereof do not otherwise prohibit the Transfer, and

(iii)   a duly executed and acknowledged counterpart of the instrument effecting the Transfer, in form and substance satisfactory to the Partnership's counsel, shall have been delivered to the General Partner, then the Transfer may be made; provided, however, that the assignee shall not have the right to be admitted as a successor Limited Partner unless

(A)   the assignor shall have indicated such intention of substitution in the instrument effecting such Transfer and the assignee shall expressly agree to be bound by the provisions of this Agreement and assume all of the obligations imposed upon a Limited Partner hereunder;

(B)   the General Partner, in its sole discretion, shall have consented in writing to such admission and,

(c)   the assignor and the assignee shall have executed or delivered such other instruments as the General Partner may deem necessary or desirable to effectuate such admission. Any assignee must also agree to pay, as the General Partner shall determine, all reasonable expenses and legal fees relating to his admission as a Limited Partner, including but not limited to the cost of any required counsel's opinion and of preparing, filing, and publishing any amendment to the Partnership's Certificate of Partnership necessary to effect such admission. Any valid Transfer shall be recognized not later than the last day of the calendar month following receipt of notice of assignment and required documentation and any required amendment of the Certificate of Partnership of the Partnership shall be filed at least once every calendar quarter to reflect the admission of a substituted Limited Partner.

(b)   The consent of the General Partner shall not be required to a Transfer of a Limited Partner's interest or to the substitution of an assignee as a successor limited partner if the Transfer is made

(i)   to another Partner or

(ii)   for no consideration to the assignor's spouse or bloodline descendant or to a trustee for the benefit of the assignor, his spouse or his bloodline descendant; provided, however, that no Transfer or substitution of a successor limited partner as provided in this Section 6.2(b) shall become effective unless the Transfer or substitution is otherwise made in accordance with and subject to the provisions of paragraph (a) of this Section 6.2.

6.3. *Limitations on Transfers of Limited Partner's Interests.* No Transfer may be made if such Transfer would result in the termination of the Partnership under the terms of any relevant section of the Code or any successor statute, and, if so attempted, such Transfer shall be void *ab initio* and shall not bind the Partnership. In making a determination of this issue, the General Partner, in its sole and absolute discretion, may require the assignee to furnish, at his expense, an opinion of counsel passing on this issue. No Transfer may be made to (and no successor limited partner shall be admitted to the Partnership who is) a person below the age of 21 years or a person who has been adjudged to be insane or incompetent, and any purported Transfer to any such person shall be void *ab initio* and shall not bind the Partnership.

As a condition of recognizing any Transfer, the General Partner may require such proof of the age and competency of the assignee as the General Partner may deem necessary.

6.4. *Disability of a Limited Partner.* The death, insanity, incompetency, bankruptcy, insolvency, dissolution or liquidation and termination (any of the foregoing being hereinafter referred to as a "Disability") of a Limited Partner shall not terminate or dissolve the Partnership. In the event of the Disability of a Limited Partner or the making of an assignment for the benefit of creditors by a Limited Partner, the executor, administrator, guardian, committee, trustee or other legal representative of such Limited Partner shall have all of the rights and liabilities of such Limited Partner, subject to the provisions of this Agreement, upon compliance with the provisions of Sections 6.2 and 6.3 hereof.

6.5. *Rights of First Refusal.* If any Limited Partner shall desire to sell his partnership interest and shall receive a bona fide offer in writing to purchase all of his interest in the Partnership ("Assigning Partner"), such Assigning Partner shall first offer, by notice in writing, to sell all of his interest to the General Partner for the same price and on the same terms and conditions of sale as set forth in the bona fide offer. Such notice or offer shall set forth the name, residence and business address of the bona fide offeror and sufficient facts concerning the bona fide offeror to enable the General Partner to arrive at an informed judgment as to the bona fides of such offer and shall contain a copy of the bona fide offer. The bona fide offer shall set forth the representations and warranties requested of the Assigning Partner and all items affecting the price. The General Partner shall have the option to purchase the interest of the Assigning Partner. The General Partner may elect to exercise the option by written notice to the Assigning Partner mailed within 30 days after the date of delivery of the notice of offer. If the option is not exercised within said 30-day period with respect to the entire interest of the Assigning Partner, then the Assigning Partner shall be free to assign his interest in the Partnership to the bona fide offeror pursuant to the terms of the bona fide offer. If such assignment to the bona fide offeror is not completed in accordance with the terms of the bona fide offer within 30 days thereafter, then the Assigning Partner shall not assign his interest in the Partnership without again offering the same to the General Partner in accordance with the procedure set forth above.

VII. Assignment by the General Partner; Disability of the General Partner; Reconstitution of the Partnership

7.1. *Assignment by the General Partner.* Except as provided in this Article VII, the General Partner may not, without the consent of the other Partners, sell, transfer, assign or otherwise dispose of all or any portion of its General Partner's interest, other than an economic interest. Any such purported sale, transfer, assignment or other disposition shall be void *ab initio* and shall not bind the Partnership unless and until a duly executed and acknowledged counterpart of the instrument effecting the same, in form and substance satisfactory to the Partnership's counsel, shall have been delivered to the Partnership's counsel; and further provided that no such sale, transfer, assignment or other disposition may be made if it would result in the termination of the Partnership under the terms of any relevant section of the Code and, if so attempted, shall be void *ab initio* and shall not bind the Partnership. No General Partner's interest shall be sold, transferred, assigned or otherwise disposed of to a person below the age of 21 years or to a person who has been adjudged to be insane or incompetent.

7.2. *Disability of the General Partner.* Upon

(a)   the death, retirement, dissolution, withdrawal, insanity, incompetency, bankruptcy, or insolvency of a General Partner or the making by a General Partner of an assignment for the benefit of creditors, or

(b)   the failure of a General Partner, at any time; in the opinion of the Partnership's counsel, to satisfy the provision of Section 7.3 hereof relating to the financial net worth requirements of a general partner of a limited partnership (each of the foregoing being hereinafter referred to as a "Disabling Event"), the Partnership may be reconstituted pursuant to Section 7.3 hereof.

7.3. *Reconstitution of the Partnership.* Upon the occurrence of a Disabling Event to the General Partner (the "Disabled General Partner"), the Partnership shall be dissolved, provided, however, within 90 days after the occurrence of a Disabling Event, the Partnership may be reconstituted pursuant to this Section 7.3 as a successor limited partnership upon the consent of such of the Limited Partners as may then be required by law. In reconstituting the Partnership pursuant to this Section 7.3, the Limited Partners may designate a successor general partner to serve in place of the Disabled General Partner; provided, however, no person shall be designated or admitted as a successor general partner unless, in the opinion of the Partnership's counsel, such person has a financial net worth to assure that he shall satisfy the financial net worth requirements of a general partner of a limited partnership. If the Limited Partners shall make such designation

     (i)   the interest of the Disabled General Partner shall be converted into and shall be deemed to be that of a limited partner with the same interest in the Partnership as the Disabled General Partner had (as a General Partner) prior to the Disabling Event (as reduced in the manner set forth below in this Section 7.3); and

     (ii)   if the successor general partner does not own a 1% interest in all material items of profit, loss, capital and cash flow of the Partnership, then each Limited Partner (including the person succeeding to the interest of the Disabled General Partner as a limited partner) shall transfer a *pro rata* portion of his interest to the successor general partner sufficient to give the successor general partner such 1% interest and the successor general partner shall pay to each Limited Partner (including the person succeeding to the interest of the Disabled General Partner as a limited partner), as the purchase price for his interest, an amount to be agreed upon by the successor general partner and the Limited Partners (including the person succeeding to the interest of the Disabled General Partner as a limited partner) and in the absence of such agreement the purchase price for such interest shall be determined by arbitration in accordance with the provisions of Article XIII hereof.

The successor general partner and all of the Limited Partners shall agree to be bound by the provisions of this Agreement; provided, however, that if this Agreement is amended by them, no amendment shall be made unless counsel to the Partnership shall issue an opinion that the Partnership shall continue to be treated as a partnership and not as an association taxable as a corporation, and, further provided, the amended agreement shall be as similar in form and substance to this Agreement as practicable and the new or successor partnership shall engage in the same business as this Partnership employing the assets and name of this Partnership to the extent possible.

For the purposes of this Agreement, unless the context otherwise requires, any successor general partner, upon compliance with the provisions of this Article VII and his admission as a general partner, shall be included within the meaning of the term "General Partner" and any interest of a General Partner or Disabled General Partner, unless the context otherwise requires, shall not include any Limited Partner's interests held by a General Partner, Disabled General Partner or successor general partner. Except for the purposes of Article VIII hereof, any person holding an interest held by a Disabled General Partner which has been converted into a limited partner's interest shall be included within the meaning of the term "limited Partner."

VIII. Distributions and Allocations

8.1. *Definitions.* As used in this Agreement:

(a)   The term "Net Cash Flow" as used herein shall mean net cash available from the operation of the Property utilizing sound accounting practices, inclusive of proceeds derived from mortgage refinancing, new financing, condemnation, sale or insurance recoveries (a "Capital Transaction") and deducting therefrom all cash expenses in connection with the operation of the Property and other cash expenses of the Partnership, provided, however, in making such computation there shall be no deduction for depreciation or for amortization of capital improvements. Expenditures for capital improvements and mortgage amortization payments shall constitute a deduction.

(b)   "Net Profits" or "Net Losses" shall mean the annual net income or net loss, respectively, of the Partnership for a fiscal year as determined by the Partnership's certified public accountant in accordance with principles applied in determining income, gains, expenses, deductions or losses, as the case may be, reported by the Partnership for Federal income tax purposes on its United States partnership tax return.

8.2. *Distributions of Net Cash Flow.* The General Partner shall, from time to time, determine the Net Cash Flow and other available proceeds which may, from time to time, be distributed to all partners, after the establishment and maintenance of appropriate reserves.

(a)   Such Net Cash Flow, except that which is attributable to a Capital Transaction, shall be divided eighty (80%) percent thereof *pro rata* to the Limited Partners, fifteen (15%) percent thereof to the General Partner, and four (4%) percent and one (1%) percent to [Name of individual] and [Name of individual], respectively, as Class B Limited Partners.

(b)   Such Net Cash Flow which is attributable to a Capital Transaction shall be devided eighty (80%) percent *pro rata* to the Limited Partners, fifteen (15%) percent thereof to the General Partner and four (4%) percent and one (1%) percent, respectively, to [Name of individual] and [Name of individual] as Class B Limited Partners. Provided, however, before any distribution may be made on account of the 20% aggregate shares to the General Partner and the Class B Limited Partners hereunder by reason of a sale or condemnation of the entire Property, the Partners* must have received in aggregate distributions under this subparagraph (b) from such transaction the net amount of their cash capital contributions to the Partnership.

8.3. *Allocation of Net Profits and Net Losses.*

(a)   All items of Net Profits or Net Losses, other than capital gain of the Partnership, shall be allocated *pro rata* in the same proportion as provided for in paragraph 8.2 (a).

(b)   All items of capital gain shall be allocated as follows: To the extent that Net Losses in excess of Net Profits have been allocated to any Partner, net capital gain shall be allocated to each such Partner whose capital account has been decreased by such excess in the proportion which such decrease bears to the total decreases until the gain allocated pursuant to this subparagraph equals the total decreases.

---

*Including the General Partner and the Class B Limited Partners with respect to any Limited Partnership interests which they may own.

(c)   To the extent that any Partner has received distributions from the Partnership in excess of Partnership Net Profits allocated to him, net capital gain shall then be allocated to such Partner in the proportion which such excess bears to the total excesses until the gain allocated pursuant to this subparagraph equals such excess. For the purpose of this subparagraph, if Net Losses allocated to any Partner exceed Net Profit allocated to such Partner, the Partnership income allocated to him shall be considered equal to zero.

(d)   After allocating the net capital gains to the extent provided in the preceding subparagraphs (b) and (c), the remaining net capital gains, if any, shall be allocated in the same proportion as set forth in paragraph 8.2(a).

8.4. *Certain Additional Allocations.*

(a)   Upon liquidation of the Partnership pursuant to Article IX hereof, Net Profits or Net Losses realized upon such liquidation shall be allocated pursuant to Section 8.3 hereof and any asset shall be distributed in cash or in kind in accordance with the provisions of Section 9.3 hereof. With respect to assets distributed in kind to the Partners,

(i)   any unrealized appreciation or unrealized depreciation in the values of such assets shall be deemed to be Net Profits or Net Losses realized by the Partnership immediately prior to the liquidation (for the purposes of this Section 8.4(a), "unrealized appreciation" or "unrealized depreciation" shall mean the difference between the appraised value of such assets (as determined pursuant to Section 9.3(b) hereof) and the Partnership's basis for such assets; and

(ii)   such Net Profits or Net Losses shall be allocated to the Capital Accounts pursuant to Section 8.3 hereof and any property so distributed shall be treated as a distribution to the Partners pursuant to Section 8.4 hereof to the extent of the aforesaid appraised value less the amount of any liability related thereto. Nothing contained in this Section 8.4 or elsewhere in this Agreement is intended to treat or cause such distributions to be treated as sales for value.

(b)   For income tax purposes, if the Partnership in any year incurs income or is allowed a deduction (including additional depreciation or amortization as a result of adding an item to its basis) as a result of the transfer of an interest in property to or from a Partner, the difference between the amount taken into account for tax purposes and the amount otherwise taken into account under this Agreement shall be allocated solely to such Partner.

8.5. *No Right to Receive Property.* No Partner shall have the right to demand and receive property other than cash in return for his capital contributions.

8.6. *Return of Distributions.* No Partner shall have any obligation to refund to the Partnership any amount which shall have been distributed to such Partner pursuant to this Agreement, subject, however, to the rights of any third party creditor.

8.7. *Allocations between Assignor and Assignee Partners.* In the case of a Transfer of a Partner's interest in the Partnership, the assignor and assignee shall each be entitled to receive distributions pursuant to Section 8.2 hereof or allocations of Net Profits or Net Losses pursuant to Sections 8.3, and 8.4 hereof, as follows:

(a)   Unless the assignor and assignee agree to the contrary and shall so provide in the instrument effecting the assignment, distributions shall be made to the person owning the Partner's interest on the date of the distribution; and

(b)   Net Profits or Net Losses shall be allocated by the number of days each person held the Partner's interest, except that if the assignor and the assignee agree to the contrary and shall so advise the General Partner in writing within ten days after the end of the fiscal year in which the assignment occurs, Net Profits or Net Losses from any Capital Transaction shall be allocated to the holder of the Partner's interest on the day the Capital Transaction occurred during such year.

IX. Liquidation and Termination of the Partnership

9.1. *General.* Upon the termination of the Partnership, the Partnership shall be liquidated in accordance with this Article IX and the Uniform Act. The liquidation shall be conducted and supervised by the General Partner or, if there be no General Partner, by a person who shall be designated for such purposes by the remaining Partners (the General Partner or such person so designated being herein referred to as the "Liquidating Agent"). The Liquidating Agent shall have all of the rights and powers with respect to the assets and liabilities of the Partnership in connection with the liquidation and termination of the Partnership that the General Partner would have with respect to the assets and liabilities of the Partnership during the term of the Partnership, and the Liquidating Agent is hereby expressly authorized and empowered to execute any and all documents necessary or desirable to effectuate the liquidation and termination of the Partnership and the transfer of any asset or liability of the Partnership. The Liquidating Agent shall have the right from time to time, by revocable powers of attorney, to delegate to one or more persons any or all of such rights and powers and such authority and power to execute documents, and, in connection therewith, to fix the reasonable compensation of each such person, which compensation shall be charged as an expense of liquidation. The Liquidating Agent is also expressly authorized to distribute the Partnership's property to the Partners subject to liens.

9.2. *Statements on Termination.* Each Partner shall be furnished with a statement prepared by the Partnership's certified public accountant which shall set forth the assets and liabilities.

9.3. *Priority on Liquidation.* The Liquidating Agent shall, to the extent feasible, liquidate the assets of the Partnership as promptly as shall be practicable. To the extent the proceeds are sufficient therefor, as the Liquidating Agent shall deem appropriate, the proceeds of such liquidation shall be applied in the following order of priority:

(a)   To the payment of matured debts and liabilities of the Partnership and the costs and expenses of the dissolution and liquidation;

(b)   To the setting up of any reserve which the Liquidating Agent may deem reasonably necessary for any contingent, unmatured or unforeseen liability of the Partnership; and

(c)   Any balance then remaining shall be allocated among the Partners and paid and distributed in the manner provided in Sections 8.2 and 8.3 hereof.

9.4. *Distribution of Non-Liquid Assets.* Subject to the provisions of Section 8.4(a) hereof, if the Liquidating Agent shall determine that it is not feasible to liquidate all of the assets of the Partnership, then the Liquidating Agent shall cause the fair market value of the assets not so liquidated to be determined by appraisal. Such assets, as so appraised, shall be retained or distributed by the Liquidating Agent as follows:

(a)   The Liquidating Agent shall retain assets having a fair market value equal to the amount by which the net proceeds of liquidated assets are insufficient to satisfy the debts and liabilities of the Partnership (other than any debt or liability for which neither the Partnership nor a Partner is personally liable), to pay the costs and expenses of the dissolution and liquidation, to establish reserves and to repay any loan or advance made by the Partners and any other debt and liability to the Partners, all subject to the provisions of Sections 8.2, 8.3 and 8.4 hereof. The foregoing notwithstanding, the Liquidating Agent shall have the right to distribute property subject to liens at the value of the Partnership's equity therein.

(b)   The remaining assets (including mortgages and other receivables) shall be distributed to the Partners *pro rata* in accordance with their respective percentage interests in the Partnership. If, in the sole and absolute judgment of the Liquidating Agent, it shall not be feasible to distribute to each Partner an aliquot share of each asset, the Liquidating Agent may allocate and distribute specific assets to one or more Partners as tenants-in-common as the Liquidating Agent shall determine to be fair and equitable, taking into consideration, *inter alia,* the basis for tax purposes of each asset distributed and after crediting or charging the Capital Accounts of the Partners for any unrealized gain or loss as provided in Section 8.4(a) hereof.

9.5. *Orderly Liquidation.* A reasonable time shall be allowed for the orderly liquidation of the assets of the Partnership and the discharge of liabilities to creditors so as to minimize the losses normally attendant upon a liquidation.

9.6. *Deficit Upon Liquidation.* Upon liquidation, no Partner shall be liable to the Partnership for any deficit in his Capital Account, except to the extent provided by law with respect to third party creditors of the Partnership, nor shall such deficit be deemed an asset of the Partnership.

9.7. *Source of Distributions.* No Partner shall be personally liable for the return of the capital contributions of any other Partners, or any portion thereof, it being expressly understood that any such return shall be made solely from Partnership assets.

X. Loans and Advances by Partners

10.1. *General.* Except as expressly provided for herein, if any Partner shall loan or advance any funds to the Partnership (other than the capital contributions provided in Article II hereof), such loan or advance shall not be deemed a contribution to the capital of the Partnership and shall not in any respect increase such Partner's interest in the Partnership. Such loan or advance shall constitute an obligation and liability of the Partnership and shall not bear any interest. Unless otherwise agreed in writing between a Partner and the Partnership, no Partner shall have personal obligation or liability for the repayment of such loans and the same shall be collectible only from Partnership assets.

XI. Power of Attorney

11.1. *General.* Each Limited Partner irrevocably constitutes and appoints the General Partner, any successor general partner and the Liquidating Agent, or any one of them, the true and lawful attorney of such Limited Partner to execute, acknowledge, swear to and file any of the following:

(a)   Any amendment to the Partnership Certificate of Limited Partnership pursuant to the Uniform Act;

(b)   Any certificate or other instrument

(i)   which may be required to be filed by the Partnership under the laws of the United States, the State of [   ] or any other State in which any of the Partners reside or

(ii)   which the General Partner shall deem advisable to file;

(c)   Any amendment to any certificate or other instrument referred to in paragraphs (a) and (b) of this Section 11.1;

(d)   Any document which may be required to effectuate the liquidation or termination of the Partnership; and

(e)   Any amendment to this Agreement or the foregoing certificates or instruments necessary to effect any change permitted under Section 14.5 hereof.

It is expressly acknowledged by each Limited Partner that the foregoing power of attorney is coupled with an interest and shall survive the death, legal incapacity or a Transfer by a Limited Partner; provided, however, if a Limited Partner shall make a Transfer of all of his interest and the assignee shall, in accordance with Section 6.3 hereof, become a successor limited partner, such power of attorney shall survive the Transfer only for the purpose of executing, acknowledging, swearing to and filing any and all instruments necessary to affectuate such substitution.

Each Limited Partner hereby agrees to execute concurrently herewith or upon 15 days' prior written notice, a special power of attorney containing the substantive provisions hereof in form satisfactory to the General Partner.

11.2. *Successor Limited Partner.* A power of attorney similar to that contained in Section 11.1 hereof shall be one of the instruments which the General Partner may require a successor limited partner to execute, acknowledge and swear to pursuant to Section 6.3 hereof.

11.3. *Additional Power of Attorney.* Upon the admission of a successor general partner pursuant to Section 7.3 hereof, or upon the liquidation or termination of the Partnership, each Limited Partner at the request of the successor general partner or the Liquidating Agent shall execute, acknowledge and swear to a new power of attorney similar to that described in Section 11.1 hereof, in favor of such successor general partner or Liquidating Agent.

XII. Indemnification

12.1. *General.*

(a)   In no event shall the General Partner, or any partners of the partnership which is the initial General Partner, be subject to any personal liability to the Partnership or to any Partner for any reason whatsoever other than the bad faith or willful misconduct of such General Partner. The conduct of the General Partner in reliance upon the advice of counsel shall in all cases be deemed to be in good faith. In the event that any action or proceeding is instituted or threatened against the General Partner, or in which he may be a party, he shall be entitled to retain legal counsel and other experts at the expense of the Partnership and to be immediately reimbursed for, indemnified against and saved harmless with respect to any liabilities, costs and expenses arising out of or in connection with such action or proceeding including, but not limited to, legal fees and expenses, as and when such liabilities, costs and expenses are incurred and whether or not such General Partner is serving in such capacity as of the date that such liabilities, costs and expenses are fixed or incurred and without awaiting determination of the action or proceeding. It is understood and agreed that the reimbursement and indemnification herein provided for shall include without limitation, to the extent permitted by law, any claim based upon fraud in the inducement of this agreement, misrepresentation, violation of any securities law, syndication law or other Blue Sky law, and it is expressly agreed that all of such matters are, to the extent permitted by law, arbitrable and subject to the arbitration provisions of Article XIII hereof.

(b)   The General Partner shall have the right and authority to require in all of the Partnership's contracts that he will not be personally liable thereon and that the person or entity contracting with the Partnership is to look solely to the Partnership and its assets for satisfaction of any claim, liability, expense, action, judgment or other matter arising thereunder or on account thereof.

XIII. Determination of Disputes

13.1. *General.* Any dispute or controversy arising under, out of, in connection with, or in relation to

(a)   this Agreement, and any amendment thereof,

(b)   the breach thereof, or

(c)   the formation, operation or termination of the Partnership shall be determined and settled by arbitration by a panel of three members in accordance with the Rules of the American Arbitration Association. Any award rendered therein shall be final and binding upon the Partners and their legal representatives and judgment may be entered in any court having jurisdiction thereof. The expenses of such arbitration shall be paid by the party against whom the award shall be entered.

XIV. Miscellaneous Provisions

14.1. *Investment Representations.* Each partner acknowledges, warrants and represents as follows:

(a)   That he is over twenty-one (21) years of age, is subject to no legal disability and has not relied and is not relying upon any information, statement or representation except those expressly set forth in this agreement.

(b)   That he has consulted with and been advised by his own attorney and accountant with respect to his participation in the Partnership.

(c)   That he and his advisors have reviewed this agreement and the exhibits annexed hereto.

(d)   That nothing in this agreement or the exhibits annexed hereto or any other document or memorandum exhibited to him has been construed or relied upon by him as a representation by the General Partner or any other person of the income, expenses or value of the Property, the tax shelter or other tax consequences to the partners or of the future distributions, income and losses of the Partnership.

(e)   That he is a sophisticated real estate investor.*

(f)   That his capital contribution does not represent more than ten (10%) percent of his net worth.*

(g)   That he has not had a professional, social and/or business relationship with the General Partners and their families prior to this transaction.*

(h)   That this is a private placement exempt from registration under § 4(2) of the Securities Act of 1933, as amended, and not a public offering and that no registration statement, prospectus or offering circular is required by law or has been used in connection with this transaction.*

(i)   That he is investing in the Partnership for his own account and for investment purposes only and not with a view to distribution or resale.*

(j)   That he is prepared to hold this partnership interest indefinitely and will not sell, pledge, transfer or otherwise dispose of his interest in the Partnership or any portion thereof without the written legal opinion of counsel satisfactory to the General Partner that such transfer is not a violation of the Securities Act of 1933, as amended, or any other applicable securities law.

14.2. *Applicable Law.* This Agreement shall be governed by, and construed in accordance with, the laws and decisions of the State of [   ].

14.3. *Creditors.* A creditor who makes a non-recourse loan to the Partnership may not have or acquire at any time as a result of making such loan any direct or indirect interest in the profits, capital or property of the Partnership other than as a secured creditor. The Partnership shall not incur any non-recourse indebtedness unless the creditor agrees to the foregoing.

14.4. *Oral Modification.* This Agreement constitutes the entire understanding among the parties hereto. No waiver or modification of the provisions hereof shall be valid unless in writing and signed by the party to be charged and then only to the extent therein set forth.

14.5. *Notices.* All notices, demands, solicitations of consent or approval, and other communications hereunder shall be in writing and shall be deemed to have been given when the same are deposited in the United States Mail and sent by postage prepaid registered or certified mail, return receipt requested, addressed as follows: If intended for the Partnership, to its principal place of business; if intended for any Partner, at the address of such Partner set forth at the foot of this Agreement; or to such other address which any Partner shall have given to the Partnership for such purpose by notice delivered as aforesaid.

14.6. *Captions.* The captions used herein are intended for convenience of reference only, shall not constitute any part of this Agreement and shall not modify or affect in any manner the meaning or interpretation of any of the provisions of this Agreement.

14.7. *Amendments.* Any amendment of this Agreement may be made only by written consent of the General Partner and Limited Partners holding not less than 51% of the total Limited Partners' interests in the Partnership.

14.8. *Binding Effect.* Except as otherwise provided herein, this Agreement shall be binding upon and shall inure to the benefit of the respective heirs, executors, administrators, legal representatives, and permitted successors and assigns of the parties hereto.

---

*Note that these representations vary somewhat from the terms of Regulation D (see Appendix H). The requirements of Regulation D must be met independently if exemption from registration is to be obtained on that basis.

14.9. *Construction.* None of the provisions of this Agreement shall be for the benefit of or enforceable by any creditor of the Partnership.

14.10. *Separability.* In case any one or more of the provisions contained in this Agreement or any application thereof shall be invalid, illegal or unenforceable in any respect, the validity, legality and enforceability of the remaining provisions contained herein and other application thereof shall not in any way be effected or impaired thereby.

14.11. *Survival of Representations, Warranties and Agreements.* All representations, warranties and agreements shall survive until the final liquidation and termination of the Partnership, except to the extent that a representation, warranty or agreement expressly provides otherwise.

14.12. *Further Assurances.* The Partners will execute and deliver such further instruments and do such further acts and things as may be required to carry out the intent and purposes of this Agreement.

14.13. *Counterparts.* This Agreement may be executed in counterparts and all such counterparts taken together shall constitute one agreement.

IN WITNESS WHEREOF, the parties have executed this

Agreement as of the date first written above.

GENERAL PARTNER

Name   Address

GP Associates

By: _____

CLASS B LIMITED PARTNERS

Name   Address

_____   _____

_____   _____

LIMITED PARTNERS

Name	Address	Capital Contribution	Percentage Interest

[Notarizations]

Appendix G

Rehab Associates Limited Partnership Pro Forma Accounting Projections with Accompanying Notes and Assumptions

---

(Projections as of September 23, 1983)

REHAB ASSOCIATES
STATEMENT OF PROJECTED

Year	1983 (3 months)	1984	1985	1986	1987	1988	1989	1990
Basic rental income (Note 3)			182500	980000	1855000	2230000	2230000	2811250
Additional rental income (Note 3)						200060	420000	590555
Interest income from capital contributions	423974	1639985	1376015	1005575	698922	418175	140849	
Total income	423974	1639985	1558515	1985575	2551922	2848235	2790849	3401805
Depreciation (Note 9)			1031073	2547293	2505793	2505793	2505793	1634293
Additional interest expense (Note 12)						200060	420000	493095
Interest on consolidated wraparound deed of trust (Note 15)			956134	3820638	3814689	3807260	3798567	3683151
Interest on supervisory services fee (Note 15)			67500	270000	270000	270000	270000	270000
Accrued interest on general partner deferred fee (Note 14)			12000	48000	48000	48000	48000	48000
Administrative fee (Note 10)			3750	15000	15000	15000	15000	16875
Monitoring fee (Note 11)			1075	7500	7500	7500	7500	8430
Other deductions (Note 5)	540780	962992	1062140	1442541	1166317	808437	401980	126073
Other administrative costs			6937	26775	25018	22958	20543	23307
Total deductions	540780	962992	3141410	8177746	7852216	7605016	7487383	6304031
Taxable income or (loss)	(116806)	676992	(1502895)	(6192172)	(5298294)	(4836781)	(4696454)	(2902226)

(Due to computer rounding, certain rows may not add.)

The notes and assumptions are an integral part of this projected statement.

LIMITED PARTNERSHIP
PARTNERSHIP TAXABLE INCOME (LOSS)

1991	1992	1993	1994	1995	1996	1997	1998	1999	2000 (9 months)
4555000	4555000	4635000	4875000	4887500	4925000	4925000	4932500	4955000	3716250
752080	923160	1104495	1296750	1500520	1716540	1945580	2188305	2445625	2038733
5307080	5478160	5739495	6171750	6388020	6641540	6870580	7120805	7400625	5754983
1634293	1634293	1634293	1634293	1634293	1634293	1634293	1634293	1634293	1225720
562320	635640	713355	795750	883080	975660	1073820	1177845	1288125	1018743
3356001	3344435	3329044	3312740	3292689	3269184	3241630	3209329	3171464	2349865
270000	270000	269918	267124	262762	257991	252772	247064	240821	176168
48000	48000	48000	48000	48000	48000	48000	48000	48000	36000
22500	22500	22500	22500	24175	30000	30000	30000	30000	22500
11250	11250	11250	11250	12100	15000	15000	15000	15000	11250
126073	126073	126073	107242	833	833	833	833	833	625
37742	35610	34621	33680	35570	55834	45894	42151	51920	30214
6069859	6128600	6190654	6232579	6193789	6286795	6342242	6404515	6480456	4871083
(762779)	(650440)	(451159)	(60829)	194231	354745	528338	716290	920169	883899

Year	1983 (3 months)	1984	1985	1986	1987	REHAB ASSOCIATES STATEMENT OF PROJECTED 1988	1989
Taxable income or (loss)	(116806)	676992	(1582895)	(6192172)	(6290294)	(4836781)	(4696454)
Add:							
Depreciation			1031073	2547293	2505793	2505793	2505793
Accrued interest–consolidated wraparound deed of trust			371826	1485548	1483465	1480574	1477185
Accrued interest–supervisory services fee			67500	270000	270000	270000	270000
Accrued interest–general partner deferred fee			12000	48000	48000	48000	48000
Salvage	300000						
Pre-opening loan to lessee			46750	187000	187000	187000	187000
Other deductions	540545	962992	1062140	1442541	1166217	808437	401900
Class A limited partners' contribution	710500	853500	3268500	2543500	2256000	2132300	2235700
Class B limited partner's contribution	30000	34000	34000	34000	34000	34000	
Short-term loan	7400000						
Construction loan	1700000	6400000	5650000				
Subtract:							
Short-term loan–principal and interest payments	271241	1085463	1085463	2166703	2893782	2493755	2337812
Amortization–consolidated wraparound deed of trust			7380	32635	38250	44840	52575
Amortization–supervisory services fee							
Accrued interest paid–consolidated wraparound deed of trust							
Accrued interest paid–supervisory services fee							
General partner fees	800000						
Personal property	700000		700000				
Land	900000						
Building	1576000						
Construction costs	1030000	7600000	5995000	675000			
Developer costs	1001000	120000	379000	150000			
Architect costs	456000	111000	108000				
Financing fees	329000						
Real estate taxes	20000	40000	40000				
Closing costs	114000						
Preopening loan to lessee		185000	750000				
Legal fees	416000						
Accounting fees	70000						
Appraisal costs	17000						
Bank fees	74000						
Printing and filing costs	30000						
Miscellaneous	12765						
Selling commissions	840000						
Real estate advisory fee	280000						
Nonaccountable expense allowance	280000						
Note guarantee fee	930510						
Monitoring fee	40000	50000	50000	50000	50000	50000	
Reserve	376723	(263979)	846052	(708629)	(329860)	40720	38737
Cash available for partners' distribution	0	0	0	0	0	0	0
Limited partners' distributions – Class A							
Limited partners' distributions – Class B							
General Partner distribution							

(Due to computer rounding, certain rows may not add.)

The notes and assumptions are an integral part of this projected statement.

LIMITED PARTNERSHIP
PARTNERSHIP CASH FLOW

1990	1991	1992	1993	1994	1995	1996	1997	1998	1999	2000 (9 months)
(2902226)	(762779)	(650440)	(451159)	(60829)	194231	354745	528338	716290	920169	803899
1634293	1634293	1634293	1634293	1634293	1634293	1634293	1634293	1634293	1634293	1225720
1367996	1043025	1045136	1040576	1035231	1028965	1021620	1013009	1002915	991083	734333
270000	270000	270000	269918	267124	262762	257991	252772	247064	240821	176168
48000	48000	48000	48000	48000	48000	48000	48000	48000	48000	36000
140250										
126873	120873	126873	126073	107242	833	833	833	833	833	625
61632	77249	84696	99206	116391	136441	159946	187500	219801	257666	221982
			10988	46499	50861	55632	60850	66550	72802	59049
265394	1061405	1059648	1057588	1055173	1052343	1049025	1045136	1040576	1035231	771724
			67500	270000	270000	270000	270000	270000	270000	202500
357960	1231759	1329518	1433139	1542998	1659439	1782879	1913760	2052460	2199499	1801489
336661	1158460	1250413	1322128	1395185	1472618	1554705	1641741	1733977	1031758	156781
17719	60972	65811	69506	73431	77506	81827	86407	91262	96408	82478
3500	12510	13294	41425	74302	109315	146347	185612	227221	271333	153930

PROJECTED INVESTMENT SUMMARY FOR A $100,000
CLASS A LIMITED PARTNER INVESTOR IN A 50% TAX BRACKET

Year	(1) Capital Contribution	(2) Interest Expense	(3) Total Investment	(4) Investment Tax Credit	(5) (A) Taxable Income (Loss)	(6) Tax Savings (Cost)	Cash Distribution (7) Annual Amount	(8) Annual Percentage Return	After Tax Benefits (9) (B) Annual	(10) (C) Net	(11) Cumulative	(12) (D) Excess Investment Interest Expense
1983 (3 months)	5075	2990	8065	6171	(3808)	1904			8076	10	10	
1984	6096	11576	17672	14281	(6825)	3413			17694	21	31	
1985	23346	9722	33068	22665	(20830)	10415			33080	12	44	
1986	18168	7106	25274		(50562)	25281			25281	7	51	3916
1987	16114	4946	21060		(42129)	21065			21065	4	55	6448
1988	15231	2972	18203		(36416)	18208			18208	6	61	6448
1989	15970	1006	16976		(33965)	16983			16983	7	68	6448
1990					(20367)	10184	2405	2.40	12588	12588	12656	6448
1991					(5353)	2677	8275	8.27	10951	10951	23608	
1992					(4565)	2282	8932	8.93	11214	11214	34821	
1993					(3166)	1583	9444	9.44	11027	11027	45848	
1994					(427)	213	9966	9.97	10179	10179	56027	
1995					1231	(616)	10519	10.52	9903	9903	65930	
1996					2210	(1105)	11105	11.11	10000	10000	75931	
1997					3237	(1619)	11727	11.73	10108	10108	80639	
1998					4322	(2161)	12386	12.39	10224	10224	96263	
1999					5474	(2737)	13084	13.08	10347	10347	106610	
2000 (9 months)					5492	(2746)	11193	11.19	8447	8447	115058	
Total	100000	40318	140318	43118	(206448)	103224	109034					

(A) Column (5) includes interest on Capital Contributions (Column 2)

(B) Column (9) equals the sum of "Investment Tax Credit" (Column 4), "Tax Savings (Cost)" (Column 6) and "Cash Distributions – Annual Amount" (Column 7)

(C) Column (10) equals "Annual After Tax Benefits" (Column 9) minus "Total Investment" (Column 3)

(D) See Note (19)

(Due to computer rounding, certain rows may not add.)

The notes and assumptions contained in this report are an integral part of this projected statement.

PROJECTED INVESTMENT SUMMARY FOR A $50,000
CLASS A LIMITED PARTNER INVESTOR IN A 50% TAX BRACKET

Year	(1) Capital Contribution	(2) Interest Expense	(3) Total Investment	(4) Investment Tax Credit	(5) (A) Taxable Income (Loss)	(6) Tax Savings (Cost)	Cash Distribution		After Tax Benefits			(12) (D) Excess Investment Interest Expense
							(7) Annual Amount	(8) Annual Percentage Return	(9) (B) Annual	(10) (C) Net	(11) Cumulative	
1983 (3 months)	2538	1495	4033	3086	(1904)	952			4038	5	5	
1984	3048	5788	8836	7141	(3413)	1706			8847	11	16	
1985	11673	4861	16534	11333	(10415)	5208			16540	6	22	1958
1986	9084	3553	12637		(25281)	12640			12640	3	25	3224
1987	8057	2473	10530		(21065)	10532			10532	2	27	3224
1988	7615	1486	9101		(18208)	9104			9104	3	30	3224
1989	7985	503	8488		(16983)	8491			8491	4	34	3224
1990					(10184)	5092	1202	2.40	6294	6294	6328	
1991					(2677)	1338	4137	8.27	5476	5476	11804	
1992					(2282)	1141	4466	8.93	5607	5607	17411	
1993					(1583)	792	4722	9.44	5513	5513	22924	
1994					(213)	107	4983	9.97	5090	5090	28014	
1995					616	(308)	5259	10.52	4952	4952	32965	
1996					1105	(552)	5553	11.11	5000	5000	37965	
1997					1619	(809)	5863	11.73	5054	5054	43019	
1998					2161	(1081)	6193	12.39	5112	5112	48131	
1999					2737	(1368)	6542	13.08	5174	5174	53305	
2000 (9 months)					2746	(1373)	5597	11.19	4224	4224	57529	
	50000	20159	70159	21559	(103224)	51612	54517					

(A) Column (5) includes interest on Capital Contributions (Column 2)

(B) Column (9) equals the sum of "Investment Tax Credit" (Column 4), "Tax Savings (Cost)" (Column 6) and "Cash Distribution – Annual Amount" (Column 7)

(C) Column (10) equals "Annual After Tax Benefits" (Column 9) minus "Total Investment" (Column 3)

(D) See Note (19)

(Due to computer rounding, certain rows may not add.)

The notes and assumptions contained in this report are an integral part of this projected statement.

PROJECTED TAX AND ECONOMIC RESULTS
OF THE ASSUMED SALE OF THE
PARTNERSHIP PROPERTY ON SEPTEMBER 1, 2000 APPLICABLE TO
A $100,000 CLASS A LIMITED PARTNERSHIP INTEREST (NOTE 18)
(000 OMITTED–EXCEPT FOR PER UNIT)

	Sale at 11% Capitalization Rate	Sale at $1 Above Existing Indebtedness	Sale at Cost
Net sales proceeds:			
Sales price	$ 80018	$32331	$33359
Less:			
Advisory fee to general partner	2401		1001
Existing indebtedness	32331	32331	32331
Net sales proceeds	$ 45286	$ -0-	$ 27
Distribution of net sales proceeds:			
Class A limited partners	$ 33244		$ 25
Class B limited partner	1750		1
General partner	10292		1
	$ 45286	$ -0-	$ 27
Gain on sale:			
Sales price (amount realized)	$ 80018	$32331	$33359
Less:			
Selling expenses	2401		1001
Adjusted basis	3165	3165	3165
Gain on sale	$ 74452	$29166	$29193
Class A limited partners' share of gain on sale:			
Ordinary	$ 3858	$ 3858	$ 3858
Capital	55593	23285	23310
Tax due (A)	$ 13119	$ 6586	$ 6590
Class A limited partners' share			
of net sales proceeds	$ 33244	$ -0-	$ 25
Less tax due	13119	6586	6590
Net after tax proceeds (cost)	$ 20125	($ 6586)	($ 6563)
Per unit (with 000)	$144000	($47000)	($47000)

(A) The projected tax does take into account any alternative minimum tax.

(Due to computer rounding, certain rows and columns may not add.)

The notes and assumptions are an integral part of this projected statement.

Accounting projections

The accounting projections have been prepared solely for purposes of illustration. They are based on estimates and assumptions which are inherently subject to uncertainty and variation depending upon evolving events. They are not represented to be, and should not be relied upon to indicate, results that will actually be obtained.

No representation or warranty of any kind is made respecting the accuracy, attainment or completeness of the accounting projections and no representation or warranty should be inferred from such projections. Prospective investors are urged to consult their own advisors with respect to the projections and the material assumptions underlying them.

Notes and assumptions

The attached statements analyzing the proposed Partnership transactions are based upon, among other things, the assumptions outlined herein. The more descriptive and informative Confidential Memorandum should be read in conjunction with the attached statements, especially materials set forth under "Fiduciary Responsibility of General Partner," "Conflicts of Interest," "Federal Tax Consequences and "Risk Factors."

The Internal Revenue Code (the Code), especially those sections enacted or amended by the Economic Recovery Tax Act of 1981 and the Tax Equity and Fiscal Responsibility Act of 1982 (TEFRA), may be affected by the adoption of further regulations. Accordingly, the reader should consult his own tax advisor with respect to the tax aspects as they may relate to his particular circumstances and tax situation.

As more fully discussed in the Confidential Memorandum, TEFRA enacted an alternative minimum tax (AMT). Limitations are imposed upon the amount of interest on the investors' notes that can be deducted in computing this AMT.

The reader is advised to consult with his tax advisor as to state and local taxes which may be payable in connection with tax preference items.

Projections are inherently subject to varying degrees of uncertainty and their achievability depends upon, among other things, the reliability of the underlying assumptions and the probability of the occurrence of a complex series of future events or transactions, many of which are not within the control of the General Partner.

THE ACCOMPANYING PROJECTIONS ARE AN ILLUSTRATION OF FINANCIAL RESULTS BASED UPON ASSUMPTIONS WHICH ARE NOT NECESSARILY THE MOST LIKELY. THE PROJECTIONS ARE PREPARED TO ILLUSTRATE WHAT WOULD HAPPEN IF THE ASSUMPTIONS CONTAINED HEREIN WOULD OCCUR.

(1)   Principles of reporting: The projections are prepared in accordance with the tax accrual method of accounting and assume that all the limited partners will be admitted on October 1, 1983. It is assumed that all transactions will be completed on October 1, 1983. If an investor is admitted after October 1, 1983, the total amounts of income and deductions allocated or available to him will vary from those projected.

(2)   Organization: The Partnership is a limited partnership organized under [state] law to acquire, substantially rehabilitate and refurnish, own and net-lease a 275-room hotel located in [   ] (the "Hotel"). The General Partner is GP Incorporated, a [   ] corporation.

(3)   Total rental income: The Partnership will lease the property as described in the "Lease" section of the "Confidential Memorandum." The Lessee will be required to pay the Partnership an Annual Basic Rent as set forth in the Lease plus Percentage Rent in an amount equal to 14% of the Lessee's Gross Income in excess of $15,000,000 per year. The amount of the Lessee's Gross Income upon which the projected Percentage Rent is based is projected in the market study and financial projections prepared in June 1982 and supplemented in April 1983 (the "Study"). The Study projects Gross Income of the Lessee from Hotel operations of $11,568,000 for the calendar year 1986, $14,160,000 for 1987, $15,951,000 for 1988, $17,863,000 for 1989, $18,936,000 for 1990 and thereafter adjusted for inflation at an assumed rate of 6%. The projections contained herein estimate that the Lessee's Gross Income from Hotel operations will begin October 1, 1985 rather than January 1, 1986. The Study numbers were adjusted accordingly. The maximum impact on the Lessee's Gross Income from Hotel operations in any particular year as a result of the adjustment will be approximately $648,000.

(4)   Syndication costs: The Partnership will incur approximately $1,542,765 of nondeductible costs in connection with the syndication of this offering. These expenditures consist of the following: $1,455,000 to [investment bank], $45,000 to Partnership counsel, $10,000 to [firm preparing market study] and $32,765 for printing, filing and other miscellaneous costs.

(5) Expenses:

It is assumed that the Partnership will deduct certain fees and expenses as ordinary and necessary business expenses. These fees and expenses are not being paid out of operating income:

REHAB ASSOCIATES LIMITED PARTNERSHIP

	1983 (3 months)	1984	1985	1986	1987	1988	1989	1990	1991	1992	1993	1994	1995-2015	Total
Partnership supervisory fee (A)			$15,625	$ 62,500	62,500	62,500	$ 46,875							$ 250,000
Loan guarantee fee (B)			3,125	12,500	12,500	12,500	9,375							50,000
Tax advice (C)	$202,000													202,000
Lease and consolidated wraparound deed of trust legal fee (D)			208	833	833	833	833	$ 833	$ 833	$ 833	833	833	$17,295	25,000
Transfer and mortgage taxes	92,065													92,065
Short-term loan costs (E)	3,800	$ 15,200	15,200	15,200	15,200	15,200	11,400							91,200
Note guarantee fee (F)	76,214	284,488	226,007	158,765	105,362	60,109	19,565							930,510
Construction loan costs (G)	6,344	25,375	19,031											50,750
Construction period costs (H)	16,505	62,020	67,146	126,040	126,040	126,040	126,040	126,040	126,040	126,040	126,040	106,409		1,260,398
Short-term loan interest (I)	143,852	575,909	703,297	1,016,703	793,782	493,755	187,812							3,915,111
Partnership monitoring fee (J)			12,500	50,000	50,000	37,500								150,000
Total	$540,780	$962,992	$1,062,140	$1,442,541	$1,166,217	$808,437	$401,900	$126,873	$126,873	$126,873	$126,873	$107,242	$17,295	

(Due to computer rounding, certain rows may not add.)

347

(A)   Partnership supervisory fee: The General Partner will receive $250,000 in 1983 and $50,000 in 2000 (see Note 14) for supervising the Partnership affairs for the first four years of operations of the Hotel. This fee will be amortized over four years commencing with the date the property is placed in service.

(B)   Loan guarantee fee: The General Partner will receive $50,000 in consideration of its obligation to make loans available to the Partnership during the rehabilitation period and for four years subsequent to the date the property is placed in service. This fee will be amortized over a four-year period commencing with the date the property is placed in service.

(C)   Tax advice: It is assumed that $202,000 will be paid by the Partnership for tax advice in regard to the structuring of the transaction. This cost is comprised of a $125,000 Partnership counsel fee, $60,000 financial consulting fee and a $17,000 appraisal fee. This cost will be deducted in 1983.

(D)   Lease and Consolidated Wraparound Deed of Trust legal fee: Legal fees of $25,000 pertaining to the Lease and Consolidated Wraparound Deed of Trust will be amortized over thirty years commencing with the date the property is placed in service.

(E)   Short-Term Loan costs:

(i)   A financing fee of $74,000 will be paid to the "Short-Term Lender" in connection with securing the loan. $66,200 of this fee will be amortized over a six-year period commencing in 1983 and the balance of $7,800 will be capitalized in accordance with Section 266 of the Code.

(ii)   A legal fee of $15,000 will be paid to the counsel of the Short-Term Lender for the review of the loan documents. This fee will be amortized over a six-year period commencing in 1983.

(iii) A legal fee of $10,000 will be paid to the Partnership counsel for review of the loan documents. This fee will be amortized over a six-year period commencing in 1983.

(F)   Note guarantee fee: A fee of $930,510 will be paid to a surety who will guarantee the payment of principal and interest on the Class Limited Partner notes. $184,207 of this fee will be amortized on a declining balance method over two years to reflect the amount of Class A Limited Partner notes assigned to the Lender and $746,303 will be amortized on a declining balance method over six years to reflect the amount of Class A Limited Partner notes assigned to the Short-Term Lender.

(G)   Construction Loan costs: The Partnership will pay $50,750 to the counsel of the Lender for review of the construction loan documents. This fee will be amortized over the construction period.

(H)   Construction period costs: The Partnership will capitalize, and thereafter amortize, $1,260,398 of interest and taxes incurred during the construction period in accordance with Section 189 of the Code.

(I)   Short-Term Loan interest: The Partnership will deduct $3,915,111, which is the portion of interest attributable to loan proceeds which are not being used for acquisition or construction of the property. The balance of the interest will be capitalized and/or amortized.

(J)   Partnership monitoring fee: Limited partners' monitoring agent will receive $150,000 for monitoring the affairs of the Partnership during the first three years of operations. This fee will be amortized over the three-year period.

(6)   General Partner fees: The General Partner will receive a total of $1,465,000 from the Partnership in connection with the organization and operation of the Partnership and the rehabilitation and operation of the Hotel. The Partnership will amortize $300,000 (see Notes 5A and 5B) of these costs over a four-year period commencing when the property is placed in service. The Partnership will capitalize and depreciate as part of the cost of the Hotel $1,160,000, consisting of a $900,000 rehabilitation supervision (see Note 14) fee and a $260,000 construction financing fee. The Partnership will also capitalize $5,000 of reimbursed expenses which the General Partner incurred in connection with the acquisition of the property. These expenses will be allocated $2,000 to land and $3,000 to the cost of the Hotel.

(7)   Source and application of initial funds:

Source of funds:

Class A Limited Partners' capital contributions	$14,000,000
Class B Limited Partner's capital contributions	200,000
Interest on Class A Limited Partners' capital contributions	5,644,589
Interest on Class B Limited Partner's capital contributions	58,905
Consolidated Wraparound Deed of Trust (A)	21,250,000
Salvage value of personal property	300,000
Services fee (A)	3,000,000
Short-Term Loan (A)	7,400,000
General Partner Deferred fee (B)	400,000
General Partner loan (C)	100,000
	$52,353,494

Application of funds:

Syndication costs (D)	$ 1,542,765
Land	1,622,274
Personal property	4,150,000
Real property	27,586,421
Transfer and mortgage taxes	92,065
General Partner fees (E)	300,000
Legal fees (F)	100,750
Tax advice (G)	202,000
Short-Term Loan fee	66,200
Note guarantee fee	930,510
Short-Term Loan interest	3,915,111
Short-Term Loan	7,400,000
Construction period costs	1,260,398
Preopening loan to Lessee	935,000
Monitoring fee	150,000
Portion of debt service short fall	2,000,000
Unallocated legal fees (C)	100,000
	$52,353,494

(A) See Note (15)	(E) See Note (6)
(B) See Note (14)	(F) See Note (5D, 5E, 5G)
(C) See Note (8)	(G) See Note (5C)
(D) See Note (4)	

(8)   Unallocated legal fees: The Partnership will pay additional counsel fees estimated to be $100,000. These fees will be funded by one of the following: (i) Partnership reserves, (ii) additional cash flow that may be available to the Partnership if the interest rate on the Short-Term Loan is less than the 14.7% average, or if (i) and (ii) are not available these fees will be paid through proceeds loaned by the General Partner to the Partnership.

(9)   Cost recovery allowances: The cost of the property, excluding certain costs that will be expensed, is assumed to be $33,358,695. For purposes of the projections, this cost has been allocated and cost recovery allowances computed as follows:

	Estimated Cost
Personal property	$ 4,150,000
Real property	27,586,421
Land	1,622,274
	$33,358,695

The real property's cost will be recovered using the straight-line method over 15 years. A 25% investment tax credit will be taken on the real property expenditures which constitute "qualified rehabilitation expenditures." The basis of the real property will be reduced by one-half of the credit taken. It has been assumed that the building will be placed in service in October 1985.

The cost of the personal property will be recovered using an accelerated method over five years.

(10)   Administrative fee: The General Partner will receive $3,750 in 1985, $15,000 each year in 1986 through 1989, $16,875 in 1990, $22,500 each year in 1991 through 1994, $24,375 in 1995, $30,000 each year in 1996 through 1999 and $22,500 in 2000 for services in the administration of the affairs of the Partnership.

(11)   Monitoring fee: Limited Partners' monitoring agent will receive $1,875 in 1985, $7,500 each year in 1986 through 1989, $8,438 in 1990, $11,250 each year in 1991 through 1994, $12,188 in 1995, $15,000 each year in 1996 through 1999 and $11,250 in 2000 for monitoring the interest of the Limited Partners.

(12)   Additional interest: Under the Consolidated Wraparound Deed of Trust, the Partnership will be obligated to pay additional interest to the Developer in an amount equal to 14% of the Lessee's Gross Income in excess of $15,000,000 but not in excess of $18,000,000 per year, and 6% of the Lessee's Gross Income in excess of $18,000,000 (see Note 3 in regard to Lessee's Gross Income).

(13)   Class A Limited Partners' capital contributions: The purchase price for each Class unit is $100,000, of which $5,075 will be payable upon admission. The following schedule illustrates the principal and interest payments for a $100,000 unit. A rate of 12.6% per annum is assumed with principal payments payable on admission and on June 30 of each year. Interest payments will be payable in quarterly installments commencing December 31, 1983.

SCHEDULE OF PAYMENTS FOR A ONE UNIT CLASS A
$100,000 INVESTMENT

Date of payment	Principal	Interest	Total payments
Upon admission	$ 5,075		$ 5,075
1983 payment		$ 2,990	2,990
1984 payment	6,096	11,576	17,672
1985 payment	23,346	9,722	33,068
1986 payment	18,168	7,106	25,274
1987 payment	16,114	4,946	21,060
1988 payment	15,231	2,972	18,203
1989 payment	15,970	1,006	16,976
	$100,000	$40,318	$140,318

(14) General Partner Deferred fee: The General Partner will defer $350,000 of the total $900,000 rehabilitation supervisory fee (see Note 6) and $50,000 of the total $250,000 Partnership supervisory fee (see Note 5A). Interest on the deferred fee will accrue at the rate of 12% per annum and will be payable along with the deferred fee subsequent to September 30, 2000. However, if the Partnership has available reserves not required to meet other obligations or if the interest rate on the Short-Term Loan is less than the 14.7% average, then available Partnership reserves and any cash flow available to Partners as a result of the interest savings on the Short-Term Loan will be applied first to reduce the accrued interest and then deferred.

(15) Financing: The following is only a summary of the financing for the property. For a full understanding, reference must be made to the "Acquisition Terms and Financing" section of the Confidential Memorandum.

GP Affiliates, an affiliate of the General Partner, has entered into an agreement to acquire the Hotel and six additional parcels of land from [   ], an unrelated limited partnership ("Developer"). GP Affiliates will assign its right to acquire the property to the Partnership. The purchase price for the property is $5,111,000. The Partnership has also agreed to bear certain additional costs that may be incurred in connection with the acquisition of the property in an amount not to exceed $50,000.

Purchase Money Note: To finance a portion of the purchase price of the property, the Partnership will issue a $1,935,000, 10-year, 16% nonrecourse Purchase Money Note to the Developer.

Construction Loan: The construction loan will be obtained from [   ] (the "Lender") in the amount of $13,750,000. This loan will mature 26 months from closing and will bear interest at a floating rate which will be 1.25% above the Lender's prime rate.

The Permanent Loan: The Partnership has obtained a permanent loan commitment from the Developer pursuant to which the Developer has agreed to:

(A) loan the Partnership $13,750,000 on the completion date or

(B)   in the event that the Developer is unable to loan the $13,750,000 to the Partnership, the Developer will then assume the Partnership obligations under the construction loan at the completion date.

The Permanent Loan will be secured by the Consolidated Wraparound Deed of Trust.

The Interim Loan: The Developer has obtained a commitment from the Lender to loan the Developer $13,750,000 upon completion of the rehabilitation. The loan will be for 5 years and will bear interest at a floating rate which will be .5% over the Lender's prime rate plus 2% of Gross Room Income of the Hotel. To secure the obligation of the Developer, the Lender will have the first deed of trust on the property (subject to a $100,000 deed of trust).

Development Agreement Obligation: At the completion date $2,200,000 of the remaining obligations to the Developer will be consolidated into the Consolidated Wraparound Deed of Trust.

FF&E Agreement Obligation: At the completion date, a $2,750,000 obligation to the Developer will be consolidated into the Consolidated Wraparound Deed of Trust.

Turnkey Rehabilitation Agreement: By the completion date, it is estimated that $3,675,000 will be owed to [service company] under the Turnkey Rehabilitation Agreement. Of this amount, $675,000 together with interest at the rate of 12% per annum (accruing from the ninety-first day following the completion date) will be paid on June 30, 1986, and the balance of $3,000,000 together with interest at 9% per annum subject to certain deferrals and capitalizations, will be paid over a 30-year period.

The Consolidated Wraparound Deed of Trust: On the completion of the Hotel the Permanent Loan ($13,750,000), the Purchase Money Note ($2,550,000, including $615,000 of interest accrued), the Development Agreement Obligation ($2,200,000) and the FF&E Agreement Obligation ($2,750,000) will be consolidated and secured by a $21,250,000 Wraparound Deed of Trust on the property. This indebtedness will bear interest at 16% per annum and will be fully amortized over a 30-year period based on the constant monthly payments. During the first 15 years there will be a 5% interest deferral for 60 months. At the end of the fifteenth year, the portion of the interest during the prior 60-month period which has accrued but which has not yet been paid will be capitalized and, together with interest at 16%, will be amortized over a 10-year period commencing in the sixteenth year. In addition, during any year in which the Partnership is not required to pay and does not pay the 5% deferred interest (anticipated to be 9/1/85 through 8/31/90), additional interest on the outstanding indebtedness will accrue at the rate of 2% per annum. The additional interest will be payable at the end of the fifteenth year. The Consolidated Wraparound Deed of Trust will bear other interest as described in note 12.

Short-Term Loan: The Partnership will also be required to borrow $7,400,000. The Partnership has received a commitment from a lender for a loan at closing, bearing interest at the rate of 1.25% above the London Interbank Offered Rate ("LIBOR") on the date of the funding of the loan, and with a partial interest deferral of up to 1.25% if, and to the extent that, such interest rate exceeds an average of 14.7% per annum. Principal payments will be made in 1986 through 1989. The Partnership may arrange for such a loan under more favorable terms.

(16) Rehabilitation investment tax credit: It is assumed that the rehabilitation of the Hotel will qualify for a 25% rehabilitation investment tax credit.

This credit is available for qualified rehabilitation expenditures attributable to a certified historic structure incurred after 1981. It is assumed for these projections that the rehabilitation investment tax credit will be taken as qualified progress expenditures. As discussed in the "Federal Tax Consequences" section of the Confidential Memorandum, it is not clear that the Internal Revenue Service will agree that the credit can be taken prior to the building being placed in service.

(17) Allocation of Income and Losses from Operations and Cash Available for Distribution:

Income and losses: Income from Operations with respect to any year or portion thereof prior to the Anniversary Date (projected to be October 1, 1990) will be allocated 98.25% to the Class A Limited Partners, .75% to the Class B Limited Partner and 1% to the General Partner. Income from Operations with respect to any year or portion thereof on or after the Anniversary Date shall be allocated to the Partners as follows: First, to the extent of and in proportion to the distributions, if any, of cash flow with respect to that year or portion thereof; however, when income allocated to any Partner would result in the creation or continuation of a positive Capital Account for that Partner while any other Partner has a negative Capital Account, Income from Operations first will be allocated to Partners having negative Capital Accounts in proportion to such negative Capital Accounts until all such negative Capital Accounts have been eliminated; Second, to the Partners in proportion to any deficits in their respective Capital Accounts and to the extent necessary to eliminate such deficits; Third, 94.05% to the Class A Limited Partners, 4.95% to the Class B Limited Partner and 1% to the General Partner until the aggregate positive balances in the Class A Limited Partners' Capital Accounts equals $14,000,000 plus a return of $1,260,000 per annum commencing on the Anniversary Date, less an amount equal to all prior distributions of net proceeds and cash flow to the Class A Limited Partners. The balance, if any, 66.5% to the Class A Limited Partners, 3.5% to the Class B Limited Partner and 30% to the General Partner.

Losses from Operations will be allocated 98.25% to the Class A Limited Partner, .75% to the Class B Limited Partner and 1% to the General Partner; if, however, when losses allocated to any Limited Partner would result in the creation or continuation of a deficit Capital Account for that Limited Partner while any other Limited Partner has a positive Capital Account Losses from Operations first will be allocated to the Limited Partners having positive Capital Accounts in proportion to such positive Capital Accounts until all such positive Capital Accounts have been eliminated.

Cash available for distribution: Any cash flow that is available for distribution with respect to any year or portion thereof prior to the Anniversary Date will be distributed 98.25% to the Class A Limited Partners, .75% to the Class B Limited Partner and 1% to the General Partner.

Any cash flow that is available for distribution with respect to any year or portion thereof on or after the Anniversary Date shall be distributed 94.05% to the Class A Limited Partners, 4.95% to the Class B Limited Partner and 1% to the General Partner until an amount equal to $1,260,000 has been distributed to the Class A Limited Partners with respect to that year, on a noncumulative basis. This would represent a distribution to the Class A Limited Partners of 9% of their capital contributions. The balance, if any, will be distributed 66.5% to the Class A Limited Partners, 3.5% to the Class B Limited Partner and 30% to the General Partner.

(18) Sale of property: For the purpose of these projections, it is assumed that the property is sold on October 1, 2000. The schedule depicting the sale of property represents the projected tax consequences if the property were to be disposed under one of the following:

    (A) At a sale price of $80,018,000 which reflects a capitalization rate of 11% on the Annual Basic Rent plus the Projected Percentage Rent in year 2001 (see Note 3).

    (B) At a price equal to the outstanding Consolidated Wraparound Deed of Trust, Services fee, General Partner Deferred fee and accrued interest at that time plus one dollar.

    (C) At a sale price of $33,358,695 which reflects all capitalized property costs upon completion of the rehabilitation of the Hotel.

Upon the sale of the property the General Partner will receive a real estate advisory fee, which is projected to be 3% of the sale price. Ordinary income rate of 50% and a capital gain rate of 20% have been applied to the taxable gain. For full details as to the distribution of net proceeds see "Summary of The Limited Partnership Agreement" section of the Confidential Memorandum.

(19) Investment interest limitation: Investment interest expense is deductible by a partner only to the extent such expense attributable to the partner's interest in the Partnership plus any other investment interest expense attributable to sources other than the Partnership does not exceed the sum of (1) $10,000 ($5,000 for a married taxpayer filing a separate return), (2) the amount of a partner's total net investment income and (3) the amount by which certain expenses incurred with respect to property subject to a net lease exceed rent received under such lease.

In determining net investment income of the Partnership, a deduction must be claimed for depreciation based on the "useful life" of the Property. The Partnership based its estimate of useful life on 45 years. The IRS may claim that the calculation of depreciation for purposes of determining the Partnership's net investment income should be based on some shorter recovery period, in which case the amount of excess investment interest per unit might be substantially increased.

(20) Other assumptions: It has been assumed that federal income tax deductions for Partnership expenses will be allowable in accordance with their treatment in the accompanying projections. There can, however, be no assurance as to the deductibility of these items upon examination by the IRS. Certain areas of possible consideration are herein set forth and should be read in conjunction with the "Federal Tax Consequences" section of the Confidential Memorandum.

The projections are based upon the assumption that the tax effects of the relationships between the Partnership and all other parties to Partnership transactions will be accepted by the IRS as reflected herein. Should the IRS successfully challenge one or more of the aforementioned relationships, the projected tax benefits to the Limited Partners could be significantly reduced or possibly eliminated.

Particular note should be given to recently issued Revenue Ruling 83-84, pending amendment to Section 267 of the Code and pending proposed changes in the regulation 704(b) of the Code.

The foregoing is not an all inclusive itemization of areas of possible challenge. It is merely intended to set forth some of the areas which require the consideration of potential investors and their tax advisors. Reference is made to the "Federal Tax Consequences" section of the Confidential Memorandum for an expanded discussion of the federal tax aspects of investing in the Partnership.

It is anticipated that the tax treatment given in the projections to specific items of Partnership income and expense will be consistent with the treatment on the Partnership returns. However, this should not be considered by potential investors as an indication of the probability that such treatment would be sustained by the IRS upon examination. Should the tax treatment of one or more items of income or expense be denied, the tax benefits from an investment in the Partnership may be significantly reduced or possibly eliminated. Potential investors should seek the advice of their own tax advisors in order to satisfy themselves as to the tax consequences of investing in the Partnership.

Potential investors should also be aware of a new penalty provision enacted by TEFRA. Newly enacted Section 6661 provides for a 10% penalty on the underpayment of tax that is attributable to a substantial understatement of income tax. This refers to an understatement in excess of 10% of the correct tax. The penalty can be avoided under certain circumstances. Potential investors should seek the advice of their own tax advisors in order to satisfy themselves as to the possibility of tax and penalty consequences of investing in the Partnership in the event of an IRS tax examination.

The IRS presently is paying increased attention to "tax shelter" partnerships. Accordingly, there may be a substantial risk of audit. An audit by the IRS of the Partnership's information return could result in an audit of an investor's individual tax return and the adjustment of nonpartnership items. The Partnership will have no obligation to pay the costs of any such audit of an investor's individual return and the investors should assume that they will have to bear any such costs.

New rules are applicable to IRS audit procedures. Under provisions enacted by TEFRA relating to 1983 and subsequent tax returns, the limited partners may be bound by certain audit determinations reached by the "tax matters partner" with the IRS.

The projections and assumptions are contingent upon the following, without limitation: (a) that the Partnership will be treated as a partnership rather than as an association taxable as a corporation, (b) that no interpretation or amendments to the Internal Revenue Code will reduce or eliminate the tax benefits projected, and (c) that the treatment given in the projections to all tax items will be accepted by the IRS.

Appendix H

Securities and Exchange Commission, Regulation D- Rules Governing the Limited Offer and Sale of Securities Without Registration Under the Securities Act of 1933

(17 C.F.R. § 230.501-§ 230.506)

Preliminary Notes

1. The following rules relate to transactions exempted from the registration requirements of section 5 of the Securities Act of 1933 (the "Act"). Such transactions are not exempt from the antifraud, civil liability, or other provisions of the federal securities laws. Issuers are reminded of their obligation to provide such further material information, if any, as may be necessary to make the information required under this regulation, in light of the circumstances under which it is furnished, not misleading.

2. Nothing in these rules obviates the need to comply with any applicable state law relating to the offer and sale of securities. Regulation D is intended to be a basic element in a uniform system of federal-state limited offering exemptions consistent with the provisions of sections 18 and 19(c) of the Act. In those states that have adopted Regulation D, or any version of Regulation D, special attention should be directed to the applicable state laws and regulations, including those relating to registration of persons who receive remuneration in connection with the offer and sale of securities, to disqualification of issuers and other persons associated with offerings based on state administrative orders or judgments, and to requirements for filings of notices of sales.

3. Attempted compliance with any rule in Regulation D does not act as an exclusive election; the issuer can also claim the availability of any other applicable exemption. For instance, an issuer's failure to satisfy all the terms and conditions of Rule 506 shall not raise any presumption that the exemption provided by section 4(2) of the Act is not available.

4. These rules are available only to the issuer of the securities and not to any affiliate of that issuer or to any other person for resales of the issuer's securities. The rules provide an exemption only for the transactions in which the securities are offered or sold by the issuer, not for the securities themselves.

5. These rules may be used for business combinations that involve sales by virtue of Rule 145(a) or otherwise.

6. In view of the objectives of these rules and the policies underlying the Act, Regulation D is not available to any issuer for any transaction or chain of transactions that, although in technical compliance with these rules, is part of a plan or scheme to evade the registration provisions of the Act. In such cases, registration under the Act is required.

7. Offers and sales of securities to foreign persons made outside the United

States effected in a manner that will result in the securities coming to rest abroad generally need not be registered under the Act. See Release No. 33-4708 (July 9, 1964). This interpretation may be relied on for such offers and sales even if coincident offers and sales are made under Regulation D inside the United States. Thus, for example, persons who are not citizens or residents of the United States would not be counted in the calculation of the number of purchasers. Similarly proceeds from sales to foreign purchasers would not be included in the aggregate offering price. The provisions of this note, however, do not apply if the issuer elects to rely solely on Regulation D for offers or sales to foreign persons.

Rule 501. Definitions and Terms Used in Regulation D

As used in Regulation D, the following terms shall have the meaning indicated:

(a)   *Accredited Investor.* "Accredited investor" shall mean any person who comes within any of the following categories, or who the issuer reasonably believes comes within any of the following categories, at the time of the sale of the securities to that person:

(1)   Any bank as defined in section 3(a) (2) of the Act whether acting in its individual or fiduciary capacity; insurance company as defined in section 2(13) of the Act; investment company registered under the Investment Company Act of 1940 or a business development company as defined in section 2(a) (48) of that Act; Small Business Investment Company licensed by the U.S. Small Business Administration under section 301(c) or (d) of the Small Business Act of 1958; employee benefit plan within the meaning of Title I of the Employee Retirement Income Security Act of 1974, if the investment decision is made by a plan fiduciary, as defined in section 3(21) of such Act, which is either a bank, insurance company, or registered investment adviser, or if the employee benefit plan has total assets in excess of $5,000,000;

(2)   Any private business development company as defined in section 202(a) (22) of the Investment Advisers Act of 1940;

(3)   Any organization described in Section 501(c) (3) of the Internal Revenue Code with total assets in excess of $5,000,000;

(4)   Any director, executive officer, or general partner of the issuer of the securities being offered or sold, or any director, executive officer, or general partner of a general partner of that issuer;

(5)   Any person who purchases at least $150,000 of the securities being offered, where the purchaser's total purchase price does not exceed 20 percent of the purchaser's net worth at the time of sale, or joint net worth with that person's spouse, for one or any combination of the following:

(i)   cash,

(ii)   securities for which market quotations are readily available,

(iii)   an unconditional obligation to pay cash or securities for which market quotations are readily available which obligation is to be discharged within five years of the sale of the securities to the purchaser, or

(iv)   the cancellation of any indebtedness owed by the issuer to the purchaser.

(6)   Any natural person whose individual net worth, or joint net worth with that person's spouse, at the time of his purchase exceeds $1,000,000.

(7)   Any natural person who had an individual income in excess of $200,000 in each of the two most recent years and who reasonably expects an income in excess of $200,000 in the current year; and

(8)   Any entity in which all of the equity owners are accredited investors under paragraphs (a) (1), (2), (3), (4), (6), or (7) of this Rule 501.

(b)   *Affiliate.* An "affiliate" of, or person "affiliated" with a specified person shall mean a person that directly, or indirectly through one or more intermediaries, controls or is controlled by, or is under common control with, the person specified.

(c)   *Aggregate Offering Price.* "Aggregate offering price" shall mean the sum of all cash, services, property, notes, cancellation of debt, or other consideration received by an issuer for issuance of its securities. When securities are being offered for both cash and noncash consideration, the aggregate offering price shall be based on the price at which the securities are offered for cash. If securities are not offered for cash, the aggregate offering price shall be based on the value of the consideration as established by bona fide sales of that consideration made within a reasonable time, or, in the absence of sales, on the fair value as determined by an accepted standard.

(d)   *Business Combination.* "Business combination" shall mean any transaction of the type specified in paragraph (a) of Rule 145 under the Act and any transaction involving the acquisition by one issuer, in exchange for all or a part of its own or its parent's stock, of stock of another issuer if, immediately after the acquisition, the acquiring issuer has control of the other issuer (whether or not it had control before the acquisition).

(e)   *Calculation of Number of Purchasers.* For purposes of calculating the number of purchasers under Rules 505(b) and 506(b) only, the following shall apply:

(1)   The following purchasers shall be excluded:

(i)   Any relative, spouse or relative of the spouse of a purchaser who has the same principal residence as the purchaser;

(ii)   Any trust or estate in which a purchaser and any of the persons related to him as specified in paragraph (e)(1)(i) or (e)(1)(iii) of this Rule 501 collectively have more than 50 percent of the beneficial interests (excluding contingent interests);

(iii)   Any corporation or other organization of which a purchaser and any of the persons related to him as specified in paragraph (e)(1)(i) or (e)(1)(ii) of this Rule 501 collectively are beneficial owners of more than 50 percent of the equity securities (excluding directors' qualifying shares) or equity interests; and

(iv)   Any accredited investor.

(2)   A corporation, partnership or other entity shall be counted as one purchaser. If, however, that entity is organized for the specific purpose of acquiring the securities offered and is not an accredited investor under paragraph (a)(8) of this Rule 501, then each beneficial owner of equity securities or equity interests in the entity shall count as a separate purchaser for all provisions of Regulation D.

Note. The issuer must satisfy all the other provisions of Regulation D for all purchasers whether or not they are included in calculating the number of purchasers.

Clients of an investment adviser or customers of a broker or dealer shall be considered the "purchasers" under Regulation D regardless of the amount of discretion given to the investment adviser or broker or dealer to act on behalf of the client or customer.

(f)   *Executive Officer.* "Executive officer" shall mean the president, any vice president in charge of a principal business unit, division or function (such as sales, administration or finance), any other officer who performs a policy making function, or any other person who performs similar policy making functions for the issuer. Executive officers of subsidiaries may be deemed executive officers of the issuer if they perform such policy making functions for the issuer.

(g)   *Issuer.* The definition of the term "issuer" in section 2(4) of the Act shall apply, except that in the case of a proceeding under the Federal Bankruptcy Code, the trustee or debtor in possession shall be considered the issuer in an offering under a plan or reorganization, if the securities are to be issued under the plan.

(h)   *Purchaser Representative.* "Purchaser representative" shall mean any person who satisfies all of the following conditions or who the issuer reasonably believes satisfies all of the following conditions:

(1)   Is not an affiliate, director, officer or other employee of the issuer, or beneficial owner of 10 percent or more of any class of the equity securities or 10 percent or more of the equity interest in the issuer, except where the purchaser is:

(i)   A relative of the purchaser representative by blood, marriage or adoption and not more remote than a first cousin;

(ii)   A trust or estate in which the purchaser representative and any persons related to him as specified in paragraph (h)(1)(i) or (h)(1)(iii) of this Rule 501 collectively have more than 50 percent of the beneficial interest (excluding contingent interest) or of which the purchaser representative serves as trustee, executor, or in any similar capacity; or

(iii)   A corporation or other organization of which the purchaser representative and any persons related to him as specified in paragraph (h)(1)(i) or (h)(1)(ii) of this Rule 501 collectively are the beneficial owners of more than 50 percent of the equity securities (excluding directors' qualifying shares) or equity interests.

(2)   Has such knowledge and experience in financial and business matters that he is capable of evaluating, alone, or together with other purchaser representatives of the purchaser, or together with the purchaser, the merits and risks of the prospective investment.

(3)   Is acknowledged by the purchaser in writing, during the course of the transaction, to be his purchaser representative in connection with evaluating the merits and risks of the prospective investment; and

(4)   Discloses to the purchaser in writing prior to the acknowledgment specified in paragraph (h)(3) of this Rule 501 any material relationship between himself or his affiliates and the issuer or its affiliates that then exists, that is mutually understood to be contemplated, or that has existed at any time during the previous two years, and any compensation received or to be received as a result of such relationship.

Note 1. A person acting as a purchaser representative should consider the applicability of the registration and antifraud provisions relating to brokers and dealers under the Securities Exchange Act of 1934 ("Exchange Act") and relating to investment advisers under the Investment Advisers Act of 1940.

Note 2. The acknowledgment required by paragraph (h)(3) and the disclosure required by paragraph (h)(4) of this Rule 501 must be made with specific reference to each prospective investment. Advance blanket acknowledgment, such as for "all securities transactions" or "all private placements," is not sufficient.

Note 3. Disclosure of any material relationships between the purchaser representative or his affiliates and the issuer or its affiliates does not relieve the purchaser representative of his obligation to act in the interest of the purchaser.

Rule 502. General Conditions to Be Met

The following conditions shall be applicable to offers and sales made under Regulation D:

(a)   *Integration.* All sales that are part of the same Regulation D offering must meet all of the terms and conditions of Regulation D. Offers and sales that are made more than six months before the start of a Regulation D offering or are made more than six months after completion of a Regulation D offering will not be considered part of that Regulation D offering, so long as during those six month periods there are no offers or sales of securities by or for the issuer that are of the same or a similar class as those offered or sold under Regulation D, other than those offers or sales of securities under an employee benefit plan as defined in Rule 405 under the Act.

Note. The term "offering" is not defined in the Act or in Regulation D. If the issuer offers or sells securities for which the safe harbor rule in paragraph (a) of this Rule 502 is unavailable, the determination as to whether separate sales of securities are part of the same offering (*i.e.,* are considered "integrated") depends on the particular facts and circumstances. Generally, transactions otherwise meeting the requirements of an exemption will not be integrated with simultaneous offerings being made outside the United States effected in a manner that will result in the securities coming to rest abroad. See Release No. 33-4708 (July 9, 1964).

The following factors should be considered in determining whether offers and sales should be integrated for purposes of the exemptions under Regulation D:

(a)   Whether the sales are part of a single plan of financing;

(b)   Whether the sales involve issuance of the same class of securities;

(c)   Whether the sales have been made at or about the same time;

(d)   Whether the same type of consideration is received; and

(e)   Whether the sales are made for the same general purpose.

See Release No. 33-4552 (November 6, 1962).

(b)   *Information Requirements.*

(1)   *When information must be furnished.*

(i)   If the issuer sells securities either under Rule 504 or only to accredited investors, paragraph (b) of this Rule 502 does not require that specific information be furnished to purchasers.

(ii)   If the issuer sells securities under Rules 505 or 506 to any purchaser that is not an accredited investor, the issuer shall furnish

the information specified in paragraph (b)(2) of this Rule 502 to all
purchasers during the course of the offering and prior to sale.

(2)   *Type of information to be furnished.*

(i)   If the issuer is not subject to the reporting requirements of
section 13 or 15(d) of the Exchange Act, the issuer shall furnish the
following information, to the extent material to an understanding of
the issuer, its business, and the securities being offered:

(A)   *Offerings up to $5,000,000.* The same kind of informa-
tion as would be required in Part I of Form S-18, except that only the
financial statements for the issuer's most recent fiscal year must be certified by an independent public or certified accountant. If Form S-18
is not available to an issuer, then the issuer shall furnish the same kind
of information as would be required in Part I of a registration statement
filed under the Act on the form that the issuer would be entitled to use,
except that only the financial statements for the most recent two fiscal
years prepared in accordance with generally accepted accounting principles shall be furnished and only the financial statements for the issuer's
most recent fiscal year shall be certified by an independent public or
certified accountant. If an issuer, other than a limited partnership, cannot obtain audited financial statements without unreasonable effort or
expense, then only the issuer's balance sheet, which shall be dated within 120 days of the start of the offering, must be audited. If the issuer
is a limited partnership and cannot obtain the required financial statements without unreasonable effort or expense, it may furnish financial
statements that have been prepared on the basis of federal income tax
requirements and examined and reported on in accordance with generally accepted auditing standards by an independent public or certified
accountant.

(B)   *Offerings over $5,000,000.* The same kind of informa-
tion as would be required in Part I of a registration statement filed
under the Act on the form that the issuer would be entitled to use. If
an issuer, other than a limited partnership, cannot obtain audited financial statements without unreasonable effort or expense, then only the
issuer's balance sheet, which shall be dated within 120 days of the start
of the offering, must be audited. If the issuer is a limited partnership
and cannot obtain the required financial statements without unreasonable effort or expense, it may furnish financial statements that have
been prepared on the basis of federal income tax requirements and
examined and reported on in accordance with generally accepted
auditing standards by an independent public or certified accountant.

(C)   If the issuer is a foreign private issuer eligible to use
Form 20-F the issuer shall disclose the same kind of information required to be included in a registration statement filed under the Act
on the form that the issuer would be entitled to use. The financial
statements need be certified only to the extent required by paragraphs
(b)(2)(i)(A) or (B) as appropriate.

(ii)   If the issuer is subject to the reporting requirements of section
13 or 15(d) of the Exchange Act, the issuer shall furnish the information
specified in paragraph (b)(2)(ii)(A) or (b)(2)(ii)(B), and in either event
the information specified in paragraph (B)(2)(ii)(C) of this Rule 502:

(A)   The issuer's annual report to shareholders for the most recent fiscal year, if such annual report meets the requirements of Rules 14a-3 or 14c-3 under the Exchange Act, the definitive proxy statement filed in connection with that annual report, and, if requested by the purchaser in writing, a copy of the issuer's most recent Form 10-K under the Exchange Act.

(B)   The information contained in an annual report on Form 10-K under the Exchange Act or in a registration statement on Form S-1 under the Act or on Form 10 under the Exchange Act, whichever filing is the most recent required to be filed.

(C)   The information contained in any reports or documents required to be filed by the issuer under sections 13(a), 14(a), 14(c), and 15(d) of the Exchange Act since the distribution or filing of the report or registration statement specified in paragraph (A) or (B), and a brief description of the securities being offered, the use of the proceeds from the offering, and any material change in the issuer's affairs that are not disclosed in the documents furnished.

(D)   If the issuer is a foreign private issuer eligible to use Form 20-F, the issuer may provide in lieu of the information specified in paragraphs (b)(2)(ii)(A) or (B), the information contained in its most recent filing on Form 20-F or Form F-1.

(iii)   Exhibits required to be filed with the Commission as part of a registration statement or report, other than an annual report of shareholders or parts of that report incorporated by reference in a Form 10-K report, need not be furnished to each purchaser if the contents of the exhibits are identified and the exhibits are made available to the purchaser, upon his written request, prior to his purchase.

(iv)   At a reasonable time prior to the purchase of securities by any purchaser that is not an accredited investor in a transaction under Rules 505 or 506, the issuer shall furnish the purchaser a brief description in writing of any written information concerning the offering that has been provided by the issuer to any accredited investor. The issuer shall furnish any portion or all of this information to the purchaser, upon his written request, prior to his purchase.

(v)   The issuer shall also make available to each purchaser at a reasonable time prior to his purchase of securities in a transaction under Rules 505 or 506 the opportunity to ask questions and receive answers concerning the terms and conditions of the offering and to obtain any additional information which the issuer possesses or can acquire without unreasonable effort or expense that is necessary to verify the accuracy of information furnished under paragraph (b)(2)(i) or (ii) of this Rule 502.

(vi)   For business combinations, in addition to information required by paragraph (b)(2) of this Rule 502, the issuer shall provide to each purchaser at the time the plan is submitted to security holders, or, with an exchange, during the course of the transaction and prior to sale, written information about any terms or arrangements of the proposed transaction that are materially different from those for all other security holders.

(c)   *Limitation on Manner of Offering.* Except as provided in Rule 504(b) (1), neither the issuer nor any person acting on its behalf shall offer or sell the securities by any form of general solicitation or general advertising, including, but not limited to, the following:

(1)   Any advertisement, article, notice or other communication published in any newspaper, magazine, or similar media or broadcast over television or radio; and

(2)   Any seminar or meeting whose attendees have been invited by any general solicitation or general advertising.

(d)   *Limitations on Resale.* Except as provided in Rule 504(b)(1), securities acquired in a transaction under Regulation D shall have the status of securities acquired in a transaction under section 4(2) of the Act and cannot be resold without registration under the Act or an exemption therefrom. The issuer shall exercise reasonable care to assure that the purchasers of the securities are not underwriters within the meaning of section 2(11) of the Act, which reasonable care shall include, but not be limited to the following:

(1)   Reasonable inquiry to determine if the purchaser is acquiring the securities for himself or for other persons;

(2)   Written disclosure to each purchaser prior to sale that the securities have not been registered under the Act and, therefore, cannot be resold unless they are registered under the Act or unless an exemption from registration is available; and

(3)   Placement of a legend on the certificate or other document that evidences the securities stating that the securities have not been registered under the Act and setting forth or referring to the restrictions on transferability and sale of the securities.

Rule 503. Filing of Notice of Sales

(a)   The issuer shall file with the Commission five copies of a notice on Form D at the following times:

(1)   No later than 15 days after the first sale of securities in an offering under Regulation D;

(2)   Every six months after the first sale of securities in an offering under Regulation D, unless the final notice required by paragraph (a)(3) of this Rule 503 has been filed; and

(3)   No later than 30 days after the last sale of securities in an offering under Regulation D.

(b)   If the offering is completed within the 15 day period described in paragraph (a)(1) of this Rule 503 and if the notice is filed no later than the end of that period but after the completion of the offering, then only one notice need be filed to comply with paragraphs (a)(1) and (3) of this Rule 503.

(c)   One copy of every notice on Form D shall be manually signed by a person duly authorized by the issuer.

(d)   If sales are made under Rule 505, the notice shall contain an undertaking by the issuer to furnish to the Commission, upon the written request of its staff, the information furnished by the issuer under Rule 502(b)(2) to any purchaser that is not an accredited investor.

(e)   If more than one notice for an offering is required to be filed under paragraph (a) of this Rule 503, notices after the first notice need only report the issuer's name and the information required by Part C and any material change in the facts from those set forth in Parts A and B of the first notice.

(f)   A notice on Form D shall be considered filed with the Commission under paragraph (a) of this Rule 503:

    (1)   As of the date on which it is received at the Commission's principal office in Washington, D.C.; or

    (2)   As of the date on which the notice is mailed by means of United States registered or certified mail to the Commission's Office of Small Business Policy, Division of Corporation Finance, at the Commission's principal office in Washington, D.C., if the notice is delivered to such office after the date on which it is required to be filed.

Rule 504. Exemption for Limited Offers and Sales of Securities not Exceeding $500,000

(a)   *Exemption.* Offers and sales of securities that satisfy the conditions in paragraph (b) of this Rule 504 by an issuer that is not subject to the reporting requirements of section 13 or 15(d) of the Exchange Act and that is not an investment company shall be exempt from the provisions of section 5 of the Act under section 3(b) of the Act.

(b)   *Conditions To Be Met*

    (1)   *General conditions.* To qualify for exemption under this Rule 504 offers and sales must satisfy the terms and conditions of Rules 501 through 503, except that the provisions of Rule 502(c) and (d) shall not apply to offers and sales of securities under this Rule 504 that are made exclusively in one or more states each of which provides for the registration of the securities and requires the delivery of a disclosure document before sale and that are made in accordance with those state provisions.

    (2)   *Specific condition.*

        (i)   *Limitation on aggregate offering price.* The aggregate offering price for an offering of securities under this Rule 504, as defined in Rule 501(c), shall not exceed $500,000, less the aggregate offering price for all securities sold within the twelve months before the start of and during the offering of securities under this Rule 504 in reliance on any exemption under section 3(b) of the Act or in violation of section 5(a) of the Act.

    Note 1. The calculation of the aggregate offering price is illustrated as follows:

    Example 1: If an issuer sold $200,000 of its securities on June 1, 1982 under this Rule 504 and an additional $100,000 on September 1, 1982, the issuer would be permitted to sell only $200,000 more under this Rule 504 until June 1, 1983. Until that date the issuer must count both prior sales toward the $500,000 limit. However, if the issuer made its third sale on June 1, 1983, the issuer could then sell $400,000 of its securities because the June 1, 1982 sale would not be within the preceding twelve months.

Example 2. If an issuer sold $100,000 of its securities on June 1, 1982 under this Rule 504 and an additional $4,500,000 on December 1, 1982 under Rule 505, the issuer could not sell any of its securities under this Rule 504 until December 1, 1983. Until then the issuer must count the December 1, 1982 sale towards the limit of $500,000 within the preceding twelve months.

Note 2. If a transaction under this Rule 504 fails to meet the limitation on the aggregate offering price, it does not affect the availability of this Rule 504 for the other transactions considered in applying such limitation. For example, if the issuer in *Example 1* made its third sale on May 31, 1983, in the amount of $250,000, this Rule 504 would not not be available for that sale, but the exemption for the prior two sales would be unaffected.

Rule 505. Exemption for Limited Offers and Sales of Securities Not Exceeding $5,000,000

(a)   *Exemption.* Offers and sales of securities that satisfy the conditions in paragraph (b) of this Rule 505 by an issuer that is not an investment company shall be exempt from the provisions of section 5 of the Act under section 3(b) of the Act.

(b)   *Conditions To Be Met*

(1)   *General conditions.* To qualify for exemption under this Rule 505, offers and sales must satisfy the terms and conditions of Rules 501 through 503.

(2)   *Specific conditions.*

(i)   *Limitation on aggregate offering price.* The aggregate offering price for an offering of securities under this Rule 505, as defined in Rule 501(c), shall not exceed $5,000,000, less the aggregate offering price for all securities sold within the twelve months before the start of and during the offering of securities under this Rule 505 in reliance on any exemption under section 3(b) of the Act or in violation of section 5(a) of the Act.

Note. The calculation of the aggregate offering price is illustrated as follows:

Example 1. If an issuer sold $2,000,000 of its securities on June 1, 1982 under this Rule 505 and an additional $1,000,000 on September 1, 1982, the issuer would be permitted to sell only $2,000,000 more under this Rule 505 until June 1, 1983. Until that date the issuer must count both prior sales towards the $5,000,000 limit. However, if the issuer made its third sale on June 1, 1983, the issuer could then sell $4,000,000 of its securities because the June 1, 1982 sale would not be within the preceding twelve months.

Example 2: If an issuer sold $500,000 of its securities on June 1, 1982 under Rule 504 and an additional $4,500,000 on December 1, 1982 under this Rule 505, then the issuer could not sell any of its securities under this Rule 505 until June 1, 1983. At that time it could sell an additional $500,000 of its securities.

(ii) *Limitation on number of purchasers.* The issuer shall reasonably believe that there are no more than 35 purchasers of securities from the issuer in any offering under this Rule 505.

Note. See Rule 501(e) for the calculation of the number of purchasers and Rule 502(a) for what may or may not constitute an offering under this Rule 505.

(iii) *Disqualifications.* No exemption under this Rule 505 shall be available for the securities of any issuer described in Rule 252(c), (d), (e) or (f) of Regulation A, except that for purposes of this Rule 505 only:

(A) The term "filing of the notification required by Rule 255" as used in Rule 252(c), (d), (e) and (f) shall mean the first sale of securities under this Rule 505;

(B) The term "underwriter" as used in Rule 252(d) and (e) shall mean a person that has been or will be paid directly or indirectly remuneration for solicitation of purchasers in connection with sales of securities under this Rule 505; and

(C) Paragraph (b)(2)(iii) of this Rule 505 shall not apply to any issuer if the Commission determines, upon a showing of good cause, that it is not necessary under the circumstances that the exemption be denied. Any such determination shall be without prejudice to any other action by the Commission in any other proceeding or matter with respect to the issuer or any other person.

Rule 506. Exemption for Limited Offers and Sales Without Regard to Dollar Amount of Offering

(a) *Exemption.* Offers and sales of securities by an issuer that satisfy the conditions in paragraph (b) of this Rule 506 shall be deemed to be transactions not involving any public offering within the meaning of section 4(2) of the Act.

(b) *Conditions To Be Met*

(1) *General conditions.* To qualify for exemption under this Rule 506, offers and sales must satisfy all the terms and conditions of Rules 501 through 503.

(2) *Specific conditions.*

(i) *Limitation on number of purchasers.* The issuer shall reasonably believe that there are no more than 35 purchasers of securities from the issuer in any offering under this Rule 506.

Note. See Rule 501(e) for the calculation of the number of purchasers and Rule 502(a) for what may or may not constitute an offering under this Rule 506.

(ii) *Nature of purchasers.* The issuer shall reasonably believe immediately prior to making any sale that each purchaser who is not an accredited investor either alone or with his purchaser representative(s) has such knowledge and experience in financial and business matters that he is capable of evaluating the merits and risks of the prospective investment.

Appendix I

National Park Service, "National Register of Historic Places"

(Final Rule, 36 C.F.R. Part 60)

Authority: National Historic Preservation Act of 1966, as amended, 16 U.S.C. 470 *et seq.,* and EO 11593.

Source: 46 FR 56187, Nov. 16, 1981, unless otherwise noted.

§ 60.1. Authorization and expansion of the National Register.

(a)   The National Historic Preservation Act of 1966, 80 Stat. 915, 16 U.S.C. 470 *et seq.,* as amended, authorizes the Secretary of the Interior to expand and maintain a National Register of districts, sites, buildings, structures, and objects significant in American history, architecture, archeology, engineering and culture. The regulations herein set forth the procedural requirements for listing properties on the National Register.

(b)   Properties are added to the National Register through the following processes.

(1)   Those Acts of Congress and Executive orders which create historic areas of the National Park System administered by the National Park Service, all or portions of which may be determined to be of historic significance consistent with the intent of Congress;

(2)   Properties declared by the Secretary of the Interior to be of National significance and designated as National Historic Landmarks;

(3)   Nominations prepared under approved State Historic Preservation Programs, submitted by the State Historic Preservation Officer and approved by the NPS;

(4)   Nominations from any person or local government (only if such property is located in a State with no approved State Historic Preservation Program) approved by the NPS and;

(5)   Nominations of Federal properties prepared by Federal agencies, submitted by the Federal Preservation Officer and approved by NPS.

§ 60.2. Effects of listing under Federal law.

The National Register is an authoritative guide to be used by Federal, State, and local governments, private groups and citizens to identify the Nation's cultural resources and to indicate what properties should be considered for protection from destruction or impairment. Listing of private property on the National Register does not prohibit under Federal law or regulation any actions which may otherwise be taken by the property owner with respect to the property.

(a)   The National Register was designed to be and is administered as a planning tool. Federal agencies undertaking a project having an effect on a listed or eligible property must provide the Advisory Council on Historic Preservation a reasonable opportunity to comment pursuant to section 106 of the National Historic Preservation Act of 1966, as amended. The Council has adopted procedures concerning, *inter alia,* their commenting responsibility in 36 CFR Part 800. Having complied with this procedural requirement the Federal agency may adopt any course of action it believes is appropriate. While the Advisory Council comments must be taken into account and integrated into the decisionmaking process, program decisions rest with the agency implementing the undertaking.

(b)   Listing in the National Register also makes property owners eligible to be considered for Federal grants-in-aid for historic preservation.

(c)   If a property is listed in the National Register, certain provisions of the Tax Reform Act of 1976 as amended by the Revenue Act of 1978 and the Tax Treatment Extension Act of 1980 may apply. These provisions encourage the preservation of depreciable historic structures by allowing favorable tax treatments for rehabilitation, and discourage destruction of historic buildings by eliminating certain otherwise available Federal tax provisions both for demolition of historic structures and for new construction on the site of demolished historic buildings. Owners of historic buildings may benefit from the investment tax credit provisions of the Revenue Act of 1978. The Economic Recovery Tax Act of 1981 generally replaces the rehabilitation tax incentives under these laws beginning January 1, 1982 with a 25% investment tax credit for rehabilitations of historic commercial, industrial and residential buildings. This can be combined with a 15-year cost recovery period for the adjusted basis of the historic building. Historic buildings with certified rehabilitations receive additional tax savings by their exemption from any requirement to reduce the basis of the building by the amount of the credit. The denial of accelerated depreciation for a building built on the site of a demolished historic building is repealed effective January 1, 1982. The Tax Treatment Extension Act of 1980 includes provisions regarding charitable contributions for conservation purposes of partial interests in historically important land areas or structures.

(d) If a property contains surface coal resources and is listed in the National Register, certain provisions of the Surface Mining and Control Act of 1977 require consideration of a property's historic values in the determination on issuance of a surface coal mining permit.

§ 60.3. Definitions.

(a) *Building.* A building is a structure created to shelter any form of human activity, such as a house, barn, church, hotel, or similar structure. Building may refer to a historically related complex such as a courthouse and jail or a house and barn.

Examples

Molly Brown House (Denver, CO)
Meek Mansion and Carriage House (Hayward, CA)
Huron County Courthouse and Jail (Norwalk, OH)
Fairntosh Plantation (Durham vicinity, NC)

(b) *Chief elected local official.* Chief elected local official means the mayor, county judge, county executive or otherwise titled chief elected administrative official who is the elected head of the local political jurisdiction in which the property is located.

(c) *Determination of eligibility.* A determination of eligibility is a decision by the Department of the Interior that a district, site, building, structure or object meets the National Register criteria for evaluation although the property is not formally listed in the National Register. A determination of eligibility does not make the property eligible for such benefits as grants, loans, or tax incentives that have listing on the National Register as a prerequisite.

(d) *District.* A district is a geographically definable area, urban or rural, possessing a significant concentration, linkage, or continuity of sites, buildings, structures, or objects united by past events or aesthetically by plan or physical development. A district may also comprise individual elements separated geographically but linked by association or history.

Examples

Georgetown Historic District (Washington, DC)
Martin Luther King Historic District (Atlanta, GA)
Durango-Silverton Narrow-Gauge Railroad (right-of-way between Durango
  and Silverton, CO)

(e)   *Federal Preservation Officer.* The Federal Preservation Officer is the official designated by the head of each Federal agency responsible for coordinating that agency's activities under the National Historic Preservation Act of 1966, as amended, and Executive Order 11593 including nominating properties under that agency's ownership or control to the National Register.

(f)   *Keeper of the National Register of Historic Places.* The Keeper is the individual who has been delegated the authority by NPS to list properties and determine their eligibility for the National Register. The Keeper may further delegate this authority as he or she deems appropriate.

(g)   *Multiple Resource Format submission.* A Multiple Resource Format submission for nominating properties to the National Register is one which includes all or a defined portion of the cultural resources identified in a specified geographical area.

(h)   *National Park Service (NPS).* The National Park Service is the bureau of the Department of Interior to which the Secretary of Interior has delegated the authority and responsibility for administering the National Register program.

(i)   *National Register Nomination Form.* National Register Nomination Form means (1) National Register Nomination Form NPS 10-900, with accompanying continuation sheets (where necessary) Form NPS 10-900a, maps and photographs or (2) for Federal nominations, Form No. 10-306, with continuation sheets (where necessary) Form No. 10-300A, maps and photographs. Such nomination forms must be "adequately documented" and "technically and professionally correct and sufficient." To meet these requirements the forms and accompanying maps and photographs must be completed in accord with requirements and guidance in the NPS publication, "How to Complete National Register Forms" and other NPS technical publications on this subject. Descriptions and statements of significance must be prepared in accord with standards generally accepted by academic historians, architectural historians and archeologists. The nomination form is a legal document and reference for historical, architectural, and archeological data upon which the protections for listed and eligible properties are founded. The nominating authority certifies that the nomination is adequately documented and technically and professionally correct and sufficient upon nomination.

(j)   *Object.* An object is a material thing of functional, aesthetic, cultural, historical or scientific value that may be, by nature or design, movable yet related to a specific setting or environment.

Examples

Delta Queen Steamboat (Cincinnati, OH)
Adams Memorial (Rock Creek Cemetery, Washington, DC)
Sumpter Valley Gold Dredge (Sumpter, OR)

(k)   *Owner or owners.* The term owner or owners means those individuals, partnerships, corporations or public agencies holding fee simple title to property. Owner or owners does not include individuals, partnerships, corporations or public agencies holding easements or less than fee interests (including leaseholds) of any nature.

(l)   *Site.* A site is the location of a significant event, a prehistoric or historic occupation or activity, or a building or structure, whether standing, ruined, or vanished, where the location itself maintains historical or archeological value regardless of the value of any existing structure.

Examples

Cabin Creek Battlefield (Pensacola vicinity, OK)
Mound Cemetery Mound (Chester vicinity, OH)
Mud Springs Pony Express Station Site (Dalton vicinity, NE)

(m) *State Historic Preservation Officer.* The State Historic Preservation Officer is the person who has been designated by the Governor or chief executive or by State statute in each State to administer the State Historic Preservation Program, including identifying and nominating eligible properties to the National Register and otherwise administering applications for listing historic properties in the National Register.

(n) *State Historic Preservation Program.* The State Historic Preservation Program is the program established by each State and approved by the Secretary of Interior for the purpose of carrying out the provisions of the National Historic Preservation Act of 1966, as amended, and related laws and regulations. Such program shall be approved by the Secretary before the State may nominate properties to the National Register. Any State Historic Preservation Program in effect under prior authority of law before December 12, 1980, shall be treated as an approved program until the Secretary approves a program submitted by the State for purposes of the Amendments or December 12, 1983, unless the Secretary chooses to rescind such approval because of program deficiencies.

(o) *State Review Board.* The State Review Board is a body whose members represent the professional fields of American history, architectural history, historic architecture, prehistoric and historic archeology, and other professional disciplines and may include citizen members. In States with approved State historic preservation programs the State Review Board reviews and approves National Register nominations concerning whether or not they meet the criteria for evaluation prior to their submittal to the NPA.

(p) *Structure.* A structure is a work made up of interdependent and interrelated parts in a definite pattern of organization. Constructed by man, it is often an engineering project large in scale.

Examples

Swanton Covered Railroad Bridge (Swanton vicinity, VT)
Old Point Loma Lighthouse (San Diego, CA)
North Point Water Tower (Milwaukee, WI)
Reber Radio Telescope (Green Bay vicinity, WI)

(q) *Thematic Group Format Submission.* A Thematic Group Format submission for nominating properties to the National Register is one which includes a finite group of resources related to one another in a clearly distinguishable way. They may be related to a single historic person, event, or developmental force; of one building type or use, or designed by a single architect; of a single archeological site form, or related to a particular set of archeological research problems.

(r) *To nominate.* To nominate is to propose that a district, site, building, structure, or object be listed in the National Register of Historic Places by preparing a nomination form, with accompanying maps and photographs which adequately document the property and are technically and professionally correct and sufficient.

§ 60.4. Criteria for evaluation.

The criteria applied to evaluate properties (other than areas of National Park System and National Historic Landmarks) for the National Register are listed below. These criteria are worded in a manner to provide for a wide diversity of resources. The following criteria shall be used in evaluating properties for nomination to the National Register, by NPS in reviewing nominations, and for evaluating National Register eligibility of properties. Guidance in applying the criteria is further discussed in the "How To" publications. Standards & Guidelines sheets and Keeper's opinions of the National Register. Such materials are available upon request.

National Register criteria for evaluation. The quality of significance in American history, architecture, archeology, engineering, and culture is present in districts, sites, buildings, structures, and objects that possess integrity of location, design, setting, materials, workmanship, feeling, and association and

(a)   that are associated with events that have made a significant contribution to the broad patterns of our history; or

(b)   that are associated with the lives of persons significant in our past; or

(c)   that embody the distinctive characteristics of a type, period, or method of construction, or that represent the work of a master, or that possess high artistic values, or that represent a significant and distinguishable entity whose components may lack individual distinction; or

(d)   that have yielded, or may be likely to yield, information important in prehistory or history.

Criteria considerations. Ordinarily cemeteries, birthplaces, or graves of historical figures, properties owned by religious institutions or used for religious purposes, structures that have been moved from their original locations, reconstructed historic buildings, properties primarily commemorative in nature, and properties that have achieved significance within the past 50 years shall not be considered eligible for the National Register. However, such properties will qualify if they are integral parts of districts that do meet the criteria or if they fall within the following categories:

(a)   A religious property deriving primary significance from architectural or artistic distinction or historical importance; or

(b)   A building or structure removed from its original location but which is significant primarily for architectural value, or which is the surviving structure most importantly associated with a historic person or event; or

(c)   A birthplace or grave of a historical figure of outstanding importance if there is no appropriate site or building directly associated with his productive life.

(d)   A cemetery which derives its primary significance from graves of persons of transcendent importance, from age, from distinctive design features, or from association with historic events; or

(e)   A reconstructed building when accurately executed in a suitable environment and presented in a dignified manner as part of a restoration master plan, and when no other building or structure with the same association has survived; or

(f)   A property primarily commemorative in intent if design, age, tradition, or symbolic value has invested it with its own exceptional significance; or

(g)   A property achieving significance within the past 50 years if it is of exceptional importance.

This exception is described further in NPS "How To", no. 2, entitled "How to Evaluate and Nominate Potential National Register Properties That Have Achieved Significance Within the Last 50 Years" which is available from the National Register of Historic Places Division, National Park Service, United States Department of the Interior, Washington, D.C. 20240.

§ 60.5. Nomination forms and information collection.

(a)   All nominations to the National Register are to be made on standard National Register forms. These forms are provided upon request to the State Historic Preservation Officer, participating Federal agencies and others by the NPS. For archival reasons, no other forms, photocopied or otherwise, will be accepted.

(b)   The information collection requirements contained in this part have been approved by the Office of Management and Budget under 44 U.S.C., 3507 and assigned clearance number *1024-0018.* The information is being collected as part of the nomination of properties to the National Register. This information will be used to evaluate the eligibility of properties for inclusion in the National Register under established criteria. The obligation to respond is required to obtain a benefit.

§ 60.6. Nominations by the State Historic Preservation Officer under approved State Historic Preservation programs.

(a)   The State Historic Preservation Officer is responsible for identifying and nominating eligible properties to the National Register. Nomination forms are prepared under the supervision of the State Historic Preservation Officer. The State Historic Preservation Officer establishes statewide priorities for preparation and submittal of nominations for all properties meeting National Register criteria for evaluation within the State. All nominations from the State shall be submitted in accord with the State priorities, which shall be consistent with an approved State historic preservation plan.

(b)   The State shall consult with local authorities in the nomination process. The State provides notice of the intent to nominate a property and solicits written comments especially on the significance of the property and whether or not it meets the National Register criteria for evaluation. The State notice also gives owners of private property an opportunity to concur in or object to listing. The notice is carried out as specified in the subsections below.

(c)   As part of the nomination process, each State is required to notify in writing the property owner(s), except as specified in paragraph (d) of this section, of the State's intent to bring the nomination before the State Review Board. The list of owners shall be obtained from either official land recordation records or tax records, whichever is more appropriate, within 90 days prior to the notification in intent to nominate. If in any State the land recordation or tax records is not the most appropriate list from which to obtain owners that State shall notify the Keeper in writing and request approval that an alternative source of owners may be used.

The State is responsible for notifying only those owners whose names appear on the list consulted. Where there is more than one owner on the list, each separate owner shall be notified. The State shall send the written notification at least 30 but not more than 75 days before the State Review Board meeting. Required notices may vary in some details of wording as the States prefer, but the content of notices must be approved by the National Register. The notice shall give the owner(s) at least 30 but not more than 75 days to submit written comments and concur in or object in writing to the nomination of such property. At least 30 but not more than 75 days before the State Review Board meeting, the States are also required to notify by the above mentioned National Register approved notice the applicable chief elected official of the county (or equivalent governmental unit) and municipal political jurisdiction in which the property is located. The National Register nomination shall be on file with the State Historic Preservation Program during the comment period and a copy made available by mail when requested by the public, or made available at a location of reasonable access to all affected property owners, such as a local library, courthouse, or other public place, prior to the State Review Board meeting so that written comments regarding the nomination can be prepared.

(d)   For a nomination with more than 50 property owners, each State is required to notify in writing at least 30 days but not more than 75 days in advance of the State Review Board meeting the chief elected local officials of the county (or equivalent governmental unit) and municipal political jurisdiction in which the property or district is located. The State shall provide general notice to property owners concerning the State's intent to nominate. The general notice shall be published at least 30 days but not more than 75 days before the State Review Board meeting and provide an opportunity for the submission of written comments and provide the owners of private property or a majority of such owners for districts an opportunity to concur in or object in writing to the nomination. Such general notice must be published in one or more local newspapers of general circulation in the area of the nomination. The content of the notices shall be approved by the National Register. If such general notice is used to notify the property owners for a nomination containing more than 50 owners, it is suggested that a public information meeting be held in the immediate area prior to the State Review Board meeting. If the State wishes to individually notify all property owners, it may do so, pursuant to procedures specified in Subsection 60.6(c), in which case, the State need not publish a general notice.

(e)   For Multiple Resource and Thematic Group Format submission, each district, site, building, structure and object included in the submission is treated as a separate nomination for the purpose of notification and to provide owners of private property the opportunity to concur in or object in writing to the nomination in accord with this section.

(f)   The commenting period following notifications can be waived only when all property owners and the chief elected local official have advised the State in writing that they agree to the waiver.

(g)   Upon notification, any owner or owners of a private property who wish to object shall submit to the State Historic Preservation Officer a notarized statement certifying that the party is the sole or partial owner of the private property, as appropriate, and objects to the listing. In nominations with multiple ownership of a single private property or of districts, the property will not be listed if a majority of the owners object to listing. Upon receipt of notarized objections respecting a district or single private property with multiple owners, it is the responsibility of the State Historic Preservation Officer to ascertain whether a majority of owners of private property have objected. If an owner whose name did not appear on the list certifies in a written notarized statement that the party is the sole or partial owner of a nominated private property such owner shall be counted by the State Historic Preservation Officer in determining whether a majority of owners has objected. Each owner of private property in a district has one vote regardless of how many properties or what part of one property that party owns and regardless of whether the property contributes to the significance of the district.

(h)   If a property has been submitted to and approved by the State Review Board for inclusion in the National Register prior to the effective date of this section, the State Historic Preservation Officer need not resubmit the property to the State Review Board; but before submitting the nomination to the NPS shall afford owners of private property the opportunity to concur in or object to the property's inclusion in the Register pursuant to applicable notification procedures described above.

(i)   (Reserved)

(j)   Completed nomination forms or the documentation proposed for submission on the nomination forms and comments concerning the significance of a property and its elibigility for the National Register are submitted to the State Review Board. The State Review Board shall review the nomination forms or documentation proposed for submission on the nomination forms and any comments concerning the property's significance and eligibility for the National Register. The State Review Board shall determine whether or not the property meets the National Register criteria for evaluation and make a recommendation to the State Historic Preservation Officer to approve or disapprove the nomination.

(k)   Nominations approved by the State Review Board and comments received are then reviewed by the State Historic Preservation Officer and if he or she finds the nominations to be adequately documented and technically, professionally, and procedurally correct and sufficient and in conformance with National Register criteria for evaluation, the nominations are submitted to the Keeper of the National Register of Historic Places, National Park Service, United States Department of the Interior, Washington, D.C. 20240. All comments received by a State and notarized statements of objection to listing are submitted with a nomination.

(l)   If the State Historic Preservation Officer and the State Review Board disagree on whether a property meets the National Register criteria for evaluation, the State Historic Preservation Officer, if he or she chooses, may submit the nomination with his or her opinion concerning whether or not the property meets the criteria for evaluation and the opinion of the State Review Board to the Keeper of the National Register for a final decision on the listing of the property. The opinion of the State Review Board may be the minutes of the Review Board meeting. The State Historic Preservation Officer shall submit such disputed nominations if so requested within 45 days of the State Review Board meeting by the State Review Board or the chief elected local official of the local, county or municipal political subdivision in which the property is located but need not otherwise do so. Such nominations will be substantively reviewed by the Keeper.

(m)   The State Historic Preservation Officer shall also submit to the Keeper nominations if so requested under the appeals process in § 60.12.

(n)   If the owner of a private property or the majority of such owners for a district or single property with multiple owners have objected to the nomination prior to the submittal of a nomination, the State Historic Preservation Officer shall submit the nomination to the Keeper only for a determination of eligibility pursuant to subsection (s) of this section.

(o)   The State Historic Preservation Officer signs block 12 of the nomination form if in his or her opinion the property meets the National Register criteria for evaluation. The State Historic Preservation Officer's signature in block 12 certifies that:

(1)   All procedural requirements have been met;

(2)   The nomination form is adequately documented;

(3)   The nomination form is technically and professional correct and sufficient;

(4)   In the opinion of the State Historic Preservation Officer, the property meets the National Register criteria for evaluation.

(p)   When a State Historic Preservation Officer submits a nomination form for a property that he or she does not believe meets the National Register criteria for evaluation, the State Historic Preservation Officer signs a continuation sheet Form NPS 10-900a explaining his/her opinions on the eligibility of the property and certifying that:

(1)   All procedural requirements have been met;

(2)   The nomination form is adequately documented;

(3)   The nomination form is technically and professionally correct and sufficient.

(q)   Notice will be provided in the Federal Reigster that the nominated property is being considered for listing in the National Register of Historic Places as specified in § 60.13.

(r)   Nominations will be included in the National Register within 45 days of receipt by the Keeper or designee unless the Keeper disapproves a nomination, an appeal is filed, or the owner of private property (or the majority of such owners for a district or single property with multiple owners) objects by notarized statements received by the Keeper prior to listing. Nominations which are technically or professionally inadequate will be returned for correction and resubmission. When a property does not appear to meet the National Register criteria for evaluation, the nomination will be returned with an explanation as to why the property does not meet the National Register criteria for evaluation.

(s)   If the owner of private property (or the majority of such owners for a district or single property with multiple owners) has objected to the nomination by notarized statement prior to listing, the Keeper shall review the nomination and make a determination of eligibility within 45 days of receipt, unless an appeal is filed. The Keeper shall list such properties determined eligible in the National

Register upon receipt of notarized statements from the owner(s) of private property that the owner(s) no longer object to listing.

(t)   Any person or organization which supports or opposes the nomination of a property by a State Historic Preservation Officer may petition the Keeper during the nomination process either to accept or reject a nomination. The petitioner must state the grounds of the petition and request in writing that the Keeper substantively review the nomination. Such petitions received by the Keeper prior to the listing of a property in the National Register or a determination of its eligibility where the private owners object to listing will be considered by the Keeper and the nomination will be substantively reviewed.

(u)   State Historic Preservation Officers are required to inform the property owners and the chief elected local official when properties are listed in the National Register. In the case of a nomination where there are more than 50 property owners, they may be notified of the entry in the National Register by the same general notice stated in § 60.6(d). States which notify all property owners individually of entries in the National Register need not publish a general notice.

(v)   In the case of nominations where the owner of private property (or the majority of such owners for a district or single property with multiple owners) has objected and the Keeper has determined the nomination eligible for the National Register, the State Historic Preservation Officer shall notify the appropriate chief elected local official and the owner(s) of such property of this determination. The general notice may be used for properties with more than 50 owners as described in § 60.6(d) or the State Historic Preservation Officer may notify the owners individually.

(w)   If subsequent to nomination a State makes major revisions to a nomination or renominates a property rejected by the Keeper, the State Historic Preservation Officer shall notify the affected property owner(s) and the chief elected local official of the revisions or renomination in the same manner as the original notification for the nomination, but need not resubmit the nomination to the State Review Board. Comments received and notarized statements of objection must be forwarded to the Keeper along with the revisions or renomination. The State Historic Preservation Officer also certifies by the resubmittal that the affected property owner(s) and the chief elected local official have been renotified. "Major revisions" as used herein means revisions of boundaries or important substantive revisions to the nomination which could be expected to change the ultimate outcome as to whether or not the property is listed in the National Register by the Keeper.

(x)   Notwithstanding any provision hereof to the contrary, the State Historic Preservation Officer in the nomination notification process or otherwise need not make available to any person or entity (except a Federal agency planning a project, the property owner, the chief elected local official of the political jurisdiction in which the property is located, and the local historic preservation commission for certified local governments) specific information relating to the location of properties proposed to be nominated to, or listed in, the National Register if he or she determines that the disclosure of specific information would create a risk of destruction or harm to such properties.

(y)   With regard to property under Federal ownership or control, completed nomination forms shall be submitted to the Federal Preservation Officer for review and comment. The Federal Preservation Officer, may approve the nomination and forward it to the Keeper of the National Register of Historic Places, National Park Service, United States Department of the Interior, Washington, D.C. 20240.

[46 FR 56187, Nov. 16, 1981, as amended at 48 FR 46308, October 12, 1983]

§§ 60.7–60.8 [Reserved]

§ 60.9. Nominations by Federal agencies.

(a)   The National Historic Preservation Act of 1966, as amended, requires that, with the advice of the Secretary and in cooperation with the State Historic Preservation Officer of the State involved, each Federal agency shall establish a program to locate, inventory and nominate to the Secretary all properties under the agency's ownership or control that appear to qualify for inclusion on the National Register. Section 2(a) of Executive Order 11593 provides that Federal agencies shall locate, inventory, and nominate to the Secretary of the Interior all sites, buildings, districts, and objects under their jurisdiction or control that appear to qualify for listing on the National Register of Historic Places. Additional responsibilities of Federal agencies are detailed in the National Historic Preservation Act of 1966, as amended, Executive Order 11593, the National Environmental Policy Act of 1969, the Archeological and Historic Preservation Act of 1974, and procedures developed pursuant to these authorities, and other related legislation.

(b)   Nomination forms are prepared under the supervision of the Federal Preservation Officer designated by the head of a Federal agency to fulfill agency responsibilities under the National Historic Preservation Act of 1966, as amended.

(c)   Completed nominations are submitted to the appropriate State Historic Preservation Officer for review and comment regarding the adequacy of the nomination, the significance of the property and its eligibility for the National Register. The chief elected local officials of the county (or equivalent governmental unit) and municipal political jurisdiction in which the property is located are notified and given 45 days in which to comment. The State Historic Preservation Officer signs block 12 of the nomination form with his/her recommendation.

(d)   After receiving the comments of the State Historic Preservation Officer, and chief elected local official, or if there has been no response within 45 days, the Federal Preservation Officer may approve the nomination and forward it to the Keeper of the National Register of Historic Places, National Park Service, United States Department of the Interior, Washington, D.C. 20240. The Federal Preservation Officer signs block 12 of the nomination form if in his or her opinion the property meets the National Register criteria for evaluation. The Federal Preservation Officer's signature in block 12 certifies that:

(1)   All procedural requirements have been met;

(2)   The nomination form is adequately documented;

(3)   The nomination form is technically and professionally correct and sufficient;

(4)   In the opinion of the Federal Preservation Officer, the property meets the National Register criteria for evaluation.

(e)   When a Federal Preservation Officer submits a nomination form for a property that he or she does not believe meets the National Register criteria for evaluation, the Federal Preservation Officer signs a continuation sheet Form NPS 10-900a explaining his/her opinions on the eligibility of the property and certifying that:

(1)   All procedural requirements have been met;

(2)   The nomination form is adequately documented;

(3)   The nomination form is technically and professionally correct and sufficient.

(f)   The comments of the State Historic Preservation Officer and chief local official are appended to the nomination, or, if there are no comments from the State Historic Preservation Officer an explanation is attached. Concurrent nominations (see § 60.10) cannot be submitted, however, until the nomination has been considered by the State in accord with Sec. 60.6 supra. Comments received by the State concerning concurrent nominations and notarized statements of objection must be submitted with the nomination.

(g)   Notice will be provided in the Federal Register that the nominated property is being considered for listing in the National Register of Historic Places in accord with § 60.13.

(h)   Nominations will be included in the National Register within 45 days of receipt by the Keeper or designee unless the Keeper disapproves such nomination or an appeal is filed. Nominations which are technically or professionally inadequate will be returned for correction and resubmission. When a property does not appear to meet the National Register criteria for evaluation, the nomination will be returned with an explanation as to why the property does not meet the National Register criteria for evaluation.

(l)   Any person or organization which supports or opposes the nomination of a property by a Federal Preservation Officer may petition the Keeper during the nomination process either to accept or reject a nomination. The petitioner must state the grounds of the petition and request in writing·that the Keeper substantively review the nomination. Such petition received by the Keeper prior to the listing of a property in the National Register or a determination of its eligibility where the private owner(s) object to listing will be considered by the Keeper and the nomination will be substantively reviewed.

§ 60.10. Concurrent State and Federal nominations.

(a)   State Historic Preservation Officers and Federal Preservation Officers are encouraged to cooperate in locating, inventorying, evaluating, and nominating all properties possessing historical, architectural archeological, or cultural value, Federal agencies may nominate properties where a portion of the property is not under Federal ownership or control.

(b)   When a portion of the area included in a Federal nomination is not located on land under the ownership or control of the Federal agency, but is an integral part of the cultural resource, the completed nomination form shall be sent to the State Historical Preservation Officer for notification to property owners, to give owners of private property an opportunity to concur in or object to the nomination, to solicit written comments and for submission to the State Review Board pursuant to the procedures in § 60.6.

(c)   If the State Historic Preservation Officer and the State Review Board agree that the nomination meets the National Register criteria for evaluation, the nomination is signed by the State Historic Preservation Officer and returned to the Federal agency initiating the nomination. If the State Historic Preservation Officer and the State Review Board disagree, the nomination shall be returned to the Federal agency with the opinions of the State Historic Preservation Officer and the State Review Board concerning the adequacy of the nomination and whether or not the property meets the criteria for evaluation. The opinion of the State Review Board may be the minutes of the State Review Board meeting. The State Historic Preservation Officer's signed opinion and comments shall confirm to the Federal agency that the State nomination procedures have been fulfilled including notification requirements. Any comments received by the State shall be included with the letter as shall any notarized statements objecting to the listing of private property.

(d)   If the owner of any privately owned property, (or a majority of the owners of such properties within a district or single property with multiple owners) objects to such inclusion by notarized statements) the Federal Historic Preservation Officer shall submit the nomination to the Keeper for review and a determination of eligibility. Comments, opinions, and notarized statements of objection shall be submitted with the nomination.

(e)   The State Historic Preservation Officer shall notify the nonfederal owners when a concurrent nomination is listed or determined eligible for the National Register as required in § 60.6.

§ 60.11. Requests for nominations.

(a)   The State Historic Preservation Officer or Federal Preservation Officer as appropriate shall respond in writing within 60 days to any person or organization submitting a completed National Register nomination form or requesting consideration for any previously prepared nomination form on record with the State or Federal agency. The response shall provide a technical opinion concerning whether or not the property is adequately documented and appears to meet the National Register criteria for evaluation in § 60.4. If the nomination form is determined to be inadequately documented, the nominating authority shall provide the applicant with an explanation of the reasons for that determination.

(b)   If the nomination form does not appear to be adequately documented, upon receiving notification, it shall be the responsibility of the applicant to provide necessary additional documentation.

(c)   If the nomination form appears to be adequately documented and if the property appears to meet the National Register criteria for evaluation, the State Historic Preservation Officer shall comply with the notification requirements in Section 60.6 and schedule the property for presentation at the earliest possible State Review Board meeting. Scheduling shall be consistent with the State's established priorities for processing nominations. If the nomination form is adequately documented, but the property does not appear to meet National Register criteria for evaluation, the State Historic Preservation Officer need not process the nomination, unless so requested by the Keeper pursuant to § 60.12.

(d)   The State Historic Preservation Officer's response shall advise the applicant of the property's position in accord with the State's priorities for processing nominations and of the approximate date the applicant can expect its consideration by the State Review Board. The State Historic Preservation Officer shall also provide notice to the applicant of the time and place of the Review Board meeting at least 30 but not more than 75 days before the meeting, as well as complying with the notification requirements in § 60.6.

(e)   Upon action on a nomination by the State Review Board, the State Historic Preservation Officer shall, within 90 days, submit the nomination to the National Park Service, or, if the State Historic Preservation Officer does not consider the property eligible for the National Register, so advise the applicant within 45 days.

(f)   If the applicant substantially revises a nomination form as a result of comments by the State or Federal agency, it may be treated by the State Historic Preservation Officer or Federal Preservation Officer as a new submittal and processed in accord with the requirements in this section.

(g)   The Federal Preservation Officer shall request the comments of the State Historic Preservation Officer and notify the applicant in writing within 90 days of receipt of an adequately documented nomination form as to whether the Federal agency will nominate the property. The Federal Preservation Officer shall submit an adequately documented nomination to the National Park Service unless in his or her opinion the property is not eligible for the National Register.

[48 FR 46308, Oct. 12, 1983]

§ 60.12.   Nomination appeals.

(a)   Any person or local government may appeal to the Keeper the failure or refusal of a nominating authority to nominate a property that the person or local government considers to meet the National Register criteria for evaluation upon decision of a nominating authority to not nominate a property for any reason when requested pursuant to § 60.11, or upon failure of a State Historic Preservation Officer to nominate a property recommended by the State Review Board. (This action differs from the procedure for appeals during the review of a nomination by the National Park Service where an individual or organization may "petition the Keeper during the nomination process," as specified in §§ 60.6(t) and 60.9(i). Upon receipt of such petition the normal 45-day review period will be extended for 30 days beyond the date of the petition to allow the petitioner to provide additional documentation for review.)

(b)   Such appeal shall include a copy of the nomination form and documentation previously submitted to the State Historic Preservation Officer or Federal Preservation Officer, an explanation of why the applicant is submitting the appeal in accord with this section and shall include pertinent correspondence from the State Historic Preservation Officer or Federal Preservation Officer.

(c)   The Keeper will respond to the appellant and the State Historic Preservation Officer or Federal Preservation Officer with a written explanation either denying or sustaining the appeal within 45 days of receipt. If the appeal is sustained, the Keeper will:

(1)   request the State Historic Preservation Officer to submit the nomination to the Keeper within 15 days if the nomination has completed the proce-

dural requirements for nomination as described in Section 60.6 or 60.9 except
that concurrence of the State Review Board, State Historic Preservation Officer
or Federal Preservation Officer is not required; or

(2)   if the nomination has not completed these procedural requirements, request the State Historic Preservation Officer or Federal Preservation
Officer to promptly process the nomination pursuant to Section 60.6 or 60.9
and submit the nomination to the Keeper without delay.

(d)   State Historic Preservation Officers and Federal Preservation Officers
shall process and submit such nominations if so requested by the Keeper pursuant
to this section. The Secretary reserves the right to list properties in the National
Register or determine properties eligible for such listing on his own motion when
necessary to assist in the preservation of historic resources and after notifying the
owner and appropriate parties and allowing for a 30-day comment period.

(e)   No person shall be considered to have exhausted administrative remedies
with respect to failure to nominate a property to the National Register until he or
she has complied with procedures set forth in this section. The decision of the
Keeper is the final administrative action on such appeals.

[48 FR 46308, Oct. 12, 1983]

§ 60.13. Publication in the "Federal Register" and other NPS notification.

(a)   When a nomination is received, NPS will publish notice in the Federal
Register that the property is being considered for listing in the National Register.
A 15-day commenting period from date of publication will be provided. When
necessary to assist in the preservation of historic properties this 15-day period may
be shortened or waived.

(b)   NPS shall notify the appropriate State Historic Preservation Officer, Federal Preservation Officer, person or local government when there is no approved
State program of the listing of the property in the National Register and will publish notice of the listing in the Federal Register.

(c)   In nominations where the owner of any privately owned property (or a
majority of the owners of such properties within a district or single property
within a district or single property with multiple owners) has objected and the
Keeper has determined the nomination eligible for the National Register, NPS
shall notify the State Historic Preservation Officer, the Federal Preservation Officer (for Federal or concurrent nominations), the person or local government where
there is no approved State Historic Preservation Program and the Advisory Council
on Historic Preservation. NPS will publish notice of the determination of eligibility
in the Federal Register.

§ 60.14. Changes and revisions to properties listed in the National Register.

(a)   *Boundary changes.*

(1)   A boundary alteration shall be considered as a new property nomination. All forms, criteria and procedures used in nominating a property to the
National Register must be used. In the case of boundary enlargements only those
owners in the newly nominated as yet unlisted area need be notified and will be

counted in determining whether a majority of private owners object to listing. In the case of a diminution of a boundary, owners shall be notified as specified in § 60.15 concerning removing properties from the National Register. A professionally justified recommendation by the State Historic Preservation Officer, Federal Preservation Officer, or person or local government where there is no approved State Historic Preservation Program shall be presented to NPS. During this process, the property is not taken off the National Register. If the Keeper or his or her designee finds the recommendation in accordance with the National Register criteria for evaluation, the change will be accepted. If the boundary change is not accepted, the old boundaries will remain. Boundary revisions may be appealed as provided for in § § 60.12 and 60.15.

(2)   Four justifications exist for altering a boundary: Professional error in the initial nomination, loss of historic integrity, recognition of additional significance, additional research documenting that a larger or smaller area should be listed. No enlargement of a boundary should be recommended unless the additional area possesses previously unrecognized significance in American history, architecture, archeology, engineering or culture. No diminution of a boundary should be recommended unless the properties being removed do not meet the National Register criteria for evaluation. Any proposal to alter a boundary has to be documented in detail including photographing the historic resources falling between the existing boundary and the other proposed boundary.

(b)   *Relocating properties listed in the National Register.*

(1)   Properties listed in the National Register should be moved only when there is no feasible alternative for preservation. When a property is moved, every effort should be made to reestablish its historic orientation, immediate setting, and general environment.

(2)   (2)   If it is proposed that a property listed in the National Register be moved and the State Historic Preservation Officer, Federal agency for a property under Federal ownership or control, or person or local government where there is no approved State Historic Preservation Program, wishes the property to remain in the National Register during and after the move, the State Historic Preservation Officer or Federal Preservation Officer having ownership or control or person or local government where there is no approved State Historic Preservation Program, shall submit documentation to NPS prior to the move. The documentation shall discuss:

(i)   The reasons for the move;

(ii)   the effect on the property's historical integrity;

(iii)   the new setting and general environment of the proposed site, including evidence that the proposed site does not possess historical or archeological significance that would be adversely affected by the intrusion of the property; and

(iv)   photographs showing the proposed location.

(3)   Any such proposal with respect to the new location shall follow the required notification procedures, shall be approved by the State Review Board if it is a State nomination and shall continue to follow normal review procedures. The Keeper shall also follow the required notification procedures for nominations. The Keeper shall respond to a properly documented request within 45 days of receipt from the State Historic Preservation Officer or Federal Preservation Officer, or within 90 days of receipt from a person or local government where there is no

approved State Historic Preservation Program, concerning whether or not the move is approved. Once the property is moved, the State Historical Preservation Officer, Federal Preservation Officer, or person or local government where there is no approved State Historic Preservation Program shall submit to the Keeper for review

      (i)   a letter notifying him or her of the date the property was moved;

      (ii)   photographs of the property on its new site; and

      (iii) revised maps, including a U.S.G.S. map;

      (iv) acreage, and

      (v)   verbal boundary description.

The Keeper shall respond to a properly documented submittal within 45 days of receipt with the final decision on whether the property will remain in the National Register. If the Keeper approves the move, the property will remain in the National Register during and after the move unless the integrity of the property is in some unforeseen manner destroyed. If the Keeper does not approve the move, the property will be automatically deleted from the National Register when moved. In cases of properties removed from the National Register, if the State, Federal agency, or person or local government where there is no approved State Historic Preservation Program has neglected to obtain prior approval for the move or has evidence that previously unrecognized significance exists, or has accrued, the State, Federal agency, person or local government may resubmit a nomination for the property.

      (4)   In the event that a property is moved, deletion from the National Register will be automatic unless the above procedures are followed prior to the move. If the property has already been moved, it is the responsibility of the State, Federal agency or person or local government which nominated the property to notify the National Park Service. Assuming that the State, Federal agency or person or local government wishes to have the structure reentered in the National Register, it must be nominated again on new forms which should discuss:

      (i)   the reasons for the move;

      (ii)   the effect on the property's historical integrity, and

      (iii) the new setting and general environment, including evidence that the new site does not possess historical or archeological significance that would be adversely affected by intrusion of the property.

In addition, new photography, acreage, verbal boundary description and a U.S.G.S. map showing the structure at its new location must be sent along with the revised nomination. Any such nomination submitted by a State must be approved by the State Review Board.

      (5)   Properties moved in a manner consistent with the comments of the Advisory Council on Historic Preservation, in accord with its procedures (36 CFR Part 800), are granted as exception to § 60.12(b). Moving of properties in accord with the Advisory Council's procedures should be dealt with individually in each memorandum of agreement. In such cases, the State Historic Preservation Officer or the Federal Preservation Officer, for properties under Federal ownership or control, shall notify the Keeper of the new location after the move including new documentation as dscribed above.

§ 60.15. Removing properties from the National Register

(a)   Grounds for removing properties from the National Register are as follows:

(1)   The property has ceased to meet the criteria for listing in the National Register because the qualities which caused it to be originally listed have been lost or destroyed, or such qualities were lost subsequent to nomination and prior to listing;

(2)   additional information shows that the property does not meet the National Register criteria for evaluation;

(3)   error in professional judgment as to whether property meets the criteria for evaluation; or

(4)   prejudicial procedural error in the nomination or listing process. Properties removed from the National Register for procedural error shall be considered for listing by the Keeper after correction of the error or errors by the State Historic Preservation Officer, Federal Preservation Officer, person or local government which originally nominated the property, or by the Keeper, as appropriate. The procedures set forth for nominations shall be followed in such reconsiderations. Any property or district removed from the National Register for procedural deficiencies in the nomination and/or listing process shall automatically be considered eligible for inclusion in the National Register without further action and will be published as such in the Federal Register.

(b)   Properties listed in the National Register prior to December 13, 1980, may only be removed from the National Register on the grounds established in paragraph (a)(1) of this section.

(c)   Any person or organization may petition in writing for removal of a property from the National Register by setting forth the reasons the property should be removed on the grounds established in paragraph (a) of this section. With respect to nominations determined eligible for the National Register because the owners of private property object to listing, anyone may petition for reconsideration of whether or not the property meets the criteria for evaluation using these procedures. Petitions for removal are submitted to the Keeper by the State Historic Preservation Officer for State nominations, the Federal Preservation Officer for Federal nominations, and directly to the Keeper from persons or local governments where there is no approved State Historic Preservation Program.

(d)   Petitions submitted by persons or local governments where there is no approved State Historic Preservation Program shall include a list of the owner(s). In such cases the Keeper shall notify the affected owner(s) and the chief elected local official and give them an opportunity to comment. For approved State programs, the State Historic Preservation Officer shall notify the affected owner(s) and chief elected local official and give them an opportunity to comment prior to submitting a petition for removal. The Federal Preservation Officer shall notify and obtain the comments of the appropriate State Historic Preservation Officer prior to forwarding an appeal to NPS. All comments and opinions shall be submitted with the petition.

(e)   The State Historic Preservation Officer or Federal Preservation Officer shall shall respond in writing within 45 days of receipt to petitioners for removal of property from the National Register. The response shall advise the petitioner of the State Historic Preservation Officer's or Federal Preservation Officer's views on the petition.

(f)   A petitioner desiring to pursue his removal request must notify the State Historic Preservation Officer or the Federal Preservation Officer in writing within 45 days of receipt of the written views on the petition.

(g)   The State Historic Preservation Officer may elect to have a property considered for removal according to the State's nomination procedures unless the petition is on procedural grounds and shall schedule it for consideration by the State Review Board as quickly as all notification requirements can be completed following procedures outlined in § 60.6, or the State Historic Preservation Officer may elect to forward the petition for removal to the Keeper with his or her comments without State Review Board consideration.

(h)   Within 15 days after receipt of the petitioner's notification of intent to pursue his removal request, the State Historic Preservation Officer shall notify the petitioner in writing either that the State Review Board will consider the petition on a specified date or that the petition will be forwarded to the Keeper after notification requirements have been completed. The State Historic Preservation Officer shall forward the petitions to the Keeper for review within 15 days after notification requirements or Review Board consideration, if applicable, have been completed.

(i)   Within 15 days after receipt of the petitioner notification of intent to pursue his petition, the Federal Preservation Officer shall forward the petition with his or her comments and those of the State Historic Preservation Officer to the Keeper.

(j)   The Keeper shall respond to a petition for removal within 45 days of receipt, except where the Keeper must notify the owners and the chief elected local official. In such cases the Keeper shall respond within 90 days of receipt. The Keeper shall notify the petitioner and the applicable State Historic Preservation Officer, Federal Preservation Officer, or person or local government where there is no approved State Historic Preservation Program, of his decision. The State Historic Preservation Officer or Federal Preservation Officer transmitting the petition shall notify the petitioner, the owner(s), and the chief elected local official in writing of the decision. The Keeper will provide such notice for petitions from persons or local governments where there is no approved State Historic Preservation Program. The general notice may be used for properties with more more than 50 owners. If the general notice is used it shall be published in one or more newspapers with general circulation in the area of the nomination.

(k)   The Keeper may remove a property from the National Register on his own motion on the grounds established in paragraph (a) of this section, except for those properties listed in the National Register prior to December 13, 1980, which may only be removed from the National Register on the grounds established in paragraph (a)(1) of this section. In such cases, the Keeper will notify the nominating authority, the affected owner(s) and the applicable chief elected local official and provide them an opportunity to comment. Upon removal, the Keeper will notify the nominating authority of the basis for the removal. The state Historic Preservation Officer, Federal Preservation Officer, or person or local government which nominated the property shall notify the owner(s) and the chief elected local official of the removal.

(l)   No person shall be considered to have exhausted administrative remedies with respect to removal of a property from the National Register until the Keeper has denied a petition for removal pursuant to this section.

Appendix J

Advisory Council on Historic Preservation Regulations Implementing Section 106 of the National Historic Preservation Act

(36 C.F.R. Part 800)

Part 800–Protection of Historic and Cultural Properties

Authority: Pub. L. 89-665, 80 Stat. 915 (16 U.S.C. 470), as amended, 84 Stat. 204 (1970), 87 Stat. 139 (1973), 90 Stat. 1320 (1976), 92 Stat. 3467 (1978); E.O. 11593, 3 CFR 1071 Comp. p. 154; President's Memorandum on Environmental Quality and Water Resources Management, July 12, 1978.

Source: 44 FR 6072, Jan. 30, 1979, unless otherwise noted.

§ 800.1. Purpose and authorities.

(a)   The National Historic Preservation Act of 1966, as amended, established the Advisory Council on Historic Preservation as an independent agency of the United States to advise the President and the Congress on historic preservation matters, recommend measures to coordinate Federal historic preservation activities, and comment on Federal actions affecting properties included in or eligible for inclusion in the National Register of Historic Places. Its members are the Secretary of the Interior, the Secretary of Housing and Urban Development, the Secretary of Commerce, the Administrator of General Services, the Secretary of the Treasury, the Attorney General, the Secretary of Agriculture, the Secretary of Transportation, the Secretary of State, the Secretary of Defense, the Secretary of Health, Education, and Welfare, the Chairman of the Council on Environmental Quality, the Chairman of the Federal Council on the Arts and Humanities, the Architect of the Capitol, the Secretary of the Smithsonian Institution, the Chairman of the National Trust for Historic Preservation, the President of the National Conference of State Historic Preservation Officers, and 12 citizen members from outside the Federal Government appointed for five-year terms by the President on the basis of their interest and experience in the matters to be considered by the Council.

(b)   The Council protects properties of historical, architectural, archeological, and cultural significance at the national, State, and local level by reviewing and commenting on Federal actions affecting National Register and eligible properties in accordance with the following authorities:

(1)   Section 106 of the National Historic Preservation Act. Section 106 requires that Federal agencies with direct or indirect jurisdiction over a Federal, federally assisted or federally licensed undertaking afford the Council a reasonable opportunity for comment on such undertakings that affect properties included in or eligible for inclusion in the National Register of Historic Places prior to the agency's approval of any such undertaking.

(2)   Section 1(3) of Executive Order 11593, May 13, 1971, "Protection and Enhancement of the Cultural Environment." Section 1(3) requires that Federal agencies, in consultation with the Council, institute procedures to assure that their plans and programs contribute to the preservation and enhancement of non-federally owned historic and cultural properties.

(3)   Section 2(b) of Executive Order 11593, May 13, 1971, "Protection and Enhancement of the Cultural Environment." Federal agencies are required by Section 2(a) of the Executive Order to locate, inventory, and nominate properties under their jurisdiction or control to the National Register. Until such processes are complete, Federal agencies must provide the Council on opportunity to comment on proposals for the transfer, sale, demolition, or substantial alteration of federally owned properties eligible for inclusion in the National Register.

(4)   The President's Memorandum on Environmental Quality and Water Resources Management. The Memorandum directs the Council to issue final regulations under the National Historic Preservation Act by March 1, 1979, and further directs Federal agencies with water resource responsibilities and programs to publish procedures implementing the Act not later than three months after promulgation of final regulations by the Council. Federal agencies' procedures are to be reviewed and, if they are consistent with the Council's regulations, approved by the Council within 60 days and published in final form.

§ 800.2. Definitions.

As used in these regulations:

(a)   "National Historic Preservation Act" means Pub. L. 89-665, approved October 15, 1966, an "Act to establish a program for the preservation of additional historic properties throughout the Nation and for other purposes" (80 Stat. 915, 16 U.S.C. 470, as amended; 84 Stat. 204 (1970), 87 Stat. 139 (1973), 90 Stat. 1320 (1976), 92 Stat. 3467 (1978)), hereinafter referred to as "the Act."

(b)   "Executive Order" means Executive Order 11593, May 13, 1971, "Protection and Enhancement of the Cultural Environment" (36 FR 8921, 16 U.S.C. 470).

(c)   "Undertaking" means any Federal, federally assisted or federally licensed action, activity, or program or the approval, sanction, assistance, or support of any non-Federal action, activity, or program. Undertakings include new and continuing projects and program activities (or elements of such activities not previously considered under Section 106 or Executive Order 11593) that are:

(1)   Directly undertaken by Federal agencies;

(2)   supported in whole or in part through Federal contracts, grants, subsidies, loans, loan guarantees, or other forms of direct and indirect funding assistance;

(3)   carried out pursuant to a Federal lease, permit, license, certificate, approval, or other form of entitlement or permission; or,

(4)   proposed by a Federal agency for Congressional authorization or appropriation. Site-specific undertakings affect areas and properties that are capable of being identified at the time of approval by the Federal agency. Non-site-specific undertakings have effects that can be anticipated on National Register and eligible properties but cannot be identified in terms of specific geographical areas or properties at the time of Federal approval. Non-site-specific undertakings include Federal approval of State plans pursuant to Federal legislation, development of comprehensive or area-wide plans, agency recommendations for legislation and the establishment or modification of regulations and planning guidelines.

(d)   "National Register" means the National Register of Historic Places. It is a register of districts, sites, buildings, structures, and objects of national, State, or local significance in American history, architecture, archeology, and culture that is expanded and maintained by the Secretary of the Interior under authority of section 2(b) of the Historic Sites Act of 1935 (49 Stat. 666, 16 U.S.C. 461) and Section 101(a)(1) of the National Historic Preservation Act implemented through 36 CFR Part 60. The National Register is published in its entirety in the Federal Register each year in February. Addenda are usually published on the first Tuesday of each month.

(e)   "National Register property" means a district, site, building, structure, or object included in the National Register.

(f)   "Eligible property" means any district, site, building, structure, or object that meets the National Register Criteria.

(g)   "National Register Criteria" means the criteria established by the Secretary of the Interior to evaluate properties to determine whether they are eligible for inclusion in the National Register. (See 36 CFR 60.6.)

(h)   "Decision" means the exercise of or the opportunity to exercise discretionary authority by a Federal agency at any state of an undertaking where alterations might be made in the undertaking to modify its impact upon National Register and eligible properties.

(i)   "Agency Official" means the head of the Federal agency having responsibility for the undertaking or a designee authorized to act for the Agency Official.

(j)   "Council" means the Advisory Council on Historic Preservation as established by Title II of the Act.

(k)   "Chairman" means the Chairman of the Advisory Council on Historic Preservation or a member designated to act for the Chairman.

(l)   "Executive Director" means the Executive Director of the Advisory Council on Historic Preservation as established by Section 205 of the Act, or a designee authorized to act for the Executive Director.

(m)   "State Historic Preservation Officer" means the official, who is responsible for administering the Act within the State or jurisdiction, or a designated representative authorized to act for the State Historic Preservation Officer. These officers are appointed pursuant to 36 CFR 61.2 by the Governors of the 50 States, Guam, American Samoa, the Commonwealth of Puerto Rico, the Virgin Islands, the Trust Territory of the Pacific Islands, the Commonwealth of the Mariana Islands, and the Mayor of the District of Columbia.

(n)   "Secretary" means the Secretary of the Interior or a designee authorized to carry·out the historic preservation responsibilities of the Secretary under the Act, Executive Order 11593, and related authorities.

(o)   "Area of the undertaking's potential environmental impact" means that geographical area within which direct and indirect effects generated by the undertaking could reasonably be expected to occur and thus cause a change in the historical, architectural, archeological, and cultural qualities possessed by a National National Register or eligible property. The boundaries of such area should be determined by the Agency Official in consultation with the State Historic Preservation Officer as early as possible in the planning of the undertaking.

(p)   "Consulting parties" means the Agency Official, the State Historic Preservation Officer, and the Executive Director.

§ 800.3.   Criteria of effect and adverse effect.

The following criteria shall be used to determine whether an undertaking has an effect or an adverse effect in accordance with these regulations.

(a)   *Criteria of Effect.* The effect of a Federal, federally assisted or federally licensed undertaking on a National Register or eligible property is evaluated in the context of the historical, architectural, archeological, or cultural significance possessed by the property. An undertaking shall be considered to have an effect whenever any condition of the undertaking causes or may cause any change, beneficial or adverse, in the quality of the historical, architectural, archeological, or cultural characteristics that qualify the property to meet the criteria of the National Register. An effect occurs when an undertaking changes the integrity of location, design, setting, materials, workmanship, feeling, or association of the property that contributes to its significance in accordance with the National Register criteria. *An* effect may be direct or indirect. Direct effects are caused by the undertaking and occur at the same time and place. Indirect effects include those caused by the undertaking that are later in time and farther removed in distance, but are still reasonably foreseeable. Such effects may include changes in the pattern of land use, population density or growth rate that may affect on properties of historical, architectural, archeological, or cultural significance.

(b)   *Criteria of Adverse Effect.* Adverse effects on National Register or eligible properties may occur under conditions which include but are not limited to:

(1)   Destruction or alteration of all or part of a property;

(2)   Isolation from or alteration of the property's surrounding environment;

(3)   Introduction of visual, audible, or atmospheric elements that are out of character with the property or alter its setting;

(4)   Neglect of a property resulting in its deterioration or destruction.

(5)   Transfer or sale of a property without adequate conditions or restrictions regarding preservation, maintenance, or use.

Review of Individual Undertakings

§ 800.4. Federal Agency responsibilities.

As early as possible before an agency makes a final decision concerning an undertaking and in any event prior to taking any action that would foreclose alternatives or the Council's ability to comment, the Agency Official shall take the following steps to comply with the requirements of Section 106 of the National Historic Preservation Act and Section 2(b) of Executive Order 11593. It is the primary responsibility of each Agency Official requesting Council comments to conduct the appropriate studies and to provide the information necessary for an adequate review of the effect a proposed undertaking may have on a National Register or eligible property, as well as the information necessary for adequate consideration of modifications or alterations to the proposed undertaking that could avoid, mitigate, or minimize any adverse effects. It is the responsibility of each Agency Official requesting consultation with a State Historic Preservation Officer under this section to provide the information that is necessary to make an informed and reasonable evaluation of whether a property meets National Register criteria and to determine the effect of a proposed undertaking on a National Register or eligible property. Although a Federal agency may require non-Federal parties to undertake certain steps required by these regulations as a prerequisite to Federal action and may authorize non-Federal participation under this section and in the consultation process under § 800.6 pursuant to approved counterpart regulations, the ultimate responsibility for compliance with these regulations remains with the Federal agency and cannot be delegated by it.

(a)   *Identification of National Register and Eligible Properties.* It is the responsibility of each Federal agency to identify or cause to be identified any National Register or eligible property that is located within the area of the undertaking's potential environmental impact and that may be affected by the undertaking.

(1)   The Agency Official shall consult the State Historic Preservation Officer, the published lists of National Register and eligible properties, public records, and other individuals or organizations with historical and cultural expertise,

as appropriate, to determine what historic and cultural properties are known to be within the area of the undertaking's potential environmental impact. The State Historic Preservation Officer should provide the Agency Official with any information available on known historic and cultural properties identified in the area (whether on the National Register or not), information on any previous surveys performed and an evaluation of their quality, a recommendation as to the need for a survey of historic and cultural properties, and recommendations as to the type of survey and/or survey methods should a survey be recommended, and recommendations on boundaries of such surveys.

(2)   The Agency Official shall, after due consideration of the information obtained pursuant to § 800.4(a)(1), determine what further actions are necessary to discharge the agency's affirmative responsibilities to locate and identify eligible properties that are within the area of the undertaking's potential environmental impact and that may be affected by the undertaking. Such actions may include a professional cultural resource survey of the environmental impact area, or parts of the area, if the area has not previously been adequately surveyed. The recommendations of the State Historic Preservation Officer should be followed in this matter.

(3)   The Agency Official, in consultation with the State Historic Preservation Officer, shall apply the National Register criteria to all properties that may possess any historical, architectural, archeological, of cultural value located within the area of the undertaking's potential environmental impact. If either the Agency Official or the State Historic Preservation Officer finds that a property meets the National Register Criteria or a question exists as to whether a property meets the Criteria, the Agency Official shall request a determination of eligibility from the Secretary of the Interior in accordance with 36 CFR Part 63. The opinion of the Secretary respecting the eligibility of a property shall be conclusive for the purposes of these regulations. If the Agency Official and the State Historic Preservation Officer agree that no identified property meets the Criteria, the Agency Official shall document this finding and, unless the Secretary has otherwise made a determination of eligibility under 36 CFR Part 63, may proceed with the undertaking.

(4)   The Agency Official shall complete the preceding steps prior to requesting the Council's comments pursuant to § 800.4(b)-(d). The Agency Official may, however, initiate a request for the Council's comments simultaneously with a request for a determination of eligibility from the Secretary when the Agency Official and the State Historic Preservation Officer agree that a property meets the National Register Criteria. Before the Council completes action pursuant to § 800.6, the Secretary must find the property eligible for inclusion in the National Register.

(b)   *Determination of Effect.* For each National Register or eligible property that is located within the area of the undertaking's potential environmental impact, the Agency Official, in consultation with the State Historic Preservation Officer, shall apply the Criteria of Effect, (§ 800.3(a)), to determine whether the undertaking will have an effect upon the historical, architectural, archeological, or cultural characteristics of the property that qualified it to meet National Register Criteria.

(1)   *No Effect.* If the Agency Official, in consultation with the State Historic Preservation Officer, finds that the undertaking will not affect these

characteristics, the undertaking may proceed. The Agency Official shall document each Determination of No Effect, which shall be available for public inspection. If the State Historic Preservation Officer objects or other timely objection is made to the Executive Director to an Agency Official's Determination of No Effect, the Executive Director may review the Determination and advise the Agency Official, the State Historic Preservation Officer and any objecting party of the findings within 15 days.

(2) *Effect determined.* If the Agency Official or the Executive Director finds that the undertaking will have an effect upon these characteristics, the Agency Official, in consultation with the State Historic Preservation Officer, shall apply the Criteria of Adverse Effect, set forth in § 800.3(b), to determine whether the effect of the undertaking may be adverse.

(c) *Determinations of no adverse effect.* If the Agency Official, in consultation with the State Historic Preservation Officer, finds the effect on the historical, architectural, archeological, or cultural characteristics of the property not to be adverse, the Agency Official shall forward adequate documentation (see § 800.13 (a)) of the Determination, including written evidence of the views of the State Historic Preservation Officer, to the Executive Director for review in accordance with § 800.6. If the State Historic Preservation Officer fails to respond to an Agency Official's request as provided in § 800.5, the Agency Official shall include evidence of having contacted the State Historical Preservation Officer.

(d) *Adverse effect determination.* If the Agency Official finds the effect on the historical, architectural, archeological, or cultural characteristics of the property to be adverse, or if the Executive Director does not accept an Agency Official's Determination of No Adverse Effect pursuant to review under § 800.6, the Agency Official shall:

(1) Prepare and submit a Preliminary Case Report requesting the comments of the Council (see § 800.13(b)),

(2) Notify the State Historic Preservation Officer of this request, and

(3) Proceed with the consultation process set forth in § 800.6.

(e) *Suspense of action.* Until the Council issues its comments under these regulations, good faith consultation shall preclude a Federal agency from taking or sanctioning any action or making any irreversible or irretrievable commitment that could result in an adverse effect on a National Register or eligible property or that would foreclose the consideration of modifications or alternatives to the proposed undertaking that could avoid, mitigate, or minimize such adverse effects.

§ 800.5. State Historic Preservation Officer responsibilities.

(a) The State Historic Preservation Officer should participate in the review process established by these regulations whenever it concerns an undertaking located within the State Historic Preservation Officer's jurisdiction.

(b) Unless a longer time is agreed to by the Agency Official, the failure of a State Historic Preservation Officer to respond to an Agency Official's request for consultation under § 800.4 within 30 days after receipt shall not prohibit the Agency Official from proceeding with the review process under these regulations.

(c) The State Historic Preservation Officer, with the Agency Official and the Executive Director, should participate in any consultation under § 800.6(b) and sign any Memorandum of Agreement developed under § 800.6(c) of these regulations. Failure of a State Historic Preservation Officer to participate in a consultation under § 800.6(b) or to sign a Memorandum of Agreement as provided in § 800.6(c) (1) within 30 days of receipt without notifying the Executive Director and the Agency Official that the State Historic Preservation Officer disagrees with the terms of the Agreement shall not prohibit the Executive Director and the Agency Official from concluding the Agreement and having it ratified by the Chairman in accordance with § 800.6(c)(2).

§ 800.6.   Council comments.

The following subsections specify how the Council will respond to Federal agency requests for the Council's comments required to satisfy an agency's responsibilities under Section 106 of the Act of Section 2(b) of the Executive Order.

(a)   *Response to determinations of no adverse effect.*

(1)   Upon receipt of a Determination of No Adverse Effect from an Agency Official, the Executive Director will review the Determination and supporting documentation. Normally, the Executive Director will concur without delay. If the documentation is not adequate, the Executive Director will so inform the Agency Official within 15 days. Unless the Executive Director objects to the Determination within 30 days after receipt of an adequately documented Determination, the Agency Official will be considered to have satisfied the agency's responsibilities under Section 106 of the Act, Section 2(b) of the Executive Order, and these regulations, and may proceed with the undertaking.

(2)   If the Executive Director objects to a Determination of No-Adverse Effect, the Executive Director shall specify the basis for the objection and may specify conditions which will eliminate the objection. As appropriate, the Executive Director may consult the Agency Official, the State Historic Preservation Officer, and other interested parties in specifying conditions. If the Agency Official accepts the conditions in writing, the conditions will be incorporated into the agency's Determination and the Executive Director's objection will be withdrawn. The Agency Official then will be considered to have satisfied the agency's responsibilities under Section 106 of the Act, Section 2(b) of the Executive Order, and these regulations, and may proceed with the undertaking.

(3)   If the Agency Official does not accept the Executive Director's conditions of if the Executive Director objects to a determination of No Adverse Effect without specifying conditions that would remove the objection, the Executive Director shall initiate the consultation process pursuant to § 800.6(t).

(b)   *Consultation Process.* The Agency Official, the State Historic Preservation Officer, and the Executive Director shall be the consulting parties to consider feasible and prudent alternatives to the undertaking that could avoid, mitigate, or minimize adverse effects on a National Register ro eligible property. When an undertaking involves more than one Federal agency, these agencies may, upon notification to the Executive Director, coordinate their consultation responsibilities through a single lead agency. Grantees, permittees, licensees, or other parties of interest, and representatives of national, State, or local units of government and public and private organizations, may be invited by the consulting parties to participate in the consultation process.

(1)   *Preliminary Case Report.* The Agency Official shall provide copies

of the report to the consulting parties at the initiation of the consultation and make
it readily available for public inspection.

(2)   *On-site inspection.* At the request of any of the consulting parties,
the Agency Official shall conduct an on-site inspection.

(3)   *Public Information Meeting.* At the request of any of the consulting
parties, the Executive Director shall conduct a meeting open to the public, where
representatives of national, State, or local units of government, representatives of
public or private organizations, and interested citizens may receive information and
express their views on the undertaking, its effects on the National Register or eligible property, and alternate courses of action that could avoid, mitigate, or minimize
any adverse effects on such properties. The Agency Official shall provide adequate
facilities for the meeting near the site of the undertaking and shall afford appropriate notice to the public, generally at least 15 days in advance of the meeting.

(4)   *Consideration of Alternatives.* Upon review of the proposed undertaking and after any on-site inspection or public information meeting, the consulting parties shall determine whether there are feasible and prudent alternatives to
avoid the adverse effects on National Register or eligible property. If the consulting
parties cannot agree on an alternative to avoid, they shall consult further to determine if there are alternatives that could satisfactorily mitigate the adverse effects.

(5)   *Avoidance or Satisfactory Mitigation of Adverse Effect.* If the consulting parties agree upon a feasible and prudent alternative to avoid or satisfactorily mitigate the adverse effects of the undertaking on the National Register or
eligible property, they shall execute a Memorandum of Agreement in accordance
with § 800.6(c) specifying how the undertaking will proceed to avoid or mitigate
the adverse effect.

(6)   *Acceptance of Adverse Effect.* If the consulting parties determine
that there are no feasible and prudent alternatives that could avoid or satisfactorily
mitigate the adverse effects and agree that it is in the public interest to proceed
with the proposed undertaking, they shall execute a Memorandum of Agreement
in accordance with § 800.6(c) acknowledging this determination and specifying
any recording, salvage, or other measures to minimize the adverse effects that shall
be taken before the undertaking proceeds.

(7)   *Failure to Agree.* Upon the failure of the consulting parties to agree
upon the terms for a Memorandum of Agreement, or upon notice of such failure
by any of the consulting parties to the Executive Director, the Executive Director
shall notify the Chairman within fifteen days and shall recommend whether or not
the matter should be scheduled for consideration at a Council meeting. The Agency
Official and the State Historic Preservation Officer shall be notified in writing of
the Executive Director's recommendation.

(c)   *Memorandum of Agreement.*

(1)   *Preparation of Memorandum of Agreement.* It shall be the responsibility of the Executive Director to prepare each Memorandum of Agreement
required under these regulations. Unless otherwise requested by the Executive
Director, the Agency Official shall prepare a proposal for inclusion in the Agreement that details the actions agreed upon by the consulting parties to be taken
to avoid, satisfactorily mitigate, or accept the adverse effects on the property.
The State Historic Preservation Officer's written concurrence shall be included
in this proposal by the Agency Official. If the Executive Director determines that
the proposal represents the agreement of the consulting parties, he shall within 10

days forward it as a Memorandum of Agreement to the Chairman for ratification pursuant to § 800.6(c)(2). If the Executive Director determines that the proposal does not adequately represent the agreement reached by the consulting parties, it may be returned to the Agency Official, or a Memorandum of Agreement revising the proposal may be submitted to the Agency Official and the State Historic Preservation Officer. As appropriate other parties in interest may be invited by the consulting parties to indicate their concurrence with the proposal or to be a signatory to the Agreement.

(2) *Review of Memorandum of Agreement.* Upon receipt of an executed Memorandum of Agreement, the Chairman shall institute a 30-day review period. Unless the Chairman notifies the Agency Official that the matter has been placed on the agenda for consideration at a Council meeting, the Agreement shall become final when ratified by the Chairman or upon the expiration of the 30-day review period with no action taken. Copies will be provided to signatories and notice of executed Memoranda of Agreement shall be published in the Federal Register. The Memorandum of Agreement should be included in the final environmental impact statement prepared pursuant to the National Environment Policy Act.

(3) *Effect of Memorandum of Agreement.* Agreements duly executed in accordance with these regulations shall constitute the comments of the Council and shall evidence satisfaction of the Federal agency's responsibilities for the proposed undertaking under Section 106 of the Act, Section 2(b) of the Executive Order, and these regulations. Failure to carry out the terms of a Memorandum of Agreement requires that the Federal agency again request the Council's comments in accordance with these regulations. In such instances, until the Council issues its comments under these regulations the Agency Official shall not take or sanction any action or make any irreversible or irretrievable commitment that could result in an adverse effect with respect to National Reigster or eligible properties covered by the Agreement or that would foreclose the Council's consideration of modifications or alternatives to the proposed undertaking that could avoid or mitigate the adverse effect.

(4) *Amendment of a Memorandum of Agreement.* If a signatory determines that the terms of the Memorandum of Agreement cannot be met or believes a change is necessary, the signatory shall immediately request the consulting parties to consider an amendment of the Agreement. Amendments will be executed in the same manner as the original Agreement.

(5) *Report on Memorandum of Agreement.* Within 90 days after carrying out the terms of the Agreement, the Agency Official shall report to all signatories on the actions taken.

(d) *Council Meetings.* The Council does not hold formal administrative hearings to develop its comments under these regulations. Reports and statements will be presented to the Council in open session in accordance with a prearranged agenda. Regular meetings of the Council generally occur quarterly.

(1) *Response to Recommendation for Consideration at Council Meeting.* Upon receipt of a notice and recommendation from the Executive Director concerning consideration of a proposed undertaking at a Council meeting, the Chairman shall determine within 15 days whether or not the undertaking will be considered and shall notify the Executive Director, the Agency Official, and the State Historic Preservation Officer of his decision. The Agency Official shall and the State

Historic Preservation Officer should provide such reports and information as may be required to assist the Chairman in this determination.

If the Chairman decides against consideration of the undertaking at a Council meeting, a written summary of the undertaking, any recommendations for action by the Federal agency, and the decision shall be sent to each member of the Council. The Chairman shall also notify the Agency Official and the State Historic Preservation Officer and other parties in interest of the decision. If three members of the Council object within 10 days of the Chairman's decision, the undertaking shall be scheduled for consideration at a Council meeting. Unless three members of the Council object, the Chairman shall notify the Agency Official, the State Historic Preservation Officer, and other parties in interest in writing that the undertaking may proceed. Such notice shall be evidence of satisfaction of the Federal agency's responsibilities for the proposed undertaking under Section 106 of the Act, Section 2(b) of the Executive Order, and these regulations.

(2) *Decision to Consider the Undertaking.* When the Council will consider an undertaking at a meeting, the Chairman shall either designate five members as a panel to hear the matter on behalf of the full Council, or schedule the matter for consideration by the full Council.

(i) A panel shall consist of three non-Federal members, one as Chairman; and two Federal members, neither of whom shall represent the Federal agency involved in the undertaking. The panel shall meet to consider the undertaking within 30 days of the Chairman's decision unless the Agency Official agrees to a longer time.

(ii) The full Council will consider an undertaking at the next regularly scheduled meeting and no less than 60 days from the date of the Chairman's decision. In exceptional cases the Chairman may schedule the matter for consideration at a special meeting of the full Council to be held less than 60 days from the date of the decision.

(iii) Prior to any panel or full Council consideration of a matter, the Chairman will notify the Agency Official and the State Historic Preservation Officer, and other parties in interest of the date on which the undertaking will be considered. The Executive Director, the Agency Official, and the State Historic Preservation Officer shall prepare reports in accordance with § 800.13. Reports required from the Agency Official and the State Historic Preservation Officer must be received by the Executive Director at least 21 days before any meeting. Failure by the Federal agency to submit its report may result in postponement of consideration of the undertaking.

(3) *Meeting Notice.* Generally, 21 days notice of all meetings involving Council review of undertakings in accordance with these regulations shall be given by publication in the Federal Register. In exceptional cases, no less than 7 days notice shall be given by publication in the Federal Register.

(4) *Statements to the Council.* An agenda shall provide for oral statements from the Executive Director; the Agency Official; other parties in interest; the Secretary of the Interior; the State Historic Preservation Officer; representatives of national, State, or local units of government, and interested public and private organizations and individuals. Parties wishing to make oral remarks should notify the Executive Director at least two days in advance of the meeting. Parties wishing to have their statements distributed to Council members prior to the meeting

should send copies of the statements to the Executive Director at least 7 days in advance.

(5) *Comments of the Council.* The written comments of the Council will be issued within 15 days after a meeting. Comments shall be made to the head of the Federal agency requesting comment or having responsibility for the undertaking. Immediately after the comments are made to the Federal agency, the comments of the Council will be forwarded to the President and Congress as a special report under authority of Section 202(b) of the Act and a notice of availability will be published in the Federal Register. The comments of the Council shall be available to the State Historic Preservation Officer, other parties in interest, and the public upon receipt of the comments by the head of the Federal Agency. The comments of the Council should be included in the final environmental impact statement prepared pursuant to the National Environmental Policy Act.

(6) *Review of Panel Decision.* Upon receipt of the panel's comments after a meeting, the head of the Federal agency shall take these comments into account in reaching a decision in regard to the proposed undertaking. If the agency determines not to follow the panel's comments, the Agency Official shall immediately provide written notice of this decision to the Council. The Chairman may convene a meeting of the full Council to consider the matter within 30 days of receipt of such notice. In the interim period the Agency Official shall not take or sanction any action or make any irreversible or irretrievable commitment that could result in an adverse effect on the National Register or eligible property or that would foreclose the Council's consideration of modifications or alternatives to the proposed undertaking that could avoid or mitigate the adverse effect. If the Chairman decides against consideration of the proposed undertaking, the consulting parties shall be immediately notified and the undertaking may proceed.

(7) *Agency Action in Response to Council Comments.* Upon receipt of the Council's comments after a meeting, the head of the Federal agency shall take these comments into account in reaching a final decision in regard to the proposed undertaking. When a final decision regarding the proposed undertaking is reached by the Federal agency, the Agency Official shall submit a written report to the Council describing the actions taken by the Federal Agency in response to the Council's comments; the actions taken by other parties pursuant to the actions of the Federal Agency; and the effect that such actions will have on the affected National Register or eligible property. Receipt of this Report by the Chairman shall be evidence that the agency has satisfied its responsibilities for the proposed undertaking under Section·106 of the Act, Section 2(b) of the Executive Order and these regulations. The Council may issue a final report to the President and Congress under authority of Section 202(b) of the Act describing the actions taken by the agency in response to the Council's comments including recommendations for changes in Federal policy and programs, as appropriate.

(8) *Continuing Review Jurisdiction.* When the Council has met and commented upon an undertaking that will require subsequent site-specific undertakings by a Federal agency, the Council's comment extends only to the undertaking as reviewed. The Agency Official shall ensure that subsequent actions related to the undertaking that have not been considered by the Council will be submitted to the Council for review in accordance with these regulations.

§ 800.7. Resources discovered during construction.

(a)   *Federal Agency Responsibilities.* If a Federal agency has previously met its responsibilities for identified National Register and eligible properties under Section 106 of the Act, Section 2(b) of the Executive Order, these regulations, and the National Environmental Policy Act (42 U.S.C. 4321 et seq.), and an Agency Official finds or is notified after construction has started that an under-taking will have an effect on a previously unidentified National Register or eligible property, the Federal agency may fulfill its responsibilities under Section 106 of the Act, Section 2(b) of the Executive Order, and these regulations, by complying with the requirements of the Archeological and Historic Preservation Act (16 U.S.C. 469(a)) as implemented by the Secretary, unless the Secretary determines that the significance of the property, the effect, and any proposed mitigation actions warrant Council consideration. If the Secretary determines the Council's comments are warranted, the Agency Official shall request the com-ments of the Council.

(b)   *Council Comments.* Within 30 days of receipt of a request for comments from an Agency Official under this section, the Executive Director, with the con-currence of the Chairman, shall transmit comments on behalf of the Council to the Agency Official or the Chairman shall convene a meeting of the Council pursu-ant to § 800.6.

Federal Program Coordination

§ 800.8. Programmatic Memorandum of Agreement.

(a)   *Application.* At the request of an Agency Official, the Council will con-sider execution of a Programmatic Memorandum of Agreement to fulfill an agen-cy's responsibilities under Section 106 of the Act and Section 2(b) of the Executive Order for a particular program or class of undertakings that would otherwise require numerous individual requests for comments under these regulations. Within 30 days after the request, the Executive Director will notify the agency official whether a Programmatic Memoranda of Agreement may be used. Generally, Programmatic Memorandum of Agreement may be used in the following types of situations.

(1)   Non-site-specific undertakings, including Federal approval of State plans pursuant to Federal legislation, development of comprehensive or area-wide plans, agency recommendations for legislation, and the establishment or modifica-tion of regulations and planning guidelines.

(2)   Undertakings that are repetitive in nature and have essentially the same effect on National Register or eligible properties.

(3)   Programs that are designed to further the preservation and enhance-ment of National Register or eligible properties.

(4)   Programs with statutory time limits for project application and approval that would not permit compliance with these regulations in the normal manner.

(b)   *Consultation Process.* Upon determination by the Executive Director that a Programmatic Memorandum of Agreement is appropriate, the Agency Official and the Executive Director shall consult to develop a Programmatic Memorandum of Agreement. When the Agreement will affect a particular State or States, the appropriate State Historic Preservation Officer may be a party to the consultation. When the Agreement involves issues national in scope, the President of the National Conference of State Historic Preservation Officers or a designated representative may be a party to the consultation. The Executive Director may invite other parties, including other Federal agencies with responsibilities which may be affected by the Agreement, to participate in the consultation and may hold a Public Information Meeting (see § 800.6(b)(3)) on the proposed Agreement.

(c)   *Preparation of the Agreement.* It shall be the responsibility of the Executive Director to prepare each Agreement. At least 30 days before executing an Agreement, the Council shall publish notice of the proposed Agreement in the Federal Register inviting comments from Federal, State, and local agencies and the public. The Council will make copies available to interested parties and to appropriate A-95 clearinghouses.

(d)   *Execution of the Agreement.* After consideration of comments received and completion of any necessary revisions, the Executive Director, the Agency Official, and other parties, if appropriate, shall sign the Agreement and it shall be sent to the Chairman for ratification.

(e)   *Chairman's Review.* Upon receipt of a signed Agreement, the Chairman shall review the Agreement and within 30 days shall take one of the following actions:

(1)   Ratify the Agreement, at which time it will take effect.

(2)   Submit the Agreement to the full Council for approval.

(3)   Disapprove the Agreement.

(f)   *Effect of the Agreement.* An approved Programmatic Memorandum of Agreement shall constitute the comments of the Council on all individual undertakings carried out pursuant to the terms of the Agreement and, unless otherwise provided by the Agreement, shall satisfy the agency's responsibilities under Section 106 of the Act, Section 2(b) of the Executive Order, and these regulations for all undertakings carried out in accordance with the Agreement.

(g)   *Notice.* Notice of an approved Programmatic Memorandum of Agreement shall be published by the Council in the Federal Register. Copies shall be distributed through appropriate A-95 clearinghouses and the consulting parties shall make copies readily available to the public. The Programmatic Memorandum of Agreement should be included in the final environmental impact statement prepared pursuant to the National Environmental Policy Act.

(h)   *Term.* Unless otherwise provided by the Agreement, duly executed Programmatic Memorandum of Agreement shall remain in effect until revoked by any one of the signatories. The Agency Official shall submit a report annually to the Executive Director and other signatories on all actions taken pursuant to the Agreement, including any recommendations for modification or termination of the Agreement. The Executive Director and other signatories shall review the report and determine whether modification or termination of the Agreement is appropriate.

§ 800.9. Coordination with agency requirements under the National Environmental Policy Act (42 U.S.C. 4321 et seq.).

Section 101(b)(4) of the National Environmental Policy Act (NEPA) declares that one objective of national environmental policy is to "preserve important historic, cultural, and natural aspects of our national heritage and maintain, wherever possible, an environment which supports diversity and variety of individual choice." In order to meet this objective, Federal agencies should coordinate NEPA compliance with the separate responsibilities of the National Historic Preservation Act and Executive Order 11593 to ensure that historic and cultural properties are given proper consideration in the preparation of environmental assessments and environmental impact statements. Agency obligations pursuant to the National Historic Preservation Act and Executive Order 11593 are independent from NEPA requirements and must be complied with even when an environmental impact statement is not required. Agencies should also be aware that the threshold for compliance with Section 106 and the Executive Order is less than that for preparation of an environmental impact statement. The former applies to any Federal, federally assisted or federally licensed undertaking having an effect on a National Register or eligible property, while the latter extends only to major Federal actions significantly affecting the human environment. Where both NEPA and the Act or Executive Order are applicable, the Council on Environmental Quality, in its National Environmental Policy Act Regulations (40 CFR 1502.25), directs that draft environmental impact statements prepared under Section 102(2)(C) of NEPA shall, to the fullest extent possible, be prepared with and integrated with other environmental impact analyses and related surveys and studies required by other authorities such as the National Historic Preservation Act and Executive Order 11593. Preparation of a draft environmental impact statement may fulfill the requirements for reports and documentation under these authorities.

Circulation of the statement for comment pursuant to Section 102(2)(C) of NEPA shall constitute a request for Council comments under § 800.4 of these regulations if Federal agencies so request in cover letters circulated with draft environmental impact statements. To coordinate the independent responsibilities of the Act and NEPA, Federal agencies should undertake compliance with these regulations whenever National Register or eligible properties may be affected by an undertaking. The following subsections indicate the appropriate means of coordinating the substance and timing of agency compliance with NEPA, Section 106, and Section 2(b). The Council will review agency environmental impact statements in accordance with this section. Adherence to these provisions will provide Federal agencies with an adequate record of the consideration of National Register and eligible properties during the planning process and will facilitate the production of a single document to meet the requirements of NEPA, Section 106, Executive Order 11593, and these regulations.

(a)   It is normally intended that the Section 106/Executive Order commenting period run concurrently with the NEPA review process. Initiation of the consideration of historic and cultural resources should coincide with the initiation of other environmental reviews. To the maximum extent possible, agencies should reflect the status of compliance with Section 106, the Executive Order, and these regulations in all documents prepared under NEPA (environmental assessments, draft environmental impact statements, and final environmental impact statements) to provide the public with the fullest and most complete information available on effects on historic and cultural resources and alternatives to reduce those effects. If the commenting process under Section 106 and the Executive Order is not completed before the final environmental impact statement is issued, as with undertakings where subsequent design stage reviews occur, agencies should include the council's comments in any supplemental statement that is prepared pursuant to NEPA.

(b)   Federal agencies should initiate compliance with Section 106 of the Act and the Executive Order in accordance with these regulations during initial environmental assessments that are undertaken to meet the requirements of NEPA and and agency environmental procedures. In any event, this should occur no later than during the preparation of the draft environmental impact statement. Identification of National Register and eligible properties should be carried out in accordance with § 800.4 of these regulations. Potential effects should then be evaluated in accordance with the Criteria of Effect and Adverse Effect in § 800.3 of these regulations. The environmental assessment and the draft environmental impact statement should fully describe any National Register or eligible properties within the area of the undertaking's potential environmental impact and the nature of the undertaking's effect on them.

(c)   If evaluation of the effect resulted in a Determination of No Effect or No Adverse Effect under § 800.4, that finding, along with supporting documentation, should be included or referenced in the environmental assessment and the draft environmental impact statement.

(d)   If evaluation of the effect resulted in a Determination of Adverse Effect, that finding and a copy of the agency's request for the Council's comments in accordance with § 800.4(d)(1) of these regulations should be included in or referenced in the environmental impact statement. Agencies should include all available relevant information on National Register and eligible properties, the effects of the undertaking and alternative courses of action so that the draft environmental impact statement can be submitted as the preliminary case report under § 800.13(b) of these regulations. In some instances, the Section 106/Executive Order commenting process will be completed prior to issuance of a draft environmental impact statement. In that event, the comments of the Council should be included in the draft.

(e)   Completion of the Council commenting process in accordance with these regulations should precede issuance of the final environmental impact statement. Comments of the Council obtained pursuant to § 800.6 or § 800.8 of these regulations should be incorporated into the final statement.

(f)   The Council, in its review of environmental impact statements for undertakings that affect National Register or eligible properties, will look for evidence of proper compliance with Section 106 of the Act, Section 2(b) of the Executive Order, and these regulations. The Council's views on the agency's compliance with those authorities will be included in its comments on environmental impact statements.

§ 800.10. Coordination with the Presidential Memorandum on Environmental Quality and Water Resources Management.

Federal Agencies with water resources responsibilities shall, not later than three months after publication of these regulations as finally adopted in the Federal Register, publish procedures to implement these regulations as required by the Presidential Memorandum on Environmental Quality and Water Resources Management. Each agency shall consult with the Council while developing its procedures and shall provide an opportunity for public review and comment on their proposed regulations. Agency procedures shall be effective when the Chairman approves them as conforming to the Presidential Memorandum and these regulations. Agency procedures must at a minimum include acceptable measures to prevent or mitigate losses of historic or cultural resources and provisions to insure that all projects not yet constructed will comply with these regulations. Additionally, such procedures shall prescribe a clear way to identify funding for environmental mitigation in an agency's appropriation requests. The procedures shall be approved by the Chairman within 60 days if they are consistent with these regulations. Once in effect they shall be filed with the Council and made readily available to the public. Agencies are also encouraged to publish explanatory guidance for the procedures.

§ 800.11. Counterpart regulations.

Individual Federal agencies may, in accordance with Section 1(3) of the Executive Order, the President's Memorandum on Environmental Quality and Water Resources Management, and these regulations, choose to adopt counterpart regulations related to their specific programs and authorities to assist in meeting their responsibilities under Section 106 of the Act and Section 2(b) of the Executive Order.

(a)   Responsibilities of individual Federal agencies pursuant to § 800.4 may be met by counterpart regulations jointly drafted by that agency and the Executive Director and approved by the Chairman. The Federal agency shall provide ample opportunity for public participation in the development of such counterpart regulations, including publication in the Federal Register as proposed and final rule making with provision for a minimum 60 day period for public comment. Once in effect such counterpart regulations may, as appropriate, supercede the requirements of § 800.4. The Federal agency shall file approved counterpart regulations with the Council and shall make them readily available to the public.

(b)   Counterpart regulations may include:

(1)   A definition of undertaking as it applies to that agency's particular activities and programs.

(2)   Methods to identify National Register and eligible properties for each class of undertakings.

(3)   Methods to evaluate effects on National Register or eligible properties.

(4)   Authorization for non-Federal participation in the consultation process, and

(5)   Standards, guidelines and other measures to ensure avoidance or

mitigation of adverse effects on National Register and eligible properties for each class of undertakings.

(c)   To the maximum extent possible, counterpart regulations developed pursuant to this section should be integrated with agency regulations for the National Environmental Policy Act.

Other Provisions

§ 800.12. Investigation of threats to National Register and eligible properties.

(a)   The Council is frequently advised by State Historic Preservation Officers and others of undertakings that threaten National Register or eligible properties and that appear to involve a Federal agency. In order to protect these properties, the Executive Director investigates these matters, generally by writing to the Federal agency that appears to be involved in the undertaking. Federal agencies should respond to these inquiries within 30 days. If there is Federal involvement in the undertaking, the agency shall fulfill its responsibilities under these regulations.

(b)   The Council will exercise its authority to comment to Federal agencies under these regulations in certain special situations even though written notice that an undertaking will have an adverse effect has not been received.

§ 800.13. Reports to the Council.

In order to meet responsibilities under these regulations, the Council prescribes that certain reports and documents be made available ot it. The content of such reports is set forth below. The purpose is to provide sufficient information for the Council to evaluate the significance of affected National Register and eligible properties, understand the objectives and requirements of the undertaking, assess the effect in terms of the criteria specified in these regulations, and analyze the feasibility and prudence of alternatives. The Council further recognizes that the Act requires that National Register and eligible properties should be preserved "as a living part of our community life and development, "and considers those elements in an undertaking that have relevance beyond historical and cultural concerns. To assist it in weighing the public interest, the Council seeks information not only bearing upon physical, esthetic, or environmental effects but also information concerning economic, social, and other benefits or detriments that will result from the undertaking. Agencies should consider these reports in the context of their compliance with the National Environmental Policy Act and incorporate their content in environmental assessments, draft environmental impact statements and final environmental impact statements as specified in § 800.9.

(a)   *Documentation for Determination of No Adverse Effect.* Adequate documentation of a Determination of No Adverse Effect pursuant to § 800.4 should include the following information:

(1)   A description of the agency's involvement with the proposed undertaking with citations of the agency's program authority and applicable implementing regulations, procedures, and guidelines.

(2)   A description of the proposed undertaking including, as appropriate, photographs, maps, drawings, and specifications;

(3)   A list of National Register and eligible properties that will be affected by the undertaking, including a description of the property's physical appearance and significance;

(4)   A brief statement explaining why each of the Criteria of Adverse Effect (see § 800.3) was found inapplicable;

(5)   Written views of the State Historic Preservation Officer concerning the Determination of No Adverse Effect, if available; and,

(6)   An estimate of the cost of the undertaking, identifying Federal and non-Federal shares.

(b)   *Preliminary Case Reports.* Preliminary Case Reports should be submitted with a request for comments purusant to § 800.4 and should include the following information:

(1)   A description of the agency's involvement with the proposed under-taking with citations of the agency's program authority and applicable implement-ing regulations, procedures, and guidelines;

(2)   The status of this project in the agency's approval process;

(3)   The status of this project in the agency's National Environmental Policy Act compliance process and the target date for completion of all environ-mental responsibilities;

(4)   A description of the proposed undertaking including, as appropriate, photographs, maps, drawings, and specifications;

(5)   A description of the National Register or eligible properties affected by the undertaking, including a description of the properties'physical appearance and significance;

(6)   A brief statement explaining why any of the Criteria of Adverse Effect (see § 800.3) apply;

(7)   Written views of the State Historic Preservation Officer concerning the effect on the property, if available.

(8)   The views of other Federal agencies, State and local governments, and the other groups or individuals, when known;

(9)   A description and analysis of alternatives that would avoid the adverse effects;

(10)   A description and analysis of alternatives that would mitigate the adverse effects; and,

(11)   An estimate of the cost of the underaking, identifying Federal and non-Federal shares.

(c)   *Reports for Council Meeting.* Consideration of an undertaking by either the full Council or a panel pursuant to § 800.6 is based on reports from the Execu-tive Director, the Agency Official, the Secretary of the Interior, the State Historic Preservation Officer, and others. The reports consist of the following:

(1)   *Secretary of the Interior's Report.* The report from the Secretary shall include a verification of the legal and historical status of the property and an assessment of the historical, architectural, archeological, or cultural significance of the property.

(2)   *Agency Official's Report.* The report from the Agency Official re-questing comments shall include a general discussion and chronology of the pro-posed undertaking; an account of the steps taken to comply with the National

Environmental Policy Act (NEPA); any relevant supporting documentation in studies that the agency has completed; an evaluation of the effect of the undertaking upon the property, with particular reference to the impact on the historical architectural, archeological, and cultural values; steps taken or proposed by the agency to avoid or mitigate adverse effects of the undertaking; a thorough discussion of alternate courses of action; and an analysis comparing the advantages resulting from the undertaking with the disadvantages resulting from the adverse effects on National Register or eligible properties. The Agency Official shall arrange for the submission and presentation of any report by a grantee, permittee, licensee, or other party receiving Federal assistance or approval to carry out the undertaking.

(3) *Other Federal Agency Reports.* A report from any other Federal agency involved in the undertaking or a related action that affects the property in question, including a general description and chronology of that agency's involvement and its relation to the undertaking being considered by the Council.

(4) *State Historic Preservation Officer's Report.* A report from the State Historic Preservation Officer should include an assessment of the significance of the property within the State preservation program; an evaluation of the effect of the undertaking upon the property and its specific components; an evaluation of known alternate courses of action; a discussion of present or proposed participation of State and local agencies or organizations in preserving or assisting in preserving the property; an indication of the support or opposition of units of government and public and private agencies and organizations within the State; and the recommendation of the State Historic Preservation Officer.

(5) *Executive Director's Report.* A report from the Executive Director shall include a description of the actions taken pursuant to these regulations, an evaluation of the effect of the undertaking on the property, a review of any known alternate courses of action, an analysis comparing the advantages resulting from the undertaking with the disadvantages resulting from the adverse effects on National Register or eligible properties and recommendations for Council action.

(6) *Other Reports.* The Council will consider other pertinent reports, statements, correspondence, transcripts, minutes, and documents received from any and all parties, public or private. Reports submitted pursuant to this section should be received by the Council at least 7 days prior to a Council meeting.

§ 800.14. Supplementary guidance.

The Executive Director may issue further guidance to interpret these regulations to assist Federal agencies and State Historic Preservation Officers in meeting their responsibilities. The guidelines are for informational purposes only and will be published in the Federal Register and will be readily available to the public.

§ 800.15. Public participation.

The Council encourages maximum public participation in the review process under these regulations. The Council, Federal agencies, and State Historic Preservation Officers should seek assistance from the public including other Federal agencies, units of local and State government, public and private organizations, individuals and federally recognized Indian tribes in evaluating National Register and eligible properties, determining effect, and developing alternatives to avoid or mitigate an adverse effect. The public has considerable information available that could assist Federal agencies, the State Historic Preservation Officer and the Council in meeting their responsibilities under these regulations. The Council especially urges that Federal agencies make every effort to involve grantees, permittees, licensees, and other parties in interest in the consultation process. To this end, the Council, the Agency Official, and the State Historic Preservation Officer should:

(a)   Make readily available, to the extent possible, documents, materials and other information and data concerning the undertaking and effects on National Register and eligible properties that may be of interest to the public. Such information should be made available within the limits of the Freedom of Information Act (5 U.S.C. 552) and need not necessarily include information on budget, financial, personnel, and other proprietary matters or the specific location of archeological sites. Material to be made available to the public by the agency and the State Historic Preservation Officer should be provided to the public at the minimum cost permissible.

(b)   Make the public aware of Public Information Meetings (§ 800.6(b)(3)), full or panel Council meetings (§ 800.6(d)), and the availability of other information related to the review process under these regulations such as a Determination of No Effect, a Determination of No Adverse Effect, a Memorandum of Agreement (see § 800.8). The purpose of such notice is to inform persons, agencies, and organizations that may be interested or affected by the proposed undertaking of the opportunity to participate in the review process under these regulations, This may include:

(1)   Mailing notice to those who have requested it on an individual undertaking or Programmatic Memorandum of Agreement;

(2)   Use of notice in local newspaper, local media, and newletters that may be expected to reach potentially interested persons;

(3)   Posting of notice on and off-site in the area where the undertaking is proposed to be located.

(c)   Solicit relevant information from the public during the identification of National Register and eligible properties, the evaluation of effects, and the consideration of alternatives.

(d)   Hold or sponsor public meetings on proposed undertakings and make diligent effort to include the public.

Appendix K

New York City Administrative Code Chapter 8-A, "Preservation of Landmarks and Historical Districts"

§ 205–1.0 Purpose and Declaration of Public Policy

(a)   The council finds that many improvements, as herein defined, and landscape features, as herein defined, having a special character or a special historical or aesthetic interest or value and many improvements representing the finest architectural products of distinct periods in the history of the city, have been uprooted, notwithstanding the feasibility of preserving and continuing the use of such improvements and landscape features, and without adequate consideration of the irreplaceable loss to the people of the city of the aesthetic, cultural and historic values represented by such improvements and landscape features. In addition distinct areas may be similarly uprooted or may have their distinctiveness destroyed, although the preservation thereof may be both feasible and desirable. It is the sense of the council that the standing of this city as a world-wide tourist center and world capital of business, culture and government cannot be maintained or enhanced by disregarding the historical and architectural heritage of the city and by countenancing the destruction of such cultural assets.

(b)   It is hereby declared as a matter of public policy that the protection, enhancement, perpetuation and use of improvements and landscape features of special character or special historical or aesthetic interest or value is a public necessity and is required in the interest of the health, prosperity, safety and welfare of the people. The purpose of this chapter is to (a) effect and accomplish the protection, enhancement and perpetuation of such improvements and landscape features and of districts which represent or reflect elements of the city's cultural, social, economic, political and architectural history; (b) safeguard the city's historic, aesthetic and cultural heritage, as embodied and reflected in such improvements, landscape features and districts; (c) stabilize and improve property values in such districts; (d) foster civic pride in the beauty and noble accomplishments of the past; (e) protect and enhance the city's attractions to tourists and visitors and the support and stimulus to business and industry thereby provided; (f) strengthen the economy of the city; and (g) promote the use of historic districts, landmarks, interior landmarks and scenic landmarks for the education, pleasure and welfare of the people of the city. (Amended by L. L. 1973, No. 71, Dec. 17.)

§ 207—1.0 Definitions

As used in this chapter, the following terms shall mean and include:

(a)   *"Alteration."* Any of the acts defined as an alteration by the building code of the city.

(b)   *"Appropriate protective interest."* Any right or interest in or title to an improvement parcel or any part thereof, including, but not limited to, fee title and scenic or other easements, the acquisition of which by the city is determined by the commission to be necessary and appropriate for the effectuation of the purpose of this chapter.

(c)   *"Capable of earning a reasonable return."* Having the capacity, under reasonably efficient and prudent management, of earning a reasonable return. For the purposes of this chapter, the net annual return, as defined in subparagraph (a) of paragraph three of subdivision v of this section, yielded by an improvement parcel during the test year, as defined in subparagraph (b) of such paragraph, shall be presumed to be the earning capacity of such improvement parcel, in the absence of substantial grounds for a contrary determination by the commission.

(d)   *"City-aided project."* Any physical betterment of real property, which:

(1)   may not be constructed or effected without the approval of one or more officers or agencies of the city; and

(2)   upon completion, will be owned in whole or in part by any person other than the city; and

(3)   is planned to be constructed or effected, in whole or in part, with any form of aid furnished by the city (other than under this chapter), including, but not limited to, any loan, grant, subsidy or other mode of financial assistance, exercise of the city's powers of eminent domain, contribution of city property, or the granting of tax exemption or tax abatement; and

(4)   will involve the construction, reconstruction, alteration or demolition of any improvement in a historic district or of a landmark.

(e)   *"Commission."* The landmarks preservation commission.

(f)   *"Day."* Any day other than a Saturday, Sunday or legal holiday; provided, however, that for the purposes of subdivision d of section 207—16.0 of this chapter, the term "day" shall mean every day in the week.

(g)   *"Exterior architectural feature."* The architectural style, design, general arrangement and components of all of the outer surfaces of an improvement, as distinguished from the interior surfaces enclosed by said exterior surfaces, including, but not limited to, the kind, color and texture of the building material and the type and style of all windows, doors, lights, signs and other fixtures appurtenant to such improvement.

(h)   *"Historic district."* Any area which:

(1)   contains improvements which:

(a)   have a special character or special historical or aesthetic interest or value; and

(b)   represent one or more periods or styles of architecture typical of one or more eras in the history of the city; and

(c)   cause such area, by reason of such factors, to constitute a distinct section of the city; and

(2)   has been designated as a historic district pursuant to the provisions of this chapter.

(i)   *"Improvement."* Any building, structure, place, work of art or other object constituting a physical betterment of real property, or any part of such betterment.

(j)   *"Improvement parcel."* The unit of real property which (1) includes a physical betterment constituting an improvement and the land embracing the site thereof, and (2) is treated as a single entity for the purpose of levying real estate taxes; provided however, that the term "improvement parcel" said* also include any unimproved area of land which is treated as a single entity for such tax purposes.

(k)   *"Interior."* The visible surfaces of the interior of an improvement.

(l)   *"Interior architectural feature."* The architectural style, design, general arrangement and components of an interior, including but not limited to, the kind, color and texture of the building material and the type and style of all windows, doors, lights, signs and other fixtures appurtenant to such interior.

(m)   *"Interior landmark."* An interior, or part thereof, any part of which is thirty years old or older, and which is customarily open or accessible to the public, or to which the public is customarily invited, and which has a special historical or aesthetic interest or value as part of the development, heritage or cultural characteristics of the city, state or nation and which has been designated as an interior landmark pursuant to the provisions of this chapter.

(n)   *"Landmark."* Any improvement, any part of which is thirty years old or older, which has a special character or special historical or aesthetic interest or value as part of the development, heritage or cultural characteristics of the city, state or nation and which has been designated as a landmark pursuant to the provisions of this chapter.

(o)   *"Landmark site."* An improvement parcel or part thereof on which is situated a landmark and any abutting improvement parcel or part thereof used as and constituting part of the premises on which the landmark is situated, and which has been designated as a landmark site pursuant to the provisions of this chapter.

(p)   *"Landscape feature."* Any grade, body of water, stream, rock, plant, shrub, tree, path, walkway, road, plaza, fountain, sculpture or other form of natural or artificial landscaping.

(q)   *"Minor work."* Any change in, addition to or removal from the parts, elements or materials comprising an improvement including, but not limited to, the exterior architectural features or interior architectural features thereof and, subject to and as prescribed by regulations of the commission if and when promulgated pursuant to section 207-18.0 of this chapter the surfacing, resurfacing, painting, renovating, restoring, or rehabilitating of the exterior architectural features or interior architectural features or the treating of the same in any manner that materially alters their appearance, where such change, addition or removal does not constitute ordinary repairs and maintenance and is of such nature that it may be lawfully effected without a permit from the department of buildings.

(r)   *"Ordinary repairs and maintenance."* Any:

(1)   work done on any improvement; or

(2)   replacement of any part of an improvement; for which a permit issued by the department of buildings is not required by law, where the purpose and effect of such work or replacement is to correct any deterioration or decay of or damage to such improvement or any part thereof and to restore same, as nearly as may be practicable, to its condition prior to the occurrence of such deterioration, decay or damage.

---

*So in original. Should probably be "shall."

(s)   *"Owner."* Any person or persons having such right to, title to or interest in any improvement so as to be legally entitled upon obtaining the required permits and approvals from the city agencies having jurisdiction over building construction, to perform with respect to such property any demolition, construction, reconstruction, alteration or other work as to which such person seeks the authorization or approval of the commission pursuant to section 207—8.0 of this chapter.

(t)   *"Person in charge."* The person or persons possessed of the freehold of an improvement or improvement parcel or a lesser estate therein, a mortgagee or vendee in possession, assignee of rents, receiver, executor, trustee, lessee, agent or any other person directly or indirectly in control of an improvement or improvement parcel.

(u)   *"Protected architectural feature."* Any exterior architectural feature of a landmark or any interior architectural feature of an interior landmark.

(v)   *"Reasonable return."*

(1)   A net annual return of six per centum of the valuation of an improvement parcel.

(2)   Such valuation shall be the current assessed valuation established by the City, which is in effect at the time of the filing of the request for a certificate of appropriateness; provided that:

(a)   The commission may make a determination that the valuation of the improvement parcel is an amount different from such assessed valuation where there has been a reduction in the assessed valuation for the year next preceding the effective date of the current assessed valuation in effect at the time of the filing of such request; and

(b)   The commission may make a determination that the value of the improvement parcel is an amount different from the assessed valuation where there has been a bona fide sale of such parcel within the period between March fifteenth, nineteen hundred fifty-eight, and the time of the filing of such request, as the result of a transaction at arms' length, on normal financing terms, at a readily ascertainable price, and unaffected by special circumstances such as, but not limited to, a forced sale, exchange of property, package deal, wash sale or sale to a cooperative. In determining whether a sale was on normal financing terms, the commission shall give due consideration to the following factors:

(1)   The ratio of the cash payment received by the seller to (a) the sales price of the improvement parcel and (b) the annual gross income from such parcel;

(2)   The total amount of the outstanding mortgages which are liens against the improvement parcel (including purchase money mortgages) as compared with the assessed valuation of such parcel;

(3)   The ratio of the sales price to the annual gross income of the improvement parcel, with consideration given, where the improvement is subject to residential rent control, to the total amount of rent adjustments previously granted, exclusive of rent adjustments because of changes in dwelling space, services, furniture, furnishings or equipment, major capital improvements, or substantial rehabilitation;

(4)   The presence of deferred amortization in purchase money mortgages, or the assignment of such mortgages at a discount;

(5)   Any other facts and circumstances surrounding such sale which, in the judgment of the commission, may have a bearing upon

the question of financing; and

    (3)   For the purposes of this subdivision v:

        (a)   Net annual return shall be the amount by which the earned income yielded by the improvement parcel during a test year exceeds the operating expenses of such parcel during such year, excluding mortgage interest and amortization, and excluding allowances for obsolescence and reserves, but including an allowance for depreciation of two per centum of the assessed value of the improvement, exclusive of the land, or the amount shown for depreciation of the improvement in the latest required federal income tax return, whichever is lower; provided, however, that no allowance for depreciation of the improvement shall be included where the improvement has been fully depreciated for federal income tax purposes or on the books of the owner; and

        (b)   Test year shall be (1) the most recent full calendar year, or (2) the owner's most recent fiscal year, or (3) any twelve consecutive months ending not more than ninety days prior to the filing (a) of the request for a certificate, or (b) of an application for a renewal of tax benefits pursuant to the provisions of section 207–8.0 of this chapter, as the case may be.

    (w)   *"Scenic landmark."* Any landscape feature or aggregate of landscape features, any part of which is thirty years or older, which has or have a special character or special historical or aesthetic interest or value as part of the development, heritage or cultural characteristics of the city, state or nation and which has been designated a scenic landmark pursuant to the provisions of this chapter. (Amended by L. L. 1973, No. 71, Dec. 17.)

§ 207–2.0 Establishment of landmarks, landmark sites, interior landmarks, scenic landmarks and historic districts

    (a)   For the purpose of effecting and furthering the protection, preservation, enhancement, perpetuation and use of landmarks, interior landmarks, scenic landmarks and historic districts, the commission shall have power, after a public hearing:

    (1)   to designate and, as herein provided in subdivision j, in order to effectuate the purposes of this chapter to make supplemental designations as additions to, a list of landmarks which are identified by a description setting forth the general characteristics and location thereof;

    (2)   to designate and, in order to effectuate the purposes of this chapter, to make supplemental designations as additions to, a list of interior landmarks, not including interiors utilized as places of religious worship, which are identified by a description setting forth the general characteristics and location thereof;

    (3)   to designate and, in order to effectuate the purposes of this chapter, to make supplemental designations as additions to a list of scenic landmarks, located on property owned by the city, which are identified by a description setting forth the general characteristics and location thereof; and

    (4)   to designate historic districts and the location and boundaries thereof, and, in order to effectuate the purposes of this chapter, to designate changes in such locations and boundaries and designate additional historic districts and the location and boundaries thereof.

    (b)   It shall be the duty of the commission, after a public hearing, to designate a landmark site for each landmark and to designate the location and boundaries of such site.

(c)   The commission shall have power, after a public hearing, to amend any designation made pursuant to the provisions of subdivisions a and b of this section.

(d)   The commission may, after a public hearing, whether at the time it designates a scenic landmark or at any time thereafter, specify the nature of any construction, reconstruction, alteration or demolition of any landscape feature which may be performed on such scenic landmark without prior issuance of a report pursuant to subdivision c of section 207—17.0. The commission shall have the power, after a public hearing, to amend any specification made pursuant to the provisions of this subdivision.

(e)   Subject to the provisions of subdivisions g and h of this section, any designation or amendment of a designation made by the commission pursuant to the provisions of subdivisions a, b and c of this section shall be in full force and effect from and after the date of the adoption thereof by the commission.

(f)   Within five days after making any such designation or amendment thereof, the commission shall file a copy of same with the secretary of the board of estimate and with the department of buildings, the city planning commission, the board of standards and appeals, the fire department and the health services administration.

(g)   (1) The secretary of the board of estimate, within five days after the filing of such copy with such secretary, shall refer such designation or amendment thereof to the city planning commission, which, within thirty days after such referral, shall submit to the board of estimate a report with respect to the relation of such designation or amendment thereof to the zoning resolution, projected public improvements and any plans for the development, growth, improvement or renewal of the area involved.

(2)   The board of estimate may modify or disapprove such designation or amendment thereof within ninety days after a copy thereof is filed with the secretary of the board provided that the planning commission has submitted the report required by this subdivision or that thirty days have elapsed since the referral of the designation or amendment to the commission by the secretary of the board. If the board shall disapprove such designation or amendment thereof, it shall cease to be in effect on the date of such action by the board. If the board shall modify such designation or amendment thereof, such modification shall be in effect on and after the date of the adoption thereof by the board.

(h)   (1) The commission shall have power, after a public hearing, to adopt a resolution proposing rescission, in whole or in part, of any designation or amendment or modification thereof mentioned in the preceding subdivisions of this section. Within five days after adopting any such resolution, the commission shall file a copy thereof with the secretary of the board of estimate, who shall, within five days after such filing, refer such resolution to the city planning commission.

(2)   Within thirty days after such referral, the city planning commission shall submit to the board of estimate a report with respect to the relation of such proposed rescission to the zoning resolution, projected public improvements and any plans for the development, growth, improvement, or renewal of the area involved.

(3)   Such board may approve, disapprove or modify such proposed rescission within ninety days after a copy of the resolution proposing same is filed with the secretary of the board. If such proposed rescission is approved or modified by the board, such rescission or modification thereof shall take effect on the date of such action by the board. If such proposed rescission is disapproved by the board, or is

not acted on by the board within such period of ninety days, it shall not take effect.

(i)   The commission may at any time make recommendations to the city planning commission with respect to amendments of the provisions of the zoning resolution applicable to improvements in historic districts.

(j)   All designations and supplemental designations of landmarks, landmark sites, interior landmarks, scenic landmarks and historic districts made pursuant to subdivision a shall be made pursuant to notices of public hearings given, as provided in section 207–12.0.

(k)   Upon its designation of any improvement parcel as a landmark and of any landmark site, interior landmark, scenic landmark or historic district or any amendment of any such designation or rescission thereof, the commission shall cause to be recorded in the office of the register of the city of New York in the county in which such landmark, interior landmark, scenic landmark or district lies, or in the case of landmarks, interior landmarks, scenic landmarks and districts in the borough and county of Richmond in the office of the clerk of said county of Richmond, a notice of such designation, amendment or rescission describing the party affected by, in the case of the county of Richmond, its land map block number or numbers, and its tax map, block and lot number or numbers, and in the case of all other counties, by its land map block and lot number or numbers. (Amended by L. L. 1977, No. 54, July 28, 1977, eff. Jan. 1, 1977.)

§ 207–3.0 Scope of commission's powers

(a)   Nothing contained in this chapter shall be construed as authorizing the commission, in acting with respect to any historic district or improvement therein, or in adopting regulations in relation thereto, to regulate or limit the height and bulk of buildings, to regulate and determine the area of yards, courts and other open spaces, to regulate density of population or to regulate and restrict the locations of trades and industries or location of buildings designed for specific uses or to create districts for any such purpose.

(b)   Except as provided in subdivision a of this section, the commission may, in exercising or performing its powers, duties or functions under this chapter with respect to any improvement in a historic district or on a landmark site or containing an interior landmark, or any landscape feature of a scenic landmark, apply or impose, with respect to the construction, reconstruction, alteration, demolition or use of such improvement, or landscape feature or the performance of minor work thereon, regulations, limitations, determinations or conditions which are more restrictive than those prescribed or made by or pursuant to other provisions of law applicable to such activities, work or use. (Subd. b amended by L. L. 1973, No. 71, Dec. 17.)

§ 207–4.0 Regulation of construction, reconstruction, alterations and demolition

(a)   (1) Except as otherwise provided in paragraph two of this subdivision a, it shall be unlawful for any person in charge of a landmark site or an improvement parcel or portion thereof located in an historic district of any part of an improvement containing an interior landmark, to alter, reconstruct or demolish any improvement constituting a part of such site or constituting a part of such parcel and located within such district or containing an interior landmark, or to construct any improvement upon land embraced within such site or such parcel and located within such district, or to cause or permit any such work to be performed on such improvement or land, unless the commission has previously issued a certificate of no effect on protected architectural features, a certificate of appropriateness or a notice to proceed authorizing such work, and it shall be unlawful for any other person to perform such work or cause same to be performed, unless such certificate or notice has been previously issued.

(2)   The provisions of paragraph one of this subdivision a shall not apply to any improvement mentioned in subdivision a of section 207−17.0 of this chapter, or to any city-aided project, or in cases subject to the provisions of section 207−11.0 of this chapter.

(3)   It shall be unlawful for the person in charge of any improvement or land mentioned in paragraph one of this subdivision a to maintain same or cause or permit same to be maintained in the condition created by any work in violation of the provisions of such paragraph one.

(b)   (1) Except in the case of any improvement mentioned in subdivision a of section 207−17.0 of this chapter and except in the case of a city-aided project, no application shall be approved and no permit or amended permit for the construction, reconstruction, alteration or demolition of any improvement located or to be located on a landmark site or in an historic district or containing an interior landmark shall be issued by the department of buildings, and no application shall be approved and no special permit or amended special permit for such construction, reconstruction or alteration, where required by article seven of the zoning resolution, shall be granted by the city planning commission or the board of standards and appeals, until the commission shall have issued either a certificate of no effect on protected architectural features, a certificate of appropriateness or a notice to proceed pursuant to the provisions of this chapter as an authorization for such work.

(c)   (1) A copy of every application or amended application for a permit to construct, reconstruct, alter or demolish any improvement located or to be located on a landmark site or in an historic district or containing an interior landmark shall, at the time of the submission of the original thereof to the department of buildings, be filed by the applicant with the commission. A copy of every application, under article seven of the zoning resolution, for a special permit for any work which includes the construction, reconstruction or alteration of any such improvement shall, at the time of the submission of such application or amended application of the city planning commission or the board of standards and appeals, as the case may be, be filed with the commission.

(2)   Every such copy of an application or amended application filed with the commission shall include plans and specifications for the work involved, or such other statement of the proposed work as would be acceptable by the department of buildings pursuant to the building code. The applicant shall furnish the commission with such other information relating to such application as the commission may from time to time require.

(3)   Together with the copies of such application, or amended application every such applicant shall file with the commission, a request for a certificate of no effect on protected architectural features or a certificate of appropriateness in relation to the proposed work specified in such application. (Amended

by L. L. 1973, No. 71, Dec. 17.)

§ 207—5.0 Determination of request for certificate of no effect on protected architectural features

(a)   (1) In any case where an applicant for a permit from the department of buildings to construct, reconstruct, alter or demolish any improvement on a landmark site or in an historic district or containing an interior landmark, or an applicant for a special permit from the city planning commission or the board of standards and appeals authorizing any such work pursuant to article seven of the zoning resolution, or amendments thereof, files, a copy of such application or amended application with the commission, together with a request for a certificate of no effect on protected architectural features, the commission shall determine: (a) whether the proposed work would change, destroy or affect any exterior architectural feature of the improvement on a landmark site or in an historic district or any interior architectural feature of the interior landmark upon which said work is to be done, and (b) in the case of construction of a new improvement, whether such construction would affect or not be in harmony with the external appearance of other, neighboring improvements on such site or in such district. If the commission determines such question in the negative, it shall grant such certificate; otherwise, it shall deny such request.

(2)   Within thirty days after the filing of such application and request, the commission shall either grant such certificate, or give notice to the applicant of a proposed denial of such request. Upon written demand of the applicant filed with the commission after the giving of notice of a proposed denial, the commission shall confer with the applicant. The commission shall determine the request for a certificate within thirty days after the filing of such demand. If a demand is not filed within ten days after the giving of notice of the proposed denial, the commission shall determine such request within five days after the expiration of such ten-day period.

(3)   In the event of a denial of such a certificate, the applicant may file with the commission a request for a certificate of appropriateness with respect to the proposed work specified in such application. (Amended by L. L. 1973, No. 71, Dec. 17.)

§ 207-6.0 Factors governing issuance of certificate of appropriateness

(a)   In any case where an applicant for a permit to construct, reconstruct, alter or demolish any improvement on a landmark site or in an historic district or containing an interior landmark, files such application with the commission together with a request for a certificate of appropriateness, and in any case where a certificate of no effect on protected architectural features is denied and the applicant thereafter, pursuant to the provisions of section 207—5.0 of this chapter, files a request for a certificate of appropriateness, the commission shall determine whether the proposed work would be appropriate for and consistent with the effectuation of the purposes of this chapter. If the commission's determination is in the affirmative on such question, it shall grant a certificate of appropriateness, and if the commission's determination is in the negative, it shall deny the applicant's request, except as otherwise provided in section 207—8.0 of this chapter.

(b)   (1) In making such determination with respect to any such application for a permit to construct, reconstruct, alter or demolish an improvement in an historic district, the commission shall consider (a) the effect of the proposed work in creating, changing, destroying or affecting the exterior architectural features of the improvement upon which such work is to be done and (b) the relationship between the results of such work and the exterior architectural features of other, neighboring improvements in such district.

(2)   In appraising such effects and relationship, the commission shall consider, in addition to any other pertinent matters, the factors of aesthetic, historical and architectural values and significance, architectural style, design, arrangement, texture, material and color.

(3)   All determinations of the commission pursuant to this subdivision b shall be made subject to the provisions of section 207—3.0 of this chapter and the commission, in making any such determination, shall not apply any regulation, limitation, determination or restriction as to the height and bulk of buildings, the area of yards, courts or other open spaces, density of population, the location of trades and industries, or location of buildings designed for specific uses, other than the regulations, limitations, determinations and restrictions as to such matters prescribed or made by or pursuant to applicable provisions of law, exclusive of this chapter; provided, however, that nothing contained in such section 207—3.0 or in this subdivision b shall be construed as limiting the power of the commission to deny a request for a certificate of appropriateness for demolition or alteration of an improvement in an historic district (whether or not such request also seeks approval, in such certificate, of construction or construction of any improvement), on the ground that such demolition or alteration would be inappropriate for and inconsistent with the effectuation of the purposes of this chapter, with due consideration for the factors hereinabove set forth in this subdivision b.

(c)   In making the determination referred to in subdivision a of this section with respect to any application for a permit to construct, reconstruct, alter or demolish any improvement on a landmark site, other than a landmark, the commission shall consider (1) the effects of the proposed work in creating, changing, destroying or affecting the exterior architectural features of the improvement upon which such work is to be done, (2) the relationship between such exterior architectural features, together with such effects, and the exterior architectural features of the landmark, and (3) the effects of the results of such work upon the protection, enhancement, perpetuation and use of the landmark on such site. In appraising such effects and relationship, the commission shall consider, in addition to any other pertinent matters, the factors mentioned in paragraph two of subdivision b of this section.

(d)   In making the determination referred to in subdivision a of this section with respect to an application for a permit to alter, reconstruct or demolish a landmark, the commission shall consider the effects of the proposed work upon the protection, enhancement, perpetuation and use of the exterior architectural features of such landmark which cause it to possess a special character or special historical or aesthetic interest or value.

(e)   In making the determination referred to in subdivision a of this section with respect to an application for a permit to alter, reconstruct or demolish an improvement containing an interior landmark, the commission shall consider the effects of the proposed work upon the protection, enhancement, perpetuation and use of the interior architectural features of such interior landmark which cause it to possess a special character or special historical or aesthetic interest or value. (Amended by L. L. 1973, No. 71, Dec. 17.)

§ 207—7.0 Procedure for determination of request for certificate of appropriateness

The commission shall hold a public hearing on each request for a certificate of appropriateness. Except as otherwise provided in section 207—8.0 of this chapter, the commission shall make its determination as to such request within ninety days after filing thereof.

§ 207—8.0 Request for certificate of appropriateness authorizing demolition, alterations or reconstruction on ground of insufficient return

(a)   (1) Except as otherwise provided in paragraph two of this subdivision a, in any case where an application for a permit to demolish any improvement located on a landmark site or in an historic district or containing an interior landmark is filed with the commission, together with a request for a certificate of appropriateness authorizing such demolition, and in any case where an application for a permit to make alterations to or reconstruct any improvement on a landmark site or containing an interior landmark is filed with the commission, and the applicant requests a certificate of appropriateness for such work, and the applicant establishes to the satisfaction of the commission that:

(a)   the improvement parcel (or parcels) which includes such improvement, as existing at the time of the filing of such request, is not capable of earning a reasonable return; and

(b)   the owner of such improvement:

(1)   in the case of an application for a permit to demolish, seeks in good faith to demolish such improvement immediately (a) for the purpose of constructing on the site thereof with reasonable promptness a new building or other income-producing facility, or (b) for the purpose of terminating the operation of the improvement at a loss; or

(2)   in the case of an application for a permit to make alterations or reconstruct, seeks in good faith to alter or reconstruct such improvement, with reasonable promptness, for the purpose of increasing the return therefrom; the commission, if it determines that the request for such certificate should be denied on the basis of the applicable standards set forth in section 207—6.0 of this chapter, shall, within ninety days after the filing of the request for such certificate of appropriateness, makes a preliminary determination of insufficient return. (Subd. a, par. 1 amended by L. L. 1973, No. 71, Dec. 17.)

(2)   In any case where any application and request for a certificate of appropriateness mentioned in paragraph one of this subdivision a is filed with the commission with respect to an improvement, the provisions of this section shall not apply to such request if the improvement parcel which includes such improvement has received, for three years next preceding the filing of such request, and at the time of such filing continues to receive, under any provision of law (other than this

chapter or sections 458, 460 or 479 of the real property tax law), exemption in whole or in part from real property taxation; provided, however, that the provisions of this section shall nevertheless apply to such request if such exemption is and has been received pursuant to Sections 420, 422, 424, 425, 426, 427, 428, 430, 432, 434, 436, 438, 440, 442, 444, 450, 452, 462, 464, 468, 470, 472 or 474 of the real property tax law and the applicant establishes to the satisfaction of the commission, in lieu of the requirements set forth in paragraph one of this subdivision a, that:

    (a)   the owner of such improvement has entered into a bona fide agreement to sell an estate of freehold or to grant a term of at least twenty years in such improvement parcel, which agreement is subject to or contingent upon the issuance of the certificate of appropriateness or a notice to proceed;

    (b)   the improvement parcel which includes such improvement, as existing at the time of the filing of such request, would not, if it were not exempt in whole or in part from real property taxation, be capable of earning a reasonable return;

    (c)   such improvement has ceased to be adequate, suitable or appropriate for use for carrying out both (1) the purposes of such owner to which it is devoted and (2) those purposes to which it had been devoted when acquired unless such owner is no longer engaged in pursuing such purposes; and

    (d)   the prospective purchasers or tenant:

        (1)   in the case of an application for a permit to demolish seeks and intends, in good faith either to demolish such improvement immediately for the purpose of constructing on the site thereof with reasonable promptness a new building or other facility; or

        (2)   in the case of an application for a permit to make alterations or reconstruct, seeks and intends in good faith to alter or reconstruct such improvement, with reasonable promptness.

    (b)   In the case of an application made pursuant to paragraph one of subdivision a of this section by an applicant not required to establish the conditions specified in paragraph two of such subdivision, as promptly as is practicable after making a preliminary determination as provided in paragraph one of such subdivision a, the commission, with the aid of such experts as it deems necessary, shall endeavor to devise, in consultation with the applicant, a plan whereby the improvement may be (1) preserved or perpetuated in such manner or form as to effectuate the purposes of this chapter, and (2) also rendered capable of earning a reasonable return.

    (c)   Any such plan may include, but shall not be limited to, (1) granting of partial or complete tax exemption, (2) remission of taxes and (3) authorization for alterations, construction or reconstruction appropriate for and not inconsistent with the effectuation of the purposes of this chapter.

    (d)   In any case where the commission formulates any such plan, it shall mail a copy thereof to the applicant promptly and in any event within sixty days after giving notice of its preliminary determination of insufficient return. The commission shall hold a public hearing upon such plan.

    (e)   (1) If the commission, after holding a public hearing pursuant to subdivision d of this section, determines that a plan which it has formulated, consisting only of tax exemption and/or remission of taxes, meets the standards set forth in subdivision b of this section, as such plan was originally formulated, or with such modifications as the commission deems necessary or appropriate, the commission shall deny the request of the applicant for a certificate of appropriateness and shall approve such plan, as originally formulated, or with such modifications.

(2)   Such plan, as so approved, shall set forth the extent of tax exemption and/or remission of taxes deemed necessary by the commission to meet such standards.

(3)   The commission shall promptly mail a certified copy of such approved plan to the applicant and shall promptly transmit a certified copy thereof to the tax commission. Upon application made by the owner of such improvement pursuant to the provisions of paragraph five of this subdivision e, the tax commission shall grant, for the fiscal year next succeeding the date of approval of such plan, the tax exemption and/or remission of taxes provided for therein.

(4)   In accordance with procedures prescribed by the regulations of the commission, it shall determine, upon application by the owner of such improvement made in advance of each succeeding fiscal year, the amount of tax exemption and/or remission of taxes, if any, which it deems necessary, as a renewal of such plan for the ensuing fiscal year, to meet the standards set forth in subdivision b of this section, and shall promptly mail a certified copy of any approved renewal of such plan to the applicant and shall promptly transmit a certified copy of such renewal to the tax commission. Upon application made by the owner of such improvement pursuant to the provisions of paragraph five of this subdivision e, the tax commission shall grant, for such fiscal year, the tax exemption and/or remission of taxes specified in such determination.

(5)   Where any such plan or a renewal thereof is approved by the commission, pursuant to the provisions of the preceding paragraphs of this subdivision e, prior to January first next preceding the fiscal year to which the tax benefits of such plan or renewal thereof are applicable, the owner shall not be entitled to such benefits for such fiscal year unless he files an application therefor with the tax commission between February first and March fifteenth, both dates inclusive, next preceding such fiscal year. Where any such plan or a renewal thereof is approved by the commission between January first and June thirtieth, both dates inclusive, next preceding the fiscal year to which the tax benefits of such plan or renewal thereof are applicable, the owner shall not be entitled to such benefits for such fiscal year unless he files an application therefor with the tax commission on or before August first of such fiscal year.

(f)   (1) In any case where the commission determines, after holding a public hearing pursuant to subdivision d of this section, that a plan which it has formulated, consisting in whole or in part of any proposal other than tax exemption and/or remission of taxes, meets the standards set forth in subdivision b of this section, as such plan was originally formulated, or with such modifications as the commission deems necessary or appropriate, the commission shall approve such plan, as originally formulated, or with such modifications, and shall promptly mail a copy of same to the applicant.

(2)   The owner of the improvement proposed to be benefited by such plan mentioned in paragraph one of this subdivision f may accept or reject such plan by written acceptance or rejection filed with the commission. If such an acceptance is filed, the commission shall deny the request of such applicant for a certificate of appropriateness. If a new application for a permit from the department of buildings and a new request for a certificate of appropriateness are filed, which application and request conform with such proposed plan, the commission shall grant such certificate as promptly as is practicable and in any event within thirty days after such filing.

(3)   If such accepted plan consists in part of tax exemption and/or remission of taxes, the provisions of paragraphs two, three, four and five of subdivision e of this section shall govern the granting of such tax exemption and/or remission of taxes.

(g)   (1) Except in a case where the applicant is required to establish the conditions set forth in paragraph two of subdivision a of this section, if

(a)   The commission does not formulate and mail a plan pursuant to the provisions of subdivisions b, c, and d of this section within the period of time prescribed by such subdivision d; or

(b)   the commission does not approve a plan pursuant to the provisions of subdivision e or f of this section within sixty days after the mailing of such plan to the applicant; or

(c)   a plan approved by the commission pursuant to the provisions of paragraph one of subdivision f of this section is rejected by the owner of such improvement pursuant to the provisions of paragraph two of such subdivision;

the commission may, within ten days after expiration of the applicable period referred to in subparagraphs (a) and (b) of this paragraph one, or within ten days after the filing of a rejection of a plan pursuant to paragraph two of subdivision f of this section, as the case may be, transmit to the mayor a written recommendation that the city acquire a specified appropriate protective interest in the improvement parcel which includes the improvement with respect to which the request for a certificate of appropriateness was filed, and shall promptly notify the applicant of such action.

(2)   If, within ninety days after transmission of such recommendation, or, if no such recommendation is transmitted, within ninety days after the expiration of the period herein prescribed for such transmission, the city does not:

(a)   give notice, pursuant to section three hundred eighty-two of the charter, of an application to condemn such interest or any other appropriate protective interest agreed upon by the mayor and the commission; or

(b)   enter into a contract with the owner of such improvement parcel to acquire such interest, as so recommended or agreed upon;

the commission shall promptly grant issue and foreward to the owner, in lieu of the certificate of appropriateness requested by the applicant, a notice to proceed.

(h)   No plan which consists in whole or in part of the granting of a partial or complete tax exemption or remission of taxes pursuant to the provisions of this chapter shall be deemed to have been approved by the commission unless it is also approved by the board of estimate within the period of time prescribed by this section for approval of such plan by the commission.

(i)   (1) In any case where the applicant is required to establish the conditions set forth in paragraph two of subdivision a of this section, as promptly as is practicable after making a preliminary determination with respect to such conditions, as provided in paragraph one of subdivision a of this section, and within one hundred and eighty days after making such preliminary determination, the commission, alone or with the aid of such persons and agencies as it deems necessary and whose aid it is able to enlist, shall endeavor to obtain a purchaser or tenant (as the case may be) of the improvement parcel or parcels with respect to which the application has been made, which purchaser or tenant will agree, without condition or contingency relating to the issuance of a certificate of appropriateness or notice to proceed and subject to the provisions of paragraph three of this subdivision i, to purchase or acquire an interest identical with that proposed to be acquired by the prospective purchaser or tenant whose agreement is the basis of the application, on reasonably equivalent terms and conditions.

(2)   The applicant shall, within a reasonable time after notice by the commission that it has obtained such a purchaser or tenant, which notice shall be served within the period of one hundred and eighty days provided by paragraph one of this subdivision i, enter into such agreement to sell or lease (as the case may be) with the purchaser or tenant so obtained. Such notice shall specify a date for the execution of such agreement, which may be postponed by the commission at the request of the applicant.

(3)   The provisions of this section shall not, after the consummation of such agreement, apply to such purchaser or tenant or to the heirs, successors or assigns of such purchaser or tenant.

(4)   (a) If, within the one-hundred eighty-day period following the commission's preliminary determination pursuant to paragraph one of subdivision a of this section, the commission shall not have succeeded in obtaining a purchaser or tenant of the improvement parcel, pursuant to paragraph one of this subdivision i, or if, having obtained such a purchaser or tenant, such purchaser or tenant fails within the time provided in paragraph two of this subdivision i, to enter into the agreement provided for by such paragraph two, the commission, within twenty days after the expiration of the one-hundred-eighty day period provided for in paragraph one of this subdivision i, or within twenty days after the date upon which a purchaser or tenant obtained by the commission pursuant to the provisions of such paragraph one fails to enter into the agreement provided for by said paragraph, whichever of said dates later occurs, may transmit to the mayor a written recommendation that the city acquire a specified appropriate protective interest in the improvement parcel or parcels which include the improvement or are part of the landmark site with respect to which the request for a certificate of appropriateness was filed, and shall promptly notify the applicant of such action.

(b)   If, within ninety days after transmission of such recommendation, or, if no such recommendation, is transmitted, within ninety days after the expiration of the period herein prescribed for such transmission, the city does not give notice, pursuant to section three hundred eighty-two of the charter, of an application to condemn such interest or any other appropriate protective interest agreed upon by the mayor and the commission, or does not enter into a contract with the owner of such improvement parcel to acquire such interest, as so recommended and agreed upon; the commission shall promptly grant, issue and forward to the owner, in lieu of the certificate of appropriateness requested by the applicant, a notice to proceed.

(5)   Such notice to proceed shall authorize the work of demolition, alteration, and/or reconstruction sought with respect to the improvement parcel or parcels concerning which the application was made, only if such work (a) is undertaken and performed by the purchaser or tenant specified pursuant to the provisions of paragraph two of subdivision a of this section, in the application, or a bona fide assignee, successor, lessee or sub-lessee of such purchaser or tenant (other than the owner who made application therefor), and (b) is undertaken and performed with reasonable promptness after the issuance of such notice to proceed.

§ 207–9.0 Regulation of minor work

(a)   (1) Except as otherwise provided in section 207–11.0 of this chapter, it shall be unlawful for any person in charge of an improvement located on a landmark site or in an historic district or containing an interior landmark to perform any minor work thereon, or to cause or permit such work to be performed, and for any other person to perform any such work thereon or cause same to be performed, unless the commission has issued a permit, pursuant to this section, authorizing such work.

(2)   It shall be unlawful for any person in charge of any such improvement to maintain same or cause or permit same to be maintained in the condition created by any work done in violation of the provisions of paragraph one of this subdivision a.

(b)   The owner of an improvement desiring to obtain such a permit, or any person authorized by the owner to perform such work, may file with the commission an application for such permit, which shall include such description of the proposed work, as the commission may prescribe. The applicant shall submit such other information with respect to the proposed work as the commission may from time to time require. The commission shall promptly transmit such application to the department of buildings, which shall, as promptly as is practicable, certify to the commission whether a permit for such proposed work, issued by such department, is required by law. If such department certifies that such a permit is required, the commission shall deny such application, and shall promptly give notice of such determination to the applicant. If such department certifies that no such permit is required, the commission shall determine such application as hereinafter provided.

(c)   (1)   The commission shall determine:

(a)   whether the proposed work would change, destroy or affect any exterior architectural feature of an improvement located on a landmark site or in an historic district or interior architectural feature of an improvement containing an interior landmark; or

(b)   If such work would have such effect, whether judged by the standards set forth in subdivisions b, c, d and e of section 207—6.0 of this chapter with respect to an improvement of similar classification hereunder, such work would be appropriate for and consistent with the effectuation of the purposes of this chapter.

(2)   If the commission determines the question set forth in subparagraph (a) of paragraph one of this subdivision e in the negative, or determines the question set forth in subparagraph (b) of such paragraph in the affirmative, it shall grant such permit, and it shall deny such permit if it determines such question set forth in subparagraph (a) in the affirmative and determines such question set forth in subparagraph (b) in the negative.

(d)   The procedure of the commission in making its determination with respect to any such application shall be as prescribed in subparagraph two of subdivision a of section 207—5.0 of this chapter, except that any period of thirty days referred to in such subparagraph shall, for the purposes of this subdivision d, be deemed to be twenty days.

(e)   The provisions of this section shall be inapplicable to any improvement mentioned in subdivision a of section 207—17.0 of this chapter and to any city-aided project. (Amended by L. L. 1973, No. 71, Dec. 17.)

§ 207—10.0 Maintenance and repair of improvements

(a)   Every person in charge of an improvement on a landmark site or in an historic district shall keep in good repair (1) all of the exterior portions of such improvement and (2) all interior portions thereof which, if not so maintained, may cause or tend to cause the exterior portions of such improvement to deteriorate, decay or become damaged or otherwise to fall into a state of disrepair.

(b)   Every person in charge of an improvement containing an interior landmark shall keep in good repair (1) all portions of such interior landmark and (2) all other portions of the improvement which, if not so maintained, may cause or tend to cause the interior landmark contained in such improvement to deteriorate, decay or become damaged or otherwise fall into a state of disrepair.

(c)   Every person in charge of a scenic landmark shall keep in good repair all portions thereof.

(d)   The provisions of this section shall be in addition to all other provisions of law requiring any such improvement to be kept in good repair. (Amended by L. L. 1973, No. 71, Dec. 17.)

§ 207—11.0 Remedying of dangerous conditions

(a)   In any case where the department of buildings, the fire department or the health service administration, or any officer or agency thereof, or any court on application or at the instance of any such department, officer or agency, shall order or direct the construction, reconstruction, alteration or demolition of any improvement on a landmark site or in an historic district or containing an interior landmark, or the performance of any minor work upon such improvement, for the purpose of remedying conditions determined to be dangerous to life, health or property, nothing contained in this chapter shall be construed as making it unlawful for any person, without prior issuance of a certificate of no effect on protected architectural features or certificates of appropriateness or permit for minor work pursuant to this chapter, to comply with such order or direction. (Subd. a amended by L. L. 1973, No. 71, Dec. 17.)

§ 207—12.0 Public hearings; conferences

(a)   The commission shall give notice of any public hearing which it is required or authorized to hold under the provisions of this chapter by publication in the City Record for at least ten days immediately prior thereto.

The owner of any improvement parcel on which a landmark or a proposed landmark is situated or which is a part of a landmark site or proposed landmark site or which contains an interior landmark or proposed interior landmark, or any property which includes a scenic landmark or proposed scenic landmark shall be given notice of any public hearing relating to the designation of such proposed landmark, landmark site, interior landmark or scenic landmark, the amendment to any designation thereof on the proposed rescission of any designation or amendment thereto. Such notice may be served by the commission by registered mail addressed to the owner or owners at his or their last known address or addresses, as the same appear in the records of the office of the city director of finance or if there is no name in such records, such notice may be served by ordinary mail addressed to "Owner" at the street address of the improvement parcel or property in question. Failure by the commission to give such notice shall not invalidate or affect any proceedings pursuant to this chapter relating to such improvement parcel or property. (Subd. a amended by L. L. 1973, No. 71, Dec. 17.)

(b)   At any such public hearing, the commission shall afford a reasonable opportunity for the presentation of facts and the expression of views by those desiring to be heard, and may, in its discretion, take the testimony of witnesses and receive evidence, provided, however, that the commission, in determining any matter as to which any such hearing is held, shall not be confined to consideration of the facts, views, testimony or evidence submitted at such hearing.

(c)   The commission may delegate to any member or members thereof the power to conduct any such public hearing and to hold any conference required to be held under the provisions of sections 207—5.0 and 207—9.0 of this chapter.

(d)   The commission, may, in its discretion, direct that notice of any such public hearing on a request for a certificate of appropriateness, or any plan formulated by the commission in relation thereto, be given by the applicant to such owners of property in the neighborhood of the improvement or improvement parcel to which such request relates, as the commission deems proper. When so directed, the applicant shall mail a notice of such hearing to such owners, at their last known addresses, as the same appear in the records of the office of the city director of finance, and shall likewise mail a notice of such hearing to persons who have filed written requests for such notice with the commission. A reasonable period of time, as prescribed by the regulations of the commission, shall be afforded the applicant for giving notice of such hearing to such owners and persons. Any failure to give or receive such notice shall not invalidate any such hearing or any determination made by the commission with respect to such request for a certificate or with respect to such plan.

§ 207—13.0 Extension of time for action by commission

Whenever, under the provisions of this chapter, the commission is required or authorized, within a prescribed period of time, to make any determination or perform any act in relation to any request for a certificate of no effect on protected architectural features, a certificate of appropriateness or a permit for minor work, the applicant may extend such period of time by his written consent filed with the commission. (Amended by L. L. 1973, No. 71, Dec. 17.)

§ 207—14.0 Determinations of the commission; notice thereof

(a)   Any determination of the commission granting or denying a certificate of no effect on protected architectural features, a certificate of appropriateness or a permit for minor work shall set forth the reasons for such determination.

(b)   The commission shall promptly give notice of any such determination and of any preliminary determination of insufficient return made pursuant to paragraph (1) of subdivision a of section 207—8.0 of this chapter to the applicant. Such notice shall include a copy of such determination.

(c)   Subject to the provisions of section 207—3.0 of this chapter, any determination of the commission granting a certificate of no effect on protected architectural features, a certificate of appropriateness or a permit for minor work may prescribe conditions under which the proposed work shall be done, in order to effectuate the purposes of this chapter, and may include recommendations by the commission as to the performance of such work, provided that the provisions of this subdivision shall not apply to any notice to proceed granted pursuant to the provisions of subdivision g of section 207–8.0 of this chapter. (Amended by L. L. 1973, No. 71, Dec. 17.)

§ 207–15.0 Transmission of certificates and applications to proper city agency

In any case where a certificate of no effect on protected architectural features, certificate of appropriateness or notice to proceed is granted by the commission to an applicant who has filed with the commission a copy of an application for a permit from the department of buildings, the commission shall transmit such certificate or a copy of such notice to the department of buildings. In any case where any such certificate or notice is granted to an applicant who has filed an application for a special permit with the city planning commission or the board of standards and appeals pursuant to article seven of the zoning resolution, the commission shall transmit such certificate or a copy of such notice to the planning commission or the board of standards and appeals, as the case may be. (Amended by L. L. 1973, No. 71, Dec. 17.)

§ 207–16.0 Penalties for violations; enforcement

(a)   Any person who violates any provision of subdivision a of section 207–4.0 of this chapter shall be guilty of a misdemeanor and shall be punished by a fine of not more than one thousand dollars and not less than one hundred dollars, or by imprisonment for not more than one year, or by both such fine and imprisonment.

(b)   Any person who violates any provision of subdivision a of section 207–9.0 of this chapter or any provision of section 207–10.0 shall be punished, for a first offense, by a fine of not more than two hundred and fifty dollars or less than twenty-five dollars or by imprisonment for not more than thirty days, or by both such fine and imprisonment, and shall be punished for a second, or any subsequent offense, by a fine of not more than five hundred dollars or less than one hundred dollars, or by imprisonment for not more than three months, or by both such fine and imprisonment.

(c)   Any person who files with the commission any application or request for a certificate or permit and who refuses to furnish, upon demand by the commission, any information relating to such application or request, or who wilfully makes any false statement in such application or request, or who, upon such demand, wilfully furnishes false information to the commission, shall be punished by a fine of not more than five hundred dollars or by imprisonment for not more than ninety days, or by both such fine and imprisonment.

(d)   For the purpose of this chapter, each day during which there exists any violation of the provisions of paragraph three of subdivision a of section 207–4.0 of this chapter or paragraph two of subdivision a of section 207–9.0 of this chapter or any violation of the provisions of section 207–10.0 of this chapter, shall constitute a separate violation of such provisions.

(e)   Whenever any person has engaged or is about to engage in any act or practice which constitutes or will constitute a violation of any provision of this chapter mentioned in subdivisions a and b of this section, the commission may make application to the supreme court for an order enjoining such act or practice, or requiring such person to remove the violation or directing the restoration, as nearly as may be practicable, of any improvement or any exterior architectural feature thereof or improvement parcel affected by or involved in such violation, and upon a showing by the commission that such person has engaged or is about to engage in any such act or practice, a permanent or temporary injunction, restraining order or other appropriate order shall be granted without bond.

§ 207—17.0 Reports by commission on plans for proposed projects

(a)   Plans for the construction, reconstruction, alteration or demolition of any improvement or proposed improvement which:

(1)   is owned by the city or is to be constructed upon property owned by the city; and

(2)   is or is to be located on a landmark site or in an historic district or contains an interior landmark; shall, prior to city action approving or otherwise authorizing the use of such plans with respect to securing the performance of such work, be referred by the agency of the city having responsibility for the preparation of such plans to the commission for a report. Such report shall be submitted to the mayor, the city council and to the agency having such responsibility and shall be published in the City Record within 45 days after such referral.

(b)   (1) No officer or agency of the city whose approval is required by law for the construction or effectuation of a city-aided project shall approve the plans or proposal for, or application for approval of, such project, unless, prior to such approval, such officer or agency has received from the commission a report on such plans, proposal or application for approval.

(2)   All such plans, proposals or applications for approval shall be referred to the commission for a report thereon before consideration of approval thereof is undertaken by any such officer or agency and the commission shall submit its report to each such officer and agency and such report shall be published in the City Record within 45 days after such referral.

(c)   Except as provided in subdivision d of section 207—2.0, where the commission so requests, plans for the construction, reconstruction* alteration or demolition of any landscape feature of a scenic landmark shall, prior to city action approving or otherwise authorizing the use of such plans with respect to securing the performance of such work, be referred by the agency of the city having responsibility for the preparation of such plans to the commission for a report. Such report shall be submitted to the mayor, the city council and to the agency having such responsibility and shall be published in the City Record within 45 days after such referral. No such report shall recommend disapproval of any such plans where land contour work or earthwork is necessary in order to conform with applicable laws concerning regulation of lots, storm water disposal and water courses. The administrator of the parks, recreation and cultural affairs administration may request

---

*So in original.

an advisory report concerning work proposed to be performed on, or in the vicinity of, a scenic landmark, and such report shall be published in the City Record. (Amended by L. L. 1973, No. 71, Dec. 17.)

§ 207—18.0 Regulations

The commission may from time to time promulgate, amend and rescind such regulations as it may deem necessary to effectuate the purposes of this chapter, including, but not limited to, regulations:

(a)   for the protection, preservation, enhancement, and perpetuation and use of landmarks, interior landmarks, scenic landmarks and historic districts, subject to the provisions of § 207—3.0 of this chapter. Such regulations may apply to one or more historic districts or to one or more portions of an historic district and may vary from area to area in their provisions. (Subd. a amended by L. L. 1973, No. 71, Dec. 17.)

(b)   relating to the determination of the earning capacity of improvement parcels by the commission pursuant to section 207—8.0 of this chapter; and

(c)   relating to the procedures of the commission in carrying out its functions, powers and duties under this chapter, including procedures for the giving of notice by the commission by mail or otherwise, where notice is required by this chapter; and

(d)   relating to forms to be used in proceedings before the commission.

§ 207—19.0 Investigations and reports

The commission may make such investigations and studies of matters relating to the protections, enhancement, perpetuation or use of landmarks, interior landmarks, scenic landmarks and historic districts, and to the restoration of landmarks, interior landmarks and scenic landmarks as the commission may from time to time, deem necessary or appropriate for the effectuation of the purposes of this chapter, and may submit reports and recommendations as to such matters to the mayor and other agencies of the city. In making such investigations and studies, the commission may hold such public hearings as it may deem necessary or appropriate. (Amended by L. L. 1973, No. 71, Dec. 17.)

§ 207—20.0 Applicability

(a)   The provisions of this chapter shall be inapplicable to the construction, reconstruction, alteration or demolition of any improvement on a landmark site or in a historic district or containing an interior landmark, or of any landscape feature of a scenic landmark, where a permit for the performance of such work was issued by the department of buildings, or, in the case of a landscape feature of a scenic landmark, where plans for such work have been approved, prior to the effective date of the designation or amended or modified designation pursuant to the provisions of section 207—2.0 of this chapter, first making the provisions of this chapter applicable to such improvement or landscape feature or to the improvement parcel or property in which such improvement or landscape feature is or is to be located.

(Amended by L. L. 1973, No. 71, Dec. 17.)

§ 207—21.0 Separability

If any provision of this chapter or the application thereof to any person or circumstances is held invalid, the remainder of this chapter and the application of such provisions to other persons or circumstances shall not be affected thereby. (Added by L. L. 1965, No. 46, April 19.)

Index